# HOW MAMMALS RUN

# P. P. Gambaryan

# HOW MAMMALS RUN

## Anatomical Adaptations

*Translated from Russian by Hilary Hardin*

A HALSTED PRESS BOOK

JOHN WILEY & SONS

New York · Toronto

ISRAEL PROGRAM FOR SCIENTIFIC TRANSLATIONS

Jerusalem · London

Sole distributors for the Western Hemisphere

HALSTED PRESS, a division of
JOHN WILEY & SONS, INC., NEW YORK

**Library of Congress Cataloging in Publication Data**

Gambarĩan, P    P
   How mammals run:  anatomical adaptations.
   "A Halsted Press book."
   Translation of *Beg mlekopitayushchikh.*
   Bibliography: p.
   1. Animal locomotion.  2. Adaptation (Biology)
3. Mammals.  I. Title.
QL739.G3513     599'.01'852     74-16190
ISBN 0-470-29059-5

Distributors for the U.K., Europe, Africa and the Middle East
JOHN WILEY & SONS LTD., CHICHESTER

Distributors for Japan, Southeast Asia and India
TOPPAN COMPANY LTD., TOKYO AND SINGAPORE

Distributed in the rest of the world by
KETTER PUBLISHING HOUSE JERUSALEM LTD.
ISBN 0 7065 1483.1
IPST cat. no. 22128

This book is a translation from Russian of
BEG MLEKOPITAYUSHCHIKH – PRISPOSOBITEL'NYE
Izdatel'stvo "Nauka," Leningradskoe Otdelenie
Leningrad, 1972

Printed and bound by Keterpress Enterprises, Jerusalem
Printed in Israel

# CONTENTS

# FOREWORD

The movements of animals have always intrigued man. From prehistoric paintings depicting man's prey in action, to the dances of primitive people mimicking familiar animals, to current anatomical and physiological studies of specific biomechanical and energetic problems, interest in how animals move continues. Some history of the scientific investigation of locomotion is included in this book, but the underlying basis for interest in the subject appears to be more fundamental than the schemes and apparatus contrived, in this century and before, to analyze locomotion. Speed, agility, grace, strength, endurance — all qualities that are universally recognized in animal movement — are ascribed only to "modern" men and women who possess special ability and training. This at once acknowledges the paralytic effect of civilization on mankind's heritage of locomotor skills, as well as a continuing fascination for animal mechanisms which machines can only mimic.

For English readers, an understanding of animal locomotion has been largely established, during the last several decades, by the original studies of A. B. Howell, James Gray and his associates, R. W. Haines, M. Hildebrand and others who have attempted to synthesize biomechanical principles from observation and experiment. These investigators were preceded by Eadweard Muybridge, W. K. Gregory and P. Magne de la Croix, who made classic contributions on photographic documentation, biomechanics, and gait analysis, respectively. Still earlier were E. J. Marey in France and E. and W. Weber in Germany whose basic approaches to biological problems of locomotion are followed by investigators even today, although with modern techniques and instrumentation. Most recently, fundamental breakthroughs in understanding basic comparative physiological aspects of locomotion have been made by McN. Alexander, K. Schmidt-Nielsen, C. R. Taylor and their associates. These developments, together with other studies employing cineradiography and electromyography, have yielded a diverse and extensive literature. As is too often the case, Western readers have been largely unaware of a Russian school of investigators which, starting from the work of Howell, Gray, and others, have developed an approach in parallel to the study of animal locomotion. Their work is seldom cited and apparently little known. Fortunately, this oversight will be corrected by the appearance of this volume.

P. P. Gambaryan, working in the Zoological Institute of the Academy of Sciences of the USSR, is the leading authority on animal locomotion in Russia. Publishing since the mid-1950's on various aspects of terrestrial locomotion, he has assembled in this monograph the most comprehensive account of the subject since A. B. Howell's "Speed in Animals." Although this book treats certain aspects of locomotion studies, such as gait analysis, in the traditional manner, students of the subject will be excited by the extraordinary diversity of Gambaryan's observations. Particularly valuable are detailed morphological data, including muscle weights and proportions, that are unavailable elsewhere. Certain of Gambaryan's hypotheses on the adaptive significance and evolutionary history of musculoskeletal structures may even startle some readers who will wish to re-examine traditional interpretations in face of these new theories and observations. Yet Gambaryan clearly perceives how mammals express adaptation by their movements, and his book, "How Mammals Run." will be the basis for further pursuit of the subject in years to come.

Farish A. Jenkins, Jr.
*Harvard University*

# INTRODUCTION

One of the chief reasons that the mammals hold sway over the animal world is their high activity, which is manifested in the immense diversity of perfected modes of locomotion in different media; on land, under ground, in water and in the air. The main trend of mammalian evolution was toward improvement of terrestrial locomotion; the other types of movement arose secondarily, on the basis of this.

Speed and maneuverability are of great importance in the life of mammals for fleeing enemies and natural disasters, for hunting, in the period of sexual activity, for covering large distances to watering places, pastures, etc. Insufficient speed and maneuverability are liable to lead to the death of individuals, populations, and even species, so that perfected means of locomotion must be considered as a vital factor in the struggle for survival. This prompted the development of diverse forms of high-speed movement over land. All these forms are characterized by a stage of free transit, and therefore such types of locomotion are called running.

Studies of the mechanics of running and the associated anatomical features of the locomotory organs are of considerable theoretical and practical interest for zoologists of different specialties, biophysicists and even technicians, as the biomechanics of movement can give them ideas for modeling fundamentally new engineering structures. This type of research is especially useful for morphologists, because it enables them to grasp the functional significance of individual organs and systems and the causes of changes occurring in them. Investigations along these lines are no less important for systematists and teriologists. Classic systematics generally uses the structural features of locomotory organs formally, as supplementary diagnostic characters, without taking account of their independent importance in evolution. Yet if we understand the processes by which the organs of movement are reorganized, we can construct a natural system of different groups of animals from a new standpoint. For instance, studies in aerodynamics and the flight mechanics of birds have allowed ornithologists not only to figure out the importance of several morphological structures but also to apply the data thus obtained to phylogenetic constructions (Kozlova, 1946; Shtegman, 1950, 1957; Yudin, 1950, 1965; and others).

It has long been a source of interest to man how mammals run, and yet there is still no single theory on terrestrial locomotion. Descriptions of the organs of movement of many species appeared at the end of the 18th century, but it was only at the end of the last century, with the development of photographic technique, that it became possible to give a clearly defined characterization of such forms of movement as the trot, rack and gallop,

known to man in the distant past. Improved methods of studying movement
made it possible at the same time to reveal the great diversity of its forms
and gave impetus to the search for new ways of studying the movements
themselves and refining their descriptions.

When the various gaits characteristic for each animal became known,
the need to classify them arose. Muybridge (1887) proposed a very convenient and simple system for determining forms of locomotion based on the
sequence of limb movements and changes of the support stage. This system
made it possible to establish the similarities and differences in the locomotion of different animals and served as a basis for describing and classifying gaits. But it proved inadequate to explain the mechanics of movement,
as the actual concept "gait" has more to do with the relative temporal links
in movement and thus characterizes only one aspect of locomotion. Other
vital features (speed of running, the distance covered during the stage of
transit, the amplitude of flexor-extensor movements in the limb joints and
in the spinal column) are another aspect which has to be understood in order
to assess the work of the skeleton and musculature.

Nevertheless, it was only when the theory of gaits was being developed
and it was realized that ideas on them were very limited did it become
clear that a distinction must be made between the main types of running,
the concept of which must include the biomechanical aspect of animal movement. One of the first and most successful classifications of types of running among mammals, for the most part ungulates and proboscidians, is that
of Gregory (1912), who distinguished four types (subcursorial, cursorial,
mediportal and graviportal) on the basis of speed and endurance and also
the influence of the body mass on the mechanics of movement. He showed
that these types make it easier to understand the process of phylogenetic
changes of the skeletal elements of the limbs of some mammals (ungulates
and proboscidians).

Several more classifications of running types (Böker, 1935; Glagolev,1941,
1952; Kas'yanenko, 1956; and others) and general theories of locomotion
(Gray, 1953, 1961, 1968; Bekker, 1955; Smith and Savadge, 1956; Rashevsky,
1960; Otoway, 1961; Aleksander, 1970; etc.) appeared later. However, these
studies suffer from one-sidedness, expressed in the former group in that
the types of movement are defined only according to the specific morphological characteristics of the structure and position of the autopodia, and
in the latter group in the certain degree of isolation from a study of the
movement of specific objects.

The recent literature contains a good number of data having a direct
bearing on the subject in hand. Mention may be made of studies on functional and ecological morphology, ecology, ethology and paleontology of
mammals which treat variously different aspects of movement or modes
of reorganization of the locomotory organs. But all these works have
different aims, and the study of movement is not their prime purpose. This
detracts from the value of the data presented, the scope of which is not broad
enough to give a full idea of the paths by which the locomotory organs became transformed in the course of adaptation to swift running.

Ecological and ethological investigations would seem to be very important for explaining those aspects of life activity and behavior which govern
the movement and anatomy of mammals. Unfortunately, the range of

questions embraced by these disciplines usually has very little to do with the problem examined here.* Paleontological studies help us understand the historical course of changes in the muscles of mammals, but the paleontologist is forced to judge the functional value of the differences he finds only by analogy with recent forms, and these have been studied insufficiently.

By applying the laws of statistics and mathematical calculations widely used in studying the strength of materials, morphologists have been able to work out many anatomical features and skeletal structures of animals. However, works of this type (Gray, 1944; Pauwels, 1948; Kummer, 1959a, 1959b, 1960; and others) give no information on the dynamic loads, which probably have decisive significance for the changes in skeletal structure.

However strange it may seem, studies of functional and ecological morphology purporting to explain the dependence between the structure and function of organs generally lack a biomechanical analysis. Such an analysis is more often made when investigating the mechanics of human sport exercises and work processes. It is used much more rarely in a study of the movements of domestic animals (as a rule, horses). It is to be regretted that practically no one has carried out comparative biomechanical research, as only this can serve as a starting point for estimating the different functions of locomotory organs in different species and groups of mammals.

In particular, comparative biomechanical characteristics of movement, quantitative data characterizing the morphology of the locomotory organs, and also explanations of those aspects of vital activity which determine the mechanics of movement and the concomitant skeletal and muscular changes are not sufficiently available. The actual approach to the problem also leaves something to be desired. In the many works devoted to the movement of mammals no true (not declarative) morphofunctional analysis has been given. What we have in essence is a set of isolated descriptions of different aspects of the phenomenon without any attempt at a synthesis. Since the publication by Gregory (1912) no examination has been made of running patterns in different groups of mammals against the background of evolution, taking into account the succession and diversity of possible biomechanical solutions to the same problems on a particular starting basis.

The present monograph proposes to fill these gaps by studying the trends of adaptive changes of locomotory organs in mammals depending on the degree of specialization for high-speed terrestrial locomotion.

The wide scope of the topics discussed called for the improvement of some points of procedure and in several cases for the elaboration of new study methods.

So as to obtain uniform and useful quantitative characteristics of the organs of movement, we made a detailed morphological analysis of the skeleton and muscles of the limbs and the spinal column of more than 100 species of seven mammalian orders. Most of the material was collected on expeditions conducted in various regions of the Soviet Union, during which we gathered the basic material on the biomechanics of running

---

* The classic range of ecological questions includes: habitat, population dynamics, reproduction, feeding, molting, enemies and parasites, and diurnal and seasonal activity. Ethology, on the other hand, is concerned mainly with describing and typifying the poses and behavior of animals in different states of stimulation.

and the ecology of the animals studied. Some of the data were gleaned from the abundant collections of the Zoological Institute of the Academy of Sciences of the USSR. Additional laboratory experiments were carried out to analyze more precisely the locomotory characteristics of a number of mammals which had been observed during the field studies.

All the comparative biomechanical data, apart from a few, specially stipulated exceptions, are original. It has not been our aim to present a full bibliography on all the topics discussed (ecology, biomechanics, morphology, etc.), all the more so that reference is made to summaries which contain a detailed bibliography of this kind.

The author was enabled to produce this work solely thanks to the support and constant encouragement of the board of directors of the Zoological Institute: Academician B. E. Bykhovskii, and Doctors of Biological Sciences A. I. Ivanov (head of the laboratory of terrestrial vertebrates) and I. M. Gromov (head of the Department of Mammals).

Invaluable assistance in compiling the work was lent by Doctor of Biological Sciences A. A. Strelkov and K. A. Yudin. Most of the figures were executed by artists E. Ya. Zakharov and V. N. Lyakhov. In completing the monograph the author was kindly aided by friends: F. Ya. Dzerzhinskii, O. V. Egorov, V. B. Sukhanov and P. P. Strelkov, and scholars K. M. Gasparyan, M. F. Zhukova, V. S. Karapetyan, S. K. Mezhlumyan, L. E. Oganesyan, R. O. Oganesyan, T. G. Protopopova and R. G. Rukhkyan.

*Chapter 1*

*MATERIAL AND METHODS*

Comparative anatomical and comparative biomechanical methods constituted the main study procedure. As regards the material, the principle of its selection is very important in order to bring out the truly objective relationship between the type of movement and structure of an animal. On the one hand, the more closely related compared animals are systematically, the greater the probability that differences in the structure of their locomotory organs will correspond to mechanical features of movement. On the other hand, in order to clarify the main paths of adaptation to running in mammals, as many forms as possible must be studied which are very different in systematic position, size, and mechanics of running. This is why we have chosen forms (Table 1) which are variously adapted to running and belong to several orders (Marsupialia, Rodentia, Lagomorpha, Carnivora, Artiodactyla, Perissodactyla, Proboscidea). Insectivora are not included in the table because we were unable to obtain forms from the family Macroscelidae which are specialized for running, although hedgehogs and shrews were studied. The largest possible number of forms was studied in each order, special attention being paid to groups which included some families and subfamilies having species which differ in the degree of cursorial specialization. Unfortunately, the selection of species was in several cases incidental, depending on the occurrence of epizooty in the zoological gardens. As a rule, both the skeleton and the muscles of the animals were subjected to morphological and biomechanical analysis (Table 1), but sometimes only the skeleton was studied (species in which only the skeleton was analyzed are not given in Table 1).

The morphological data obtained on domestic animals will not be presented except in rare cases, as domestication may have brought about marked distortions of the adaptive mechanics of movement.

Although our material far from covers all forms specialized for running, almost all the main trends of this specialization have been investigated, and a clear picture is given of the great diversity of pathways along which mammals have developed high-speed locomotion over land. Cursorial specialization is expressed most differently in ungulates and rodents, and therefore these groups were studied in greater detail than the Carnivora and Lagomorpha.

The degree of cursorial specialization was estimated by comparing indexes of speed and endurance, the size of jumps, and so on. The data were obtained by various methods of field observations which made it possible to assess these aspects of biology objectively. Data on running speeds were taken partly from the literature (Andrews, 1924; Zverev, 1948; Rakov, 1955; Grzimek and Grzimek, 1960; Sludskii, 1962; Solomatin, 1965; and others) and are partly original. Original data were obtained by measuring speeds with the aid of speedometers from automobiles and motorcycles, and in some cases from a helicopter; other data were yielded by an analysis of motion pictures and, finally, visually in the field with stopwatches. The accuracy of the speedometer readings was checked on areas of known extent with the use of a stopwatch. Apart from this, a detailed description of the experiment was made with respect to both the running order of the automobile and the behavior of the animal. It was especially noted whether the wheels of the vehicle skidded and whether the animal kept up with the car, fell behind, outpaced it, etc. Despite the relative inaccuracy of experiments using automobile speedometers, we were able to collect numerous data (both our own and from published sources) on the speed and stamina of large plain-dwelling animals. When studying the running speed of animals inhabiting very broken terrain, speedometer readings could naturally not be used, and in this case the speed was determined by other methods. In the analysis of the motion pictures the running speed was established according to the displacement of the animal in relation to the reference point marked on the frame. Knowing the filming speed, expressed in frames per second, and the average length of the animal's body, it was possible to calculate the running speed with an accuracy of $\pm 10\%$.

The most precise determinations of speed during observations in nature were carried out in areas where a trail remained (shifting sands, loose solonchaks, gray takyr soils, fresh powder, etc.). The stopwatch was started the moment the animal reached some reference point and was stopped either when the animal switched to a different gait or when it reached the next reference point. Measurements according to tracks, with a description of the nature of these, give the most accurate information on running speeds at different gaits. It is much more complicated to determine the speed of running in a place where a trail is not discovered. In this case successful results may be obtained by working in twos. Starting the stopwatch at the time the animal crosses over from one reference point to another, the leader directs the measurements taken by his assistant from the spot. If the work is done single-handed, the reference points often become unrecognizable close to, and it is very difficult to take measurements. It is sometimes convenient to calculate the distance geometrically. The following of a series of geometric means was used. A peg was driven in at the point of observation (Figure 1, O) and from this threads about a meter long (Oa and Ob) were stretched in the direction of the reference points (A and B). From point O at a right angle to Oa and Ob threads Oc and Od were stretched (these threads should preferably be at least 5 m long). From point c to reference point A and from point d to reference point B short threads are stretched (not more than a meter long). The exact value of legs Oc and Od and of angles aOb, OcA and OdB is measured. From the leg and angle of each right-angled triangle the distances OA and OB are found and then the unknown distance AB is calculated from OA, OB, and the angle between them.

TABLE 1.  List of species subjected to morphological and biomechanical analysis

| Order, family, species | Number of specimens analyzed morpho- logically | Description of biomechanical analysis | |
|---|---|---|---|
| | | according to motion-picture photography, frames/sec | according to photographs and tracks |
| Order *Marsupialia* | | | |
| Great gray kangaroo (Macropus giganteus).... | 1 | — | + |
| Red kangaroo (M. rufus) ........................ | 2 | — | + |
| Order *Rodentia* | | | |
| Family Sciuridae | | | |
| Persian squirrel (Sciurus persicus)............ | 3 | — | + |
| Red squirrel (S. vulgaris ) ..................... | 3 | 300 | + |
| Long-clawed ground squirrel (Spermophilopsis leptodactylus) ........................... | 3 | 40, 500 | + |
| Long-tailed Siberian souslik (Citellus undulatus) | 3 | 64 | + |
| European souslik (C. citellus) ................. | 5 | 96, 180 | + |
| Altai marmot (Marmota baibacina).......... | 2 | — | — |
| Mongolian bobak (M. sibirica) ................ | 2 | — | — |
| Long-tailed marmot (M. caudata) .............. | 1 | 96 | — |
| Family Hystricidae | | | |
| Indian crested porcupine (Hystrix leucura) .... | 2 | — | — |
| Family Dasyproctidae | | | |
| Agouti (Dasyprocta agouti)................. | 2 | 120 | + |
| Family Myocastoridae | | | |
| Coypu (Myocastor coypus)................... | 3 | — | + |
| Family Dipodidae | | | |
| Northern birch mouse (Sicista betulina) ....... | 2 | — | — |
| Caucasian birch mouse (S. caucasica)........... | 1 | — | — |
| Far Eastern birch mouse (S. caudata) ............ | 1 | — | — |
| Great jerboa (Allactaga jaculus)............. | 2 | — | — |
| William's jerboa (A. williamsi) ............... | 4 | 120 | + |
| A. sibirica saltator ........................ | 1 | 120 | — |
| Severtsov's jerboa (A. severtzovi) ............. | 2 | — | — |
| Greater fat-tailed jerboa (Pygerethmus platyurus)................................. | 2 | — | — |
| Lichtenstein's jerboa (Eremodipus lichten- steini)..................................... | 1 | — | — |
| Family Muridae | | | |
| Bandicoot rat (Nesokia indica) .............. | 2 | — | — |
| Norway rat (Rattus norvegicus).............. | 6 | 120, 64 | + |
| Turkestan rat (R. turkestanicus) .............. | 2 | — | — |
| Conilurus sp. ............................. | 1 | — | — |

TABLE 1 (continued)

| Order, family, species | Number of specimens analyzed morpho- logically | Description of biomechanical analysis | |
|---|---|---|---|
| | | according to motion-picture photography, frames/sec | according to photographs and tracks |
| Family Cricetidae | | | |
| Mouselike hamster (Calomyscus bailwardi)..... | 6 | 64, 120, 300 | + |
| Striped hairy-footed hamster (Phodopus sungorus) | 5 | 96, 300 | + |
| Migratory hamster (Cricetulus migratorius).... | 6 | 96, 1080 | + |
| Common hamster (Cricetus cricetus)........... | 2 | — | — |
| Mesocricetus brandti ................. | 5 | 120, 300 | + |
| Gerbil (Meriones blackleri) ................. | 12 | 64, 120, 1,000 | + |
| Persian jird (M. persicus)........................ | 10 | 96, 300 | + |
| Midday gerbil (M. meridianus) .................. | 10 | 96, 300 | + |
| Red-tailed Libyan jird (M. erythrourus).......... | 2 | 300 | + |
| Vinogradov's gerbil (M. vinogradovi) ............ | 10 | 96, 300 | + |
| Great gerbil (Rhombomys opimus) ............. | 2 | 40 | + |
| Order *Lagomorpha* | | | |
| Family Leporidae | | | |
| Blue hare (Lepus timidus) ..................... | 5 | — | + |
| European hare (L. europaeus) .................... | 6 | — | + |
| Tolai hare (L. tolai) .............................. | 1 | — | — |
| Family Lagomyidae | | | |
| Northern pika (Ochotona alpina) .............. | 3 | 64 | + |
| Pallas' pika (O. pricei) ........................ | 1 | 64 | — |
| Order *Carnivora* | | | |
| Family Canidae    ........ | | | |
| Wolf (Canus lupus) .......................... | 4 | 120 | + |
| Jackal (C. aureus) ............................. | 1 | — | — |
| Common fox (Vulpes vulpes) .................. | 3 | — | — |
| Corsac fox (V. corsak) ......................... | 1 | — | + |
| Raccoon-dog (Nyctereutes procyonoides) ..... | 2 | 350 | + |
| Cape hunting dog (Lycaon pictus)  ............ | 2 | — | — |
| Family Ursidae | | | |
| European brown bear (Ursus arctos) .............. | 5 | 120 | + |
| Black bear (U. tibetanus)........................ | 2 | — | — |
| Malayan sun bear (U. malajanus) .............. | 1 | — | — |
| Family Mustelidae | | | |
| Weasel (Mustela nivalis)..................... | 2 | 96 | ⊥ |
| Polecat (M. putorius) ........................ | 1 | — | — |
| Siberian polecat (M. eversmanni)................ | 1 | — | — |

TABLE 1 (continued)

| Order, family, species | Number of specimens analyzed morpho- logically | Description of biomechanical analysis | |
|---|---|---|---|
| | | according to motion-picture photography, frames/sec | according to photographs and tracks |
| Tiger weasel (Vormela peregusna) .......... | 4 | 40 | + |
| Stone marten (Martes foina) ................. | 2 | 300 | + |
| Wolverine (Gulo gulo) ....................... | 1 | — | — |
| Indian honey-badger (Mellivora indica) ....... | 1 | — | — |
| Badger (Meles meles) ....................... | 6 | 120 | — |
| Common otter (Lutra lutra) .................. | 3 | — | + |
| Sea otter (Enhydra lutris) .................. | 1 | — | — |
| Family Felidae | | | |
| Jungle cat (Felis chaus) ..................... | 2 | — | — |
| Lynx (F. lynx)................................ | 2 | — | + |
| Caracal lynx (F. caracal) .................... | 1 | — | — |
| F. manul ................................... | 1 | — | — |
| Leopard (F. pardus)........................... | 4 | — | + |
| Snow leopard (F. uncia) ..................... | 2 | — | — |
| Panther (F. pantera)......................... | 1 | — | — |
| Puma (F. concolor) ......................... | 2 | — | — |
| Jaguar (F. onza) ............................ | 1 | — | — |
| Lion (F. leo) ................................ | 3 | — | + |
| Cheetah (Acinonyx jubatus) ................. | 3 | — | + |
| Order Perissodactyla | | | |
| Family Tapiridae | | | |
| Tapirus americanus ....................... | 1 | 40 | + |
| Family Equidae | | | |
| Asiatic wild ass (Equus hemionus onager) .... | 3 | 40 | + |
| Zebra (E. quagga chapmani) ................. | 2 | — | + |
| Order Artiodactyla | | | |
| Family Cervidae | | | |
| Roe deer (Capreolus capreolus).............. | 2 | 120 | + |
| Fallow deer (Cervus dama) ................... | 2 | — | + |
| Axis deer (C. nippon) ......................... | 2 | 40, 64 | + |
| Red deer (C. elaphus)......................... | 3 | — | + |
| Reindeer (Rangifer tarandus)................. | 8 | 64 | + |
| Family Giraffidae | | | |
| Giraffe (Giraffa camelopardalis) ........... | 2 | 64 | + |

TABLE 1 (continued)

| Order, family, species | Number of specimens analyzed morpho- logically | Description of biomechanical analysis | |
|---|---|---|---|
| | | according to motion-picture photography, frames/sec | according to photographs and tracks |
| Family Moschidae | | | |
| Musk deer (Moschus moschiferus) .......... | 1 | — | — |
| Family Bovidae | | | |
| Goitered gazelle (Gazella subgutturosa) ..... | 2 | 40 | + |
| Saiga (Saiga tatarica) ...................... | 8 | 64 | + |
| Chamois (Rupicapra rupicapra)............. | 5 | 64 | + |
| Wild goat (Capra aegagrus) ................. | 2 | — | + |
| Siberian ibex (C. sibirica) ................... | 4 | 64 | + |
| Markhor (C. falconeri) ...................... | 5 | 64 | + |
| Caucasian tur (C. caucasica)................. | 5 | 64 | + |
| Asiatic mouflon (Ovis orientalis) ........... | 2 | — | + |
| Pamir argali (O. ammon) ..................... | 1 | — | — |
| European bison (Bison bonasus) .............. | 2 | 64 | + |
| North American bison (B. americanus)......... | 1 | 64 | + |
| Zebu (Bos indicus) ......................... | 1 | — | — |
| Gnu (Connochaetes gnou) .................. | 3 | — | + |
| Eland (Taurotragus oryx) .................. | 3 | — | + |
| Kudu (Strepsiceros cudu) .................. | 1 | — | — |
| Anoa (Anoa depressicornis) ................ | 1 | — | — |
| Yak (Phoëphagus mutus) .................. | 1 | — | — |
| Order *Proboscidea* | | | |
| Indian elephant (Elephas indicus)............ | 1 | 120 | + |
| African elephant (Loxodonta africana) ...... | 1 | — | + |

Note. A plus sign indicates that analysis was done according to photographs and tracks, a minus sign that it was not.

For accurate measurements the following rules must be used. The greater the distance OA, the longer must be the threads forming the legs of the right-angled triangles (Oa and Ob). The whole system of straight lines (Oa, Ob, Oc, ce, Od, df) should be distributed in a horizontal plane as far as possible. This is done by hammering in the pegs vertically and using a protractor when stringing the threads. If observations are conducted on animals running along mountain slopes, the entire system should be laid out in the plane of the triangle OAB. With the aid of one assistant these geometric constructions take 3—5 minutes on flat surfaces and more than 10 minutes on mountain slopes.

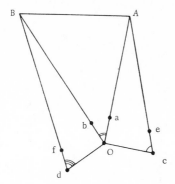

FIGURE 1. Geometric scheme for determining the
distance between two points A and B.

Explanation in the text.

Speed is determined from a helicopter only in cases when the animal is
running at about the same speed as the helicopter and is covering a large
distance which can be measured on a map.

In order to characterize the conditions for running, it is often necessary
to calculate the slope of the mountain. If a geological compass is not at
hand, the angle is measured as follows: a thread is stretched between two
points along the slope. A perpendicular is dropped from some point along
the thread, and the angle between the line and the thread is measured.

In measuring the size of jumps the difference in the levels of their beginning and end must be taken into account. This is easily done by stringing a
thread along the line of the jump and determining the corresponding angle.

For the comparative biomechanical studies, which we considered especially important, a great variety of methods was used, from analyses of
tracks, individual photographs and motion pictures to work with special
instruments determining the force of the footfalls.

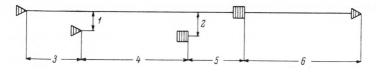

FIGURE 2. Procedure for measuring footfalls:

1) width of base of hind feet; 2) of forefeet; 3) hind pace; 4) extended flight;
5) fore pace; 6) crossed flight. Triangles show prints of hind feet, squares tracks
of forefeet.

A study of tracks helps elucidate several characteristics of running. We therefore developed a procedure for measuring tracks and made up a key to gaits according to them (see p. 46). For the running cycle to be assessed, a section with at least five consecutive footfalls has to be sought, in other words, a stretch of a full pace between the prints of the same foot. A straight line is drawn from the first to the fifth footfall (Figure 2) and this is taken to be the axis of the cycle or of the pace. A perpendicular is dropped onto the axis of the cycle from the front edge of each imprint. The size of the perpendiculars (Figure 2, 1, 2) characterizes the width of the base of the limbs. The size of the intervals between the perpendiculars determines the nature of the different stages of a full pace (Figure 2, 3—6).

Clear photographs of a moving animal can often enable one to judge the gait (for ways of determining gaits from photographs see p. 48), the angles in the leg joints and the flexor-extensor movements of the spine. When we had a number of such photographs we sketched the position of the skeleton in the outline of the animal, so that we were able to get an idea of the amplitudes of movement in the leg joints, etc. But the most useful material for studying the comparative biomechanics of movement was obtained by motion-picture photography. Every fifth frame is marked for an analysis of the duration of support and transfer of the limbs. If less than 30 frames correspond to a full pace, the work of each limb is analyzed separately, that is, the strip of film with a full pace is examined four times. If more than 30 frames correspond to a full pace, the analysis is more conveniently done as the film is being run through. Data on the support and transfer of each limb are written down in the form of a table (Table 2). A count is made of the frames where the four limbs show an identical position with respect to the ground. The transitional moments when the legs are leaving the ground or landing on it are scrutinized especially carefully, and a frame-by-frame record is then entered in the table (Table 2). The intervals of support and transfer of each limb obtained in this manner and the sequence of support stages in the cycle serve to define the animal's gait.

TABLE 2. Motion-picture analysis of the movements of a giraffe

| Serial No. of frames | rf | lf | rh | lh | Serial No. of frames | rf | lf | rh | lh | Serial No. of frames | rf | lf | rh | lh | Serial No. of frames | rf | lf | rh | lh |
|---|---|---|---|---|---|---|---|---|---|---|---|---|---|---|---|---|---|---|---|
| 1—6 | + | + | + | + | 30 | − | + | − | + | 61 | − | + | + | + | 92 | + | − | + | − |
| 7 | + | + | + | + | 31 | − | + | − | + | 62 | + | + | + | + | 93—101 | + | − | + | − |
| 8 | + | + | + | + | 32—38 | − | + | − | + | 63—69 | + | + | + | + | 102 | + | − | + | + |
| 9 | + | + | − | + | 39 | − | + | − | + | 70 | + | + | + | − | 103—120 | + | − | + | + |
| 10 | + | + | − | + | 40 | − | + | + | + | 71 | + | + | + | − | 121 | + | + | + | + |
| 11—28 | + | + | − | + | 41 | − | + | + | + | 72—90 | + | + | + | − | | | | | |
| 29 | + | + | − | + | 42—60 | − | + | + | + | 91 | + | + | + | − | | | | | |

Note. rf — right fore, lf — left fore, rh — right hind, lh — left hind leg. A plus sign denotes support of the limb, a minus sign free transit.

Each frame is then enlarged and the skeleton of the limbs is traced in. The amplitude of movement in the joints can be determined after calculating the change of the angles in the limb joints according to the skeleton. Knowing the filming speed in addition, an idea can be gained of the angular velocities of movement in the limb joints.

The patterns of movement of the skeleton obtained in this way are of great value in studying the work of the leg muscles. They are especially important for an analysis of the work of double-jointed muscles, the contraction and extension of which depend both on the flexor-extensor movements in these joints simultaneously and on the distance of their point of insertion from the center of each joint. The extreme variants of limb position can be selected from a number of frames with outlined skeletons. A comparison of these schemes with similar forms of locomotion in different species of animals presents a graphic picture of the different nature of muscle work in different animals.

An analysis was carried out on several small mammals to determine the force of pushing off and landing during jumps of different size. For this a special device was constructed consisting of a racing corridor with an exit and an entrance cage. In the middle of the corridor there was a section 1.5 m long with a transparent glass wall. The bottom of this section was extensible. A vessel containing water was placed in the unoccupied space. The animal raced from one end of the corridor to the other. After it had fallen into the water a few times, it became used to jumping over the obstacle and learned not to hurl itself onto the glass. During the experiment the glass was removed and all the films were shot without it. After the animal had been taught to jump a set distance, mobile platforms were set up at the edges of the vessel to record the force of the jumps upon pushing off and landing.

Of the many mechanical-electrical transducer systems tried out, we finally used two.* In the first, each of the platforms was connected by nonelastic thread to the lever of the potentiometer, in which the terminals were connected through an amplifier to an oscilloscope. The lever of the potentiometer moved by the pull of the threads proportionally to the pressure on the platform and returned to its original position owing to the pull on the rubber band (Figure 3, 1). The movement of the lever altered the resistance, leading to a change in the character of the oscillogram. The force of jumps was ascertained after experiments to compare the oscillograms obtained with reference oscillograms from small loads placed on the jumping-off and landing platforms. In the second system the change in resistance arose on account of the compression of a piece of leather the inner surface of which was sprinkled with powdered graphite. The leather was placed on a sheet of vinyl plastic covered with copper foil which was etched to a certain form (Figure 3, 9). When the piece of leather was squeezed, the graphite particles were pressed closer together, causing a change in the resistance. This change of resistance, as in the first system, was recorded on an oscillogram.

* The errors associated with the method are described on p. 90.

A reference oscillogram was obtained after each experiment by placing small loads on the platforms. In the second system the horizontal indexes of force were recorded as well as the vertical. The free play of the platforms was limited by the thickness of the leather and did not exceed 0.5 mm. A 16-mm movie camera (Pentafleks-16) was switched on at the same time as the oscilloscope. A 500 cps time marker was plotted on the oscillogram and a 50 cps time marker on the film. Analysis of the oscillograms proceeded in parallel with monitoring of the picture frames. Such experiments were conducted only on small animals (gerbils, hamsters, sousliks, pikas, weasels, etc.).

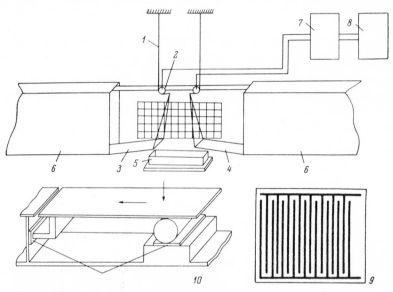

FIGURE 3. Diagram of a device for biomechanical studies:

1) rubber band drawing the lever of the potentiometer back to its original position; 2) potentiometer; 3) jumping-off platform; 4) landing platform; 5) vessel containing water; 6) racing corridor; 7) amplifier; 8) oscilloscope; 9) apparatus for etching foil for the experimental platforms; 10) experimental platforms of the second type. Arrows show the direction of the pressure studied.

After completion of the series of experiments with successive frames of the most typical jump, photographic prints of the animals were prepared natural size. Stencils of the animals were made from these photographs. An anesthetized animal was fixed with adhesive plaster in the corresponding pose on the stencil, and in a Borelli stand (Figure 4) the frame-by-frame position of the animal's center of gravity was determined, which served as a control for a study of the dynamics of shifting in certain phases of the cycle of movement.* The dynamics of movement of the center of gravity

---

* The Borelli stand is a prism embedded in wax. One corner of the prism fits into a groove in the middle of a Plexiglas sheet. This keeps the sheet horizontal. The animal fixed on the stencil is placed on the sheet, and by pulling the stencil in different directions, the sheet is brought into balance, and two points corresponding to the line of the prism's corner are drawn on the stencil. The animal is then turned around and balanced anew. A second straight line is thus plotted. From the intersection of these lines the position of the animal's center of gravity is found and is plotted on the stencil.

itself was established from frame-by-frame cardboard models of the
moving animal.* The models were placed on a background analogous to
that used when the animals were photographed, and the main indexes of
the change of position (speed, acceleration, angles of displacement relative
to the horizontal plane, etc.) were determined from the projection of the
centers of gravity. This done, X-ray photographs were taken of the animal
fixed in the position of each frame. The skeleton was then sketched in on
all the stencils, and its position corrected according to the series of X-ray
photographs. All the above-mentioned means of obtaining initial data for
a biomechanical analysis made it possible to gain a fairly clear idea in
quantitative terms of the different mechanics of movement of different
species of mammals. The mechanics of movement in turn can be a basis
for understanding the structure of animals, and therefore for our morphological studies we also had to find ways of obtaining quantitative characteristics of the organs of movement, enabling us to discern the relationship
between the structure of the skeleton and muscles and the mechanics of
movement.

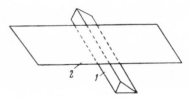

FIGURE 4. Borelli stand:

1) Duralex prism; 2) Plexiglas sheet.

The procedure for comparing the characteristics of muscle work in
different species of animals consists in determining their relative weight
and the lever arms corresponding to the application of the force. For this
purpose all the leg muscles and some of the trunk muscles were weighed
and their percentage weight to the total weight of the leg muscles was calculated. In cases where marked differences were observed in the structure
of the muscles in compared animals, the appropriate corrections were
introduced (all such cases are mentioned in the text). To determine the
levers of the application of the muscle force, stencils of the bones were
made and the outlines of the insertion of each muscle were plotted on them.
A perpendicular was then dropped from the center of movement in the joint
onto the line of pull of the muscle. Because of the linear insertion of the
muscle two perpendiculars were usually dropped. Measurement of their
length yielded two values of the lever of the application of the muscle force;
these were averaged or, if the muscle was of uneven thickness, additional
corrections were introduced for the mean.

* As a rule, the center of gravity of cardboard models, determined by the usual method, proves to be within
  the margin of error obtained when determining the center of gravity on the animal; this error depends on
  the difficulty of holding it in the position matching the outlines of the frame.

In muscles showing clear differences of a structural nature the angles
of pinnation were measured. For this a protractor was placed along a
tendon and a thread attached in the middle of the protractor was drawn
along the muscle fiber. The angle at which the force was applied to the
bone was also determined in a number of muscles. Here the protractor
was laid alongside the bone and a thread was drawn along the muscle fiber.

In all cases where more than three specimens of muscles were analyzed
(Table 1) the indexes (the percent weight of the muscle to the total weight
of the leg muscles) were processed statistically. It was found that for all
muscles the index of which exceeded 0.5, the coefficient of variation was not
more than 15% and was generally less than 10%. Therefore, when comparing the indexes of the muscles of different species, differences of more than
a factor of 1.2 were considered sufficiently objective. Less pronounced
differences were not used.

Various measurements of the bones were made and indexes calculated
for a study of the structural features of the skeleton. In measurements of
the limb segments the maximum length of the bones was determined in all
cases except those specified. Apart from this, the length of the lumbar and
thoracic regions of the spine was measured in all species to obtain relative
values for the limb segments. On selected skeletons all the bones were
preliminarily joined together with modeling clay. The procedure was as
follows. A dab of clay was stuck on the head of each corpus vertebrae and
the vertebrae were then stuck together so that the articular processes
formed natural-looking joints. This increased the measurable distance by
approximately the thickness of the intervertebral cartilage. If modeling clay
interlayers were not inserted, the natural position of the joints could not be
restored.

To illustrate the changes in the length of the limb segments occurring
upon specialization for different types of running it seemed best to use the
ratio of the length of the segments to the sum of the lengths of the thoracic
and lumbar regions of the spine. The length of the spine was taken to be
relatively stable in the phylogeny of these animals, that is, it becomes longer
and shorter in proportion to the increase in the size of the animals, irre-
spective of the increase in the length of the limbs.

Specialization for swift running involves the expenditure of a great amount
of energy, and in order to supply this energy the rate of metabolism has to
be stepped up. This probably dictated the need for increased size of lungs,
heart and digestive tract. All this must have brought about an increase in
the length of the spine. On the other hand, to increase the length of jumps
and speed of running, it proved more advantageous for the center of gravity
to be situated nearer the hind, jumping limbs, and this necessarily called
for a shortening of the spine. These two contradictory influences apparently
determine the relative stability of the size of the spine. Calculation of the
indexes to the sum of the lengths of segments, as is adopted in many works,
seems a less convenient method and, which is more important, a less com-
petent one, as it is impossible to detect the general trend of elongation of
the limbs. Furthermore, the increase in size of any segment leads with this
method to a decrease in the relative dimensions of the other two segments,
and this in turn incurs additional errors in the calculations. For example,

in connection with what is known as the arithmetic mirage,* a comparison of the lengths of the segments in the fore and hind limbs led to the conclusion that the forelimbs undergo a process of reduction in animals adapted to bipedal ricocheting. The general trend of shortening or lengthening of segments is more easily perceived by calculating the indexes to the length of the trunk or the trunk region of the spine, but some useful relationships are ascertained better by calculating the indexes inside the limbs. Such ratios are especially useful when comparing one's data with published findings. For instance, numerous paleontological studies of the length ratios of segments are generally based on calculation of the indexes to the sum of the lengths of all three segments of the limbs or to the ratios of the segments (Gregory, 1912; Osborn, 1936, 1942; and others). A similar calculation procedure has been used in studying recent animals (Howell, 1944; Egorov, 1955; and others). Apart from this system of calculating indexes, the indexes of the ratio of the segments to the length of the trunk or the ratio of individual segments to each other, etc., are often used (Böker, 1935; Bryant, 1945; Sokolov, Klebanova and Sokolov, 1964; Polyakova, 1965; and others).

---

* An arithmetic mirage is produced when unreal numerical data are obtained due to the initial methodical error.

*Chapter 2*

## GAITS OF MAMMALS

### HISTORY OF THE STUDY OF LOCOMOTION

In the study of locomotion knowledge of gaits has reached an especially high level. By gaits we mean the way animals move, determined by the sequences of both the leg movements and the successive stages in the cycle. In theory, six sequences of limb movements can be visualized: 1) symmetrical diagonal (Figure 5, A), when movement of a forelimb is followed by that of the contralateral hind limb; 2) symmetrical lateral (Figure 5, B), when movement of a forelimb is followed by that of the ipselateral hind limb*; 3) asymmetrical diagonal direct (Figure 5, C); 4) asymmetrical diagonal converse (Figure 5, $C_1$), when movement of the two forelimbs is followed by that of the hind limb contralateral to the last forelimb; 5) asymmetrical lateral direct (Figure 5, D); 6) asymmetrical lateral converse (Figure 5, $D_1$), when movement of the two forelimbs is followed by that of the hind limb on the same side as the last forelimb.** In addition, there are bipedal gaits which, as will be shown below, can arise on the basis of symmetrical and asymmetrical gaits.

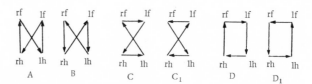

FIGURE 5. Six possible sequences of leg movements:

A, B) symmetrical; A) diagonal, B) lateral; C, D) asymmetrical: C, $C_1$) diagonal; D, $D_1$) lateral; C, D) direct, $C_1$, $D_1$) converse; rf) right foreleg; lf) left foreleg; rh) right hind leg; lh) left hind leg.

---

* The terms "diagonal and lateral symmetrical sequences" were coined long ago, although they do not actually convey the differences between them. The definition "lateral" may even lead one astray, as it stresses a special connection of movement of the limbs on one side.
** Gaits based on the first two sequences are called symmetrical and on the other four asymmetrical. This classification of gaits was adopted by Croix (1929), Bourdelle (1934), Howell (1944), and Sukhanov (1963, 1967, 1968).

Locomotion of terrestrial quadrupeds has a long history of study. Concepts such as trot, gallop and rack existed much earlier than their scientific characterization. It may be said that a precise recording of these movements began with an aural study of the pounding of horses' hooves, on the end of each one of which a bell with a specific ring was attached (Goiffon and Vincent, 1779). With the aid of this device the first concepts on gaits of different speeds (two- and four-time trot and rack, three- and four-time gallop, etc.) were developed. Goiffon and Vincent presented a graphic scheme of gaits in the form of four lines each of which corresponds to one limb. The incidence of footfalls and their rhythm, even or uneven, are marked on the lines.

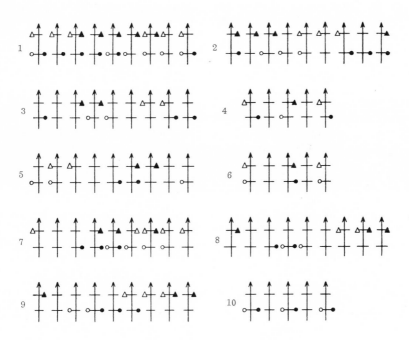

FIGURE 6. Graphic representation of gaits (after Muybridge, 1887):

1) walk; 2) amble; 3—4) trot; 5—6) rack; 7) canter; 8) rotatory-gallop; 9) transverso-gallop; 10) ricochet. Other explanations in the text.

Marei (Marei, 1875; Marey, 1894) devised a more perfect way of recording movements with the aid of the automatic recording apparatus he invented (Marei's pneumatic automatic recorder), which enabled him not only to determine the sequence of the incidence of footfalls but also to estimate the duration of support and free transit. He specified the characteristics of the walk, trot, rack, three- and four-time gallop and established the duration of support and suspension of the limbs with these forms of locomotion and also the dynamics of the pressure forces of the hooves on the ground. His graphic representation of gaits is a considerable improvement over the line method used widely before him (Goiffon and Vincent, 1779). All this permitted Marei to revise his predecessors' views on gaits.

Gaits were studied at the same time as by Marei, but by another method, by Muybridge (1887), who analyzed types of animal movements by snapshots. By drawing threads connected to the shutters of a series of cameras across the line of movement of an animal, he achieved successive snappings performed by the animal itself. This study of series of shots of animals in motion enabled him to give a detailed description of various gaits. Muybridge paid particular attention to the stages of support. By stage of support he meant the part of the cycle characterized by a certain combination of limbs present on the ground at that moment. Muybridge suggested that each stage be depicted in the form of an arrow (the trunk) which the lines of the limbs meet perpendicularly. If a line ends in a circle (for the hind legs) or a triangle (for the forelegs), it means that the limb is on the ground at this stage (Figure 6). The absence of a circle or triangle at any stage indicates that the relevant limb is in the air. He also demonstrated that in the cycle of quadrupedal four-time movement there must be eight stages of support, arising in connection with the fact that each limb has phases of support and free transit in the cycle of movement. During the work of any pair of limbs in unison, the number of stages of support in the cycle is also reduced.

Muybridge described eight gaits: walk, amble, trot, rack (or pace), canter, transverso-gallop, rotatory-gallop and ricochet. He considered all other types of movement as divergences from the norm. With the walk there is a sequence of three-point and two-point (lateral and diagonal) support stages (Figure 6, 1). A four-point support stage sometimes appears or the two-point stage of lateral support is omitted in the walk. With the amble (fast pace in Sukhanov's terminology — 1963, 1968) the three-point support stages are replaced with one-point support stages (2). Muybridge described two variants of the trot and rack (pace). In the first there is a switch from one- and two-point support stages back to one-point support with a stage of flight in the air.* In the second variant two-point support stages alternate with flight in the air. The rack (5, 6) differs from the trot (3, 4) in the two-point support stages, which are diagonal for the trot and lateral for the rack. With the canter, there is a sequence of stages of support in the first half of the cycle as with the walk, and in the second half of the cycle as with the amble. In the second half of the cycle one of the two-point support stages is sometimes replaced with a stage of flight in the air (7).

The gallop is characterized by alternate propulsion with the hind legs, after which there is a stage of flight in the air, ending in landing on the forefeet, the thrust of which leads again to flight in the air. If after the first stage of flight landing is achieved on the foot opposite the last one to push off, the gallop is called diagonal (transverso-gallop; 9), but if landing is on the foot of the same side, it is lateral (rotatory-gallop; 8). The first stage of flight is often superseded by support on one hind and one forefoot. Muybridge also described a bipedal gait, the ricochet (10).

---

* The concept of "flight" incorporates the physical meaning of free transit of the body, thrown at an angle to the horizontal. This has nothing to do with the active flight typical for birds.

Thus, the end of the last century marked the arrival at a definite stage in the study of gaits, during which time accurate descriptions of a whole series of gaits appeared and two main trends of graphic representation were worked out: the system of recording on bars (Goiffon and Vincent, 1779; Marei, 1875) and the recording of the system of support in the form of eight stages (Muybridge, 1887). Our notions on gaits were greatly enhanced by the work of Howell (1944), who perfected the characterizations of gaits. Like Muybridge, Howell devoted special attention to their eight-measure graphic representation. He also simplified Muybridge's method; instead of plotting nine (the ninth repeats the first) vertical arrows with light and dark triangles and circles, he drew eight horizontal lines with circles above and below, corresponding to the left and right limbs, respectively, which are at that moment on the ground (Figure 7). Apart from the above-mentioned footfall formulas he introduced support formulas characterizing the change of the number of supporting limbs in the consecutive stages of support in one cycle (e. g., for the walk the sequence of two- and three-point support stages is expressed as 3—2—3—2—3—2—3—2). Howell was the first to show clearly that acceleration of movement leads to a decrease in the number of limbs supporting the body at the same time. He was the father of the concept on symmetrical and asymmetrical gaits and described many of them.

For further progress in classifying and elucidating the fundamentals of symmetrical gaits we are indebted to Sukhanov (1963, 1967, 1968), who introduced two very important concepts, the rhythm of locomotion and the rhythm of limb work, and also made some other improvements in gait classification (see below). A very similar system of symmetrical gaits was constructed independently by Hildebrand (1963, 1966), and his scheme found prompt application in the analysis of symmetrical gaits in various groups of mammals (Hildebrand, 1967, 1968; Dagg and Vos, 1968).

In parallel with this trend of research into the characteristics of gaits another system of describing movements developed (Hesse, 1935; Böker and Pfaff, 1931; Croix, 1932, 1934, 1936; Hatt, 1932; Böker, 1935; Krüger, 1958; Casamiquela, 1964; and others). None of these authors gave graphic systems of recording the cycle of movement. They proposed terms which should, it seems, have brought some clarity to the presentation, but because each author interpreted these terms differently, the picture became even more confused. In view of the great similarity of the systems mentioned above in their basic principles, we will discuss in detail the points in the two extreme variants which have been developed the most fully (Hatt, 1932; Böker, 1935).

Böker suggested dividing all terrestrial species into three groups: Schreiten, Laufen and Rennen. The first group comprises all the slow-moving species and the second and third the swift-moving forms. Running (Laufen) differs from the full gallop (Rennen) in being slower. For instance, a trotting movement is running, while a gallop is running at full speed. In addition to this, in another functional grouping, Böker returns to saltatory movements over land, and here too he singles out three groups: Laufspringen, Hindernisspringen and Hakenspringen; in his characterization of these he

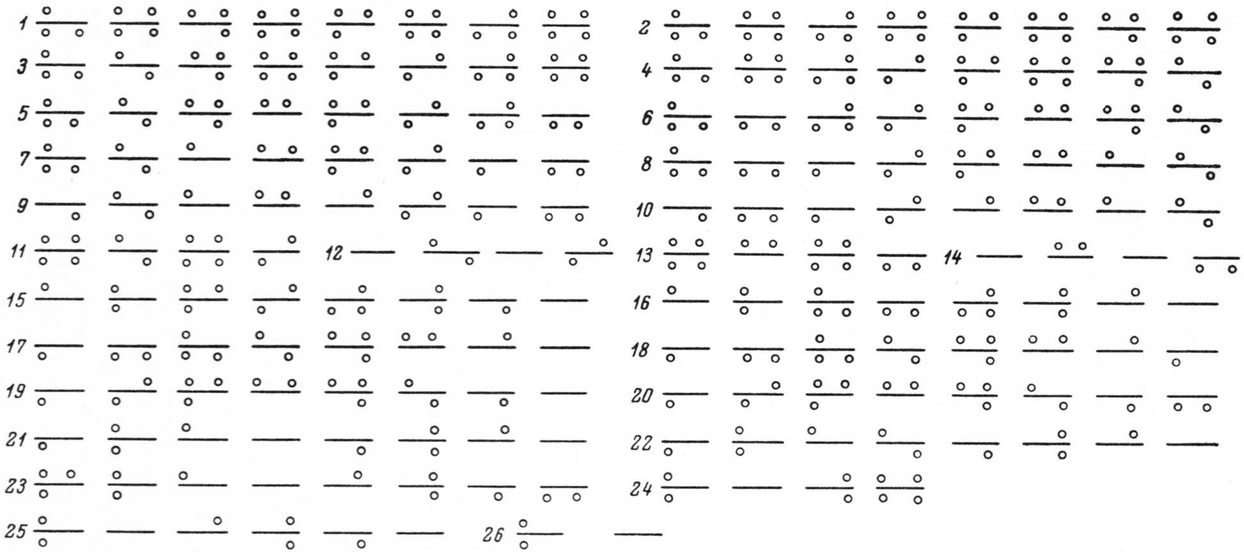

FIGURE 7.  Graphic representation of gaits (after Howell, 1944):

1) diagonal crawl; 2) lateral crawl; 3) slow diagonal walk; 4) slow lateral walk; 5) diagonal walk; 6) lateral walk; 7) diagonal slow running walk; 8) lateral slow running walk; 9) diagonal fast running walk; 10) lateral fast running walk; 11) slow trot; 12) fast trot; 13) slow pace; 14) fast pace; 15) transverse four-time canter in typical gallop sequence; 16) lateral four-time canter in typical gallop sequence; 17) transverse four-time canter in typical gallop sequence; 18) transverse four-time canter having no period of suspension; 19) lateral four-time canter in atypical gallop sequence; 20) lateral four-time canter having no period of suspension; 21) transverse fast gallop; 22) transverse slow gallop; 23) lateral fast gallop; 24) bound; 25) half-bound gallop; 26) ricochet.

intermingles morphological features of the structure of the animals with
a superficial description of movement.  The description of movement is
in practice generally substituted by photographs or a scheme which contributes little.

Hatt pointed out the considerable muddle in the terminology describing
movement and proposed his own set of terms.  He places them in two
groups:  saltation — spring, ricochet and hop, and bipedalism — walk, run,
hop and ricochet.  Saltation is achieved in two ways:  quadrupedal (spring)
and bipedal (ricochet and hop).  Bipedal movement may alternate between
slow walking and fast running or it may be paired:  slow (hop) and fast (ricochet).  In animals with weak forelimbs, the forefeet and then the hind feet
land on the ground on one spot after flight in the air, and a second stage of
flight is not observed.  Here the hind feet often touch down before the forefeet leave the ground, which leads to the appearance of a quadrupedal stage
of support.

The spring is quadrupedal saltation, in which the hind legs work to push
off while the forelegs take the weight of the body.  Hatt takes one of the
varieties of the spring to be the rotatory gallop, which has two variants:
the gallop of animals with strong forelimbs and that of animals with weak
forelimbs.  Both Böker's scheme (1935) and Hatt's (1932) suffer from lack
of clarity in the descriptions of movements, making it hard to define the
forms of locomotion of mammals.  For this reason we believe that these
schemes do not show any advantages over the classifications of gaits proposed by preceding investigators.

SYMMETRICAL GAITS

Symmetrical gaits are characterized first by the sequence of movement
of each limb, which can be defined by the fact that movement of a forelimb
is necessarily followed by that of either of the hind limbs, after which the
next fore and the next hind limb move;  secondly, typical for most symmetrical gaits (excluding the canters) is symmetry of the two halves of the
cycle, i. e., the right side of their graphic representation is the mirror
image of the left side (Figure 8).  On an animal in motion the picture obtained is that of the movement of the legs in the first half of the cycle being
repeated by the symmetrically arranged limbs in the second half.

The most comprehensive scheme of symmetrical gaits is that propounded
by Sukhanov (1963, 1967, 1968).  Retaining the best aspects of the graphic
systems of recording gaits, Sukhanov (1966, 1968) supplemented the support
schemes of Muybridge and Howell with a system of crosshatched bars taken
from Marei.  These four bars are distributed in twos at the sides of the
footfall formulas.  The outer bars correspond to the hind limbs and the
inner ones to the forelimbs.  The crosshatched part of the bar denotes the
phase of support of the limb, the nonhatched part the phase of free transit.
During a switch from one gait to another, the duration of some support
stages is shortened or lengthened until these stages are omitted from the
cycle altogether.  This indicates the transition to another gait.

Sukhanov distinguishes the following symmetrical diagonal gaits: very slow walk, normal walk, fast walk, very fast walk, slow trotlike walk, slow trot, fast trotlike walk, fast trot, slow racklike walk, slow rack, fast racklike walk, fast rack, slow canter and fast canter. In all these gaits the time of support of the fore and hind limbs is equal, but if it is not, exterior modifications arise. If support on the hind limbs proves to last longer in the cycle than support on the forelimbs, symmetrical gaits arise with the hind limbs dominant, while if, on the other hand, support on the forelimbs lasts longer, symmetrical gaits appear with the forelimbs dominant. In addition, there is the very rare phenomenon of symmetrical lateral gaits. Theoretically, all the above diagonal gaits can be visualized in lateral sequence, but in fact a lateral sequence of movement occurs extremely rarely with symmetrical gaits.

Another of Sukhanov's achievements was that he showed the mutual relation between all the symmetrical gaits and established new concepts on the rhythm of limb work and the rhythm of locomotion. The rhythm of limb work is defined as the relation between the phase of support and the period of free movement of each limb, in other words, in the graphic form, the ratio between the hatched and nonhatched areas of the bar corresponding to a given limb. The rhythm of locomotion characterizes the temporal relationships in the joint work of all four limbs. It is defined as the ratio of the time intervals between the moments at which the feet land consecutively. However, the presence of gaits with either the fore or the hind limbs dominant makes it more convenient to give a graphic definition of the rhythm of symmetrical locomotion according to the ratio between the two sections in the graph of the cycle. The first is equal to the distance between the centers of support (or of free movement) of the limbs on one side. The second is equal to the distance from the end of the first section to the middle of free movement (or support) of the limb on the other side (Figures 8 and 9).

Sukhanov (1967) considers the basic symmetrical gait to be the very slow diagonal walk (Figure 8, *1*; Figure 9, *1*), identical with the crawl (Howell, 1944), which is characterized by a sequence of four-point and three-point support stages only, that is, the support formula is 4−3−4−3−4−3−4−3. With the norm* of this gait the rhythm of limb work is equal to 7:1 and the rhythm of locomotion is 1:1. In practice we can derive from this gait all the gaits associated with simple acceleration of movement and also all their modifications toward trots and racks.

Acceleration is reflected in the first place in the relation between the time of support and that of free transit of the limbs: the shorter the former in relation to the latter, the quicker the movement. When the situation is reversed, and a foot takes off sooner and lands later, which is graphically expressed in an increase in size of the nonhatched parts of the bars backward and forward, the rhythm of limb work also changes. When this rhythm

---

* The arbitrary norm of a gait is the term attributed to a gait in which all the support stages of the cycle are equal to each other, which allows us, when calculating the rhythms of limb work and of locomotion, to use the relationship of the simple number of stages, not the ratio of the values expressed in absolute units (e.g. in seconds).

becomes equal to 3:1, four-point support stages disappear (Figure 9, 2).
Further acceleration leads to the appearance of normal walking (Figure 8, 2;
Figure 9, 3, 4;  Figure 10) (walk, after Howell, 1944), the norm of which
(Figure 9, 4) arises with a rhythm of limb work equal to 1.66:1. With this
gait a sequence now takes place with two- and three-point support stages
according to the formula 2−3−2−3−2−3−2−3. Acceleration of this gait
leads, with a rhythm of limb work equal to 1:1, to the disappearance of
three-point support stages in the cycle and then to the transition to fast
walking (amble, after Muybridge, 1887;  fast running walk after Howell, 1944).
with a rhythm of limb work equal to 0.66:1 (Figure 8, 3;  Figure 11). The
support formula of this gait is 2−1−2−1−2−1−2−1. Further acceleration
of this gait might lead to the appearance of very fast walking, a hypothetical
gait which has not been recorded in nature and in which there should occur
an alternation of support stages on one leg with flight in the air (Figure 8, 4).

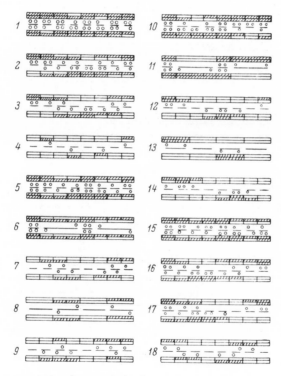

FIGURE 8. Graphic representation of symmetrical gaits:

1) very slow diagonal walk;  2) normal walk;  3) fast walk;  4) very fast walk;  5) slow trotlike walk;
6) slow trot;  7) fast trotlike walk;  8) fast trot;  9) fast trot with the hind limbs dominant;  10) slow
racklike walk;  11) slow rack;  12) fast racklike walk;  13) fast rack;  14) fast rack with the hind limbs
dominant;  15) slow trot with the hind limbs dominant;  16) slow canter;  17) fast canter;  18) lateral
fast trotlike walk. Progression is from left to right in all the graphs.

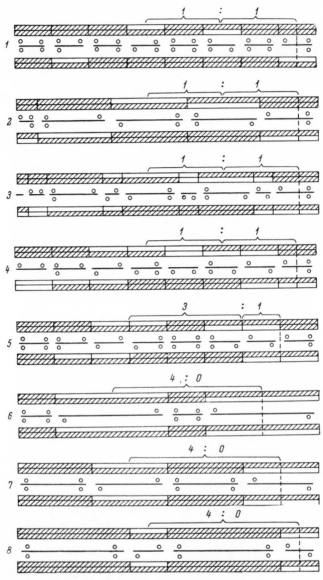

FIGURE 9. Diagram showing the changes of symmetrical gaits depending on the increase of the speed of movement (1–5) or a change in the rhythm of locomotion (6–8):

1) very slow diagonal walk; 2) switch from very slow diagonal walk to normal walk; 3) initial stages of normal walk; 4) norm of normal walk; 5) slow trotlike walk; 6–8) slow trot: 6) with rhythm of limb work 1.66:1, 7) with rhythm of limb work 3:1; 8) with rhythm of limb work 7:1. Figures on top of the diagrams denote the rhythms of locomotion; progression is from left to right in all the graphs.

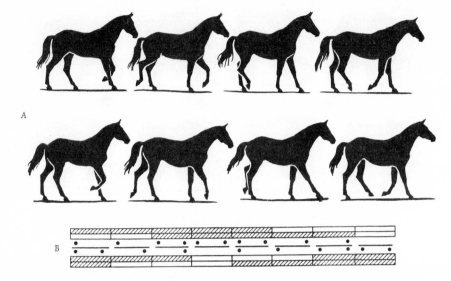

FIGURE 10. Normal walk of the horse (A) and its support graph (B)

All the changes of gaits discussed above take place with the maintenance of the same rhythm of locomotion, 1:1. However, acceleration may occur with a simultaneous change in the rhythm of locomotion. Such an eventuality may be graphically expressed in an increase in the nonhatched parts of the bars of the hind limbs on one side and of the forelimbs on the other side. Thus, if the hind feet begin to leave the ground earlier and the forefeet land later, which is expressed in the graph by an increase in size of the nonhatched parts of the bars of the hind limbs backward and of the forelimbs forward, even with a rhythm of limb work equal to 3:1 the norm of slow trotlike walk arises (Figure 12). But if the hind feet land later and the forefeet leave the ground earlier, represented by an increase in size of the nonhatched parts of the bars of the forelimbs backward and of the hind limbs forward, with a rhythm of limb work of 3:1 a slow racklike walk appears (Figure 13). In both these gaits with the same rhythm of limb work (3:1) and outwardly almost the same support formula (4—3—2—3—4—3—2—3 and 2—3—4—3—2—3—4—3)* the rhythm of locomotion and the nature of the two-point support stages change in different directions. With the slow trotlike walk the rhythm of locomotion becomes 3:1 and the two-point support stage is diagonal. With the slow racklike walk the rhythm of locomotion is 1:3 and the two-point support is lateral. Further acceleration with a change in the rhythm of locomotion in the same directions leads to the appearance of the slow trot (Figure 14) and to the slow rack with a rhythm of limb work equal to 1.66:1. In other words, the slow trot and the slow rack are in terms of speed more or less the same as the normal walk.

* It is readily shown that the two-point support stages of the slow racklike walk replace the stages of quadrupedal support of the slow trotlike walk.

FIGURE 11. Fast walk of the elephant (A) and its support graph (B)

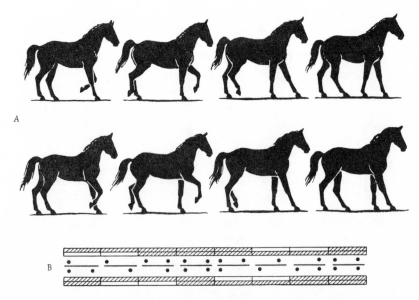

FIGURE 12. Slow trotlike walk of the horse (A) and its support graph (B)

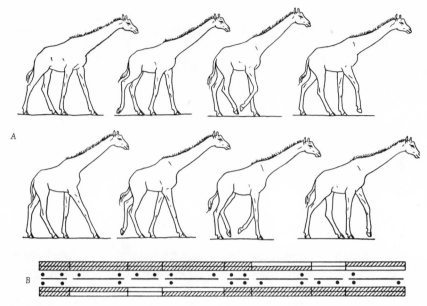

FIGURE 13.  Slow racklike walk of the giraffe (A) and its support graph (B)

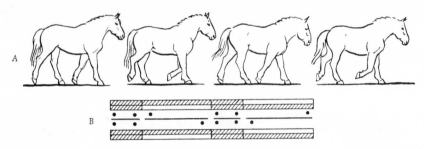

FIGURE 14.  Slow trot of the horse (A) and its support graph (B)

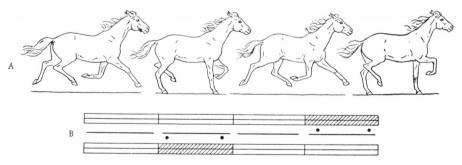

FIGURE 15.  Fast rack of the horse (A) and its support graph (B)

FIGURE 16. Fast trot of the horse (A) and its support graph (B)

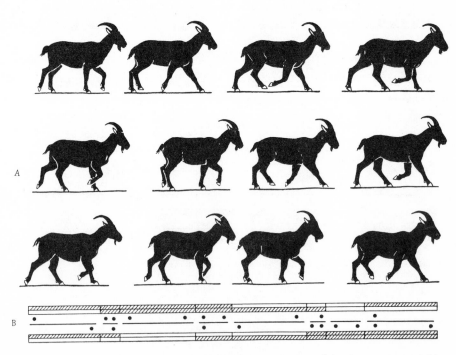

FIGURE 17. Normal walk of the goat with predominance of the hind limbs (A) and its support graph (B) (after Muybridge, 1887)

An analogous change in the rhythm of locomotion with an acceleration of the fast walk leads to the appearance of the fast racklike and fast trot-like walk and then to the fast rack (Figure 15) and fast trot (Figure 16). As in the case with the change from the very slow walk to the slow trotlike and slow racklike walk, a change from the fast walk to the fast trotlike and racklike walk can be imagined without any acceleration of movement. Nevertheless, when the rhythm of locomotion of the slow walk changes, there is generally a simultaneous speeding up of movement, while with a change in the rhythm of locomotion of the fast walk the speed of movement frequently remains unchanged.

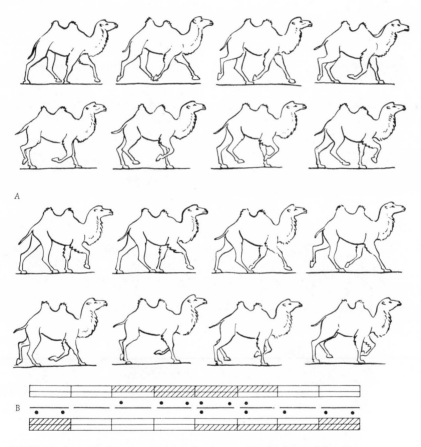

FIGURE 18. Fast rack of the camel with predominance of the hind limbs (A) and its support graph (B) (after Muybridge, 1887)

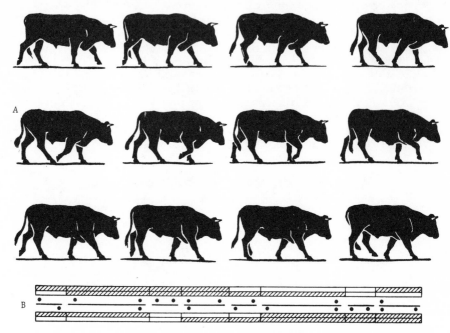

FIGURE 19. Normal walk of the ox with predominance of the forelimbs (A) and its support graph (B) (after Muybridge, 1887)

FIGURE 20. Slow canter of the horse (A) and its support graph (B)

A peculiar form of symmetrical gaits are the exterior gaits, the essence of which lies in the uneven work of the fore and hind limbs. If the support stages of the hind limbs have a longer duration than those of the forelimbs, the result are gaits with the predominance of the hind limbs (Figure 8, *9, 14, 15;* Figures 17 and 18). Similarly, there may be exterior gaits with the predominance of the forelimbs (Figure 19).

Unlike Howell (1944), Sukhanov (1967) classifies the canter also with the symmetrical gaits, noting rightly that slow canters are something of a combination of two gaits, the left part of the formula and of the footfall pattern corresponding to normal walk and the right part to the fast walk (Figure 8, *16;* Figure 20). This shift of gaits in the cycle leads to asymmetry of its two sides. Howell's attempt to bring together in his concept of the canter two gaits with a sequence of limb movements typical and atypical of the gallop should also be considered unsuccessful. The only thing common to these two gaits is the support formula $(3-2-3-2-1-0-1-2)$.

We should point out it is very rare to find in nature symmetrical gaits with a lateral symmetrical sequence of limb movements. A lateral trotlike walk* is known for primates (Figure 8, *18;* Figure 21). Gray (1944) made a detailed study of the stability of an animal's body during various sequences of limb work and showed that with a lateral symmetrical sequence of limb movement all the three-point support stages prove unstable; this is apparently why a lateral sequence of limb movement with symmetrical gaits is hardly ever met in nature.

The patterns governing the ratios of the rhythms of limb work and the rhythms of locomotion are represented in the graph of symmetrical gaits (Sukhanov, 1968). We see that each diagonal symmetrical gait can be represented by a point in the coordinate system, along one axis of which (the horizontal) the values of the rhythms of locomotion are plotted (Figure 22 shows the rhythm of the rack 0, of the half-rack 1:3, of the walk 1:1, of the half-trot 3:1 and of the trot 00) and on the second axis the values of the rhythms of limb work from 7:1 to 1:7). The area of the triangles shown in the graph characterizes the possible deviations from the norm for the given gait. Along the lines dividing the triangles the number of stages in the cycle decreases due to the deletion of some of them; this occurs during movement of a transitional nature between any two kinds of gait. The geometric center of each triangle characterizes the arbitrary norm of the gait. Let us take one case of rapid motion-picture photography of a giraffe. The full cycle of the animal's movement occupied 120 frames (filming lasted around two seconds at a speed of 64 frames/sec; the camera was a 16-SP). In the first eight frames the giraffe stands on its four legs, and then the right hind leg lifts up (Figure 13, second frame) and for the next 21 frames the animal stands on three legs, after which also the right foreleg rises (Figure 13, third frame) and for the next ten frames the giraffe stands on its two left legs. The right hind leg then comes down and again during 21 frames the animal has a three-point support stage, and then the right foreleg also lands on the ground, and for eight frames the four-point support stage is repeated;

---

* In this case the rhythm of locomotion is calculated from the fore to the hind leg on one side and from the latter to the diagonal foreleg.

then the whole sequence is repeated on the left side (Figure 13). Hence
the individual stages of support in the cycle of movement are laid down in
the following sequence: 8—21—10—21—8—21—10—21. Taking the number of
frames for each stage of support as the basis of the scale, we can map the
bars of reference for each limb (Figure 13, B), and then work out the rhythm
of work of each limb, equal in this case to 2.8:1, and also according to the
scheme presented above, the rhythm of locomotion, which is seen to be 1:1.8.
The rhythm of the limbs and the rhythm of locomotion intersect on the graph
of symmetrical gaits (Figure 22) in the area of the slow racklike walk, that is,
the same gait which was used by the giraffe. Actually, this motion diverges
somewhat from the norm of the gait in the direction of the walk.

FIGURE 21. Lateral fast trotlike walk of the baboon (A) and its support graph (B)(after Muybridge, 1887).
The last stage in the graph is the probable one.

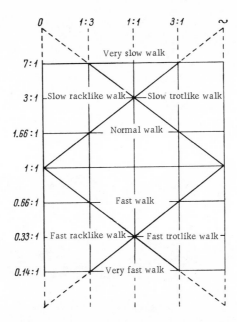

FIGURE 22. Diagram showing the changes in symmetrical gaits as a factor of the rhythm of limb work and the rhythm of loco-motion (after Sukhanov, 1968).

Top figures show the rhythm of locomotion, figures on the side the rhythm of limb work. Other explanations in the text.

## ASYMMETRICAL GAITS

Asymmetrical gaits differ from symmetrical gaits in the sequence of limb movement, the two forelimbs moving first and then the hind limbs. There is thus no symmetry in the cycle, because the movement of the limbs in the first half is never repeated by symmetrically arranged limbs in the second half.

An asymmetrical sequence of leg movement is sometimes also observed during the very slow walk (Figure 23). It would seem by analogy with the symmetrical gaits that we should have to analyze all the transitions from the very slow walk to the swift types of asymmetrical movement. Numerous observations in nature and analyses of motion pictures and published data show, however, that mammals almost always go over to symmetrical gaits when they slow down, and it did not seem worthwhile to discuss the exceptional cases. Apart from this, for clarity's sake it is more convenient to begin by describing the faster typical quadrupedal asymmetrical gaits: primitive ricocheting saltation (Gambaryan, 1955, 1960) and the gallop.

These two gaits differ both in the sequence of support stages and in the type of motion of the limbs in the stage of flight, which sets in after pushing off with the hind legs.   Such a stage is absent with some slow forms of primitive ricocheting saltation and gallop, making it rather difficult to distinguish these gaits.   As with the symmetrical gaits, the sequence of change of different forms of asymmetrical gaits depends on the speed of locomotion; during the gallop a decrease in the number of simultaneously supporting limbs is observed in the consecutive stages of the cycle as motion is accelerated, whereas with ricocheting saltation the number of supporting limbs is likewise reduced at the beginning but then increases again.

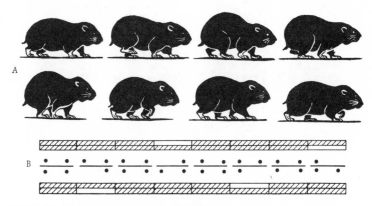

FIGURE 23.  Very slow asymmetrical walk of the hamster (M e s o c r i c e t u s
b r a n d t i) (A) and its support graph (B)

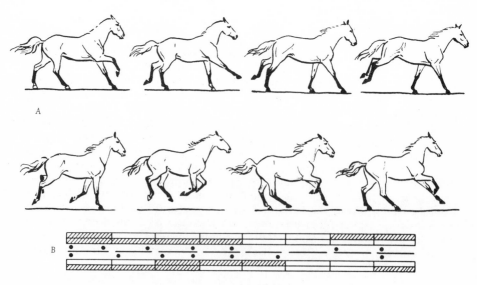

FIGURE 24.  Slow gallop of the horse (A) and its support graph (B)

FIGURE 25.  Slow lateral gallop of the bear (Ursus arctos) (A) and its support graph (B)

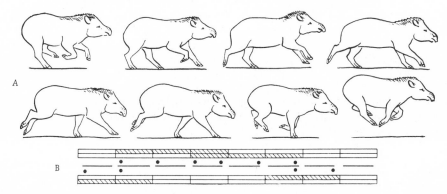

FIGURE 26.  Heavy lateral gallop of the tapir (Tapirus americanus) (A) and its support graph (B)

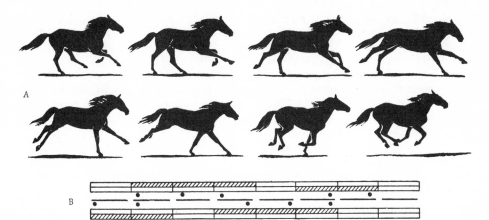

FIGURE 27.  Heavy diagonal gallop of the horse (A) and its support graph (B)

All forms of the gallop can be classed as alternate, when each of the fore and hind limbs rises and lands independently of each other, or paired, when the hind limbs and sometimes also the forelimbs work in unison.

There are five known forms of alternate gallop, depending on the speed of movement.* The slowest form, the slow gallop (Figures 24 and 25) (according to Howell, 1944, this is a canter with the sequence of leg movement typical for the gallop), is often called the manege gallop in horse breeding. Its support formula is $1-2-3-2-3-2-1-0$. After the stage of flight landing is achieved on one, two and three feet successively. If the order is hind foot, hind foot and then the contralateral forefoot, the gallop is called diagonal (Figure 24) and when it is hind foot, hind foot and then the ipselateral forefoot, it is called lateral (Figure 25). This three-point support stage ends with the lifting of the first hind foot and then the second forefoot touches down. After the second three-point support stage the hind foot and then both forefeet push off one after the other, after which a stage of flight ensues. This moment is called the stage of crossed flight, as the animal moves with its legs very close together or even crossed. With acceleration of movement the first hind foot manages to lift off before the first forefoot lands, and the second hind foot lifts before the second forefoot lands, which leads to the disappearance of three-point support stages. This is called a heavy gallop (Figures 26 and 27) and its support formula is $1-2-1-2-1-2-1-0$. It has been found that for horses the speed of movement during the slow (manege) gallop is appreciably slower than record speeds for the trot and rack. Record speeds over a distance of 1 km have been set by horses running at a heavy gallop.

Further acceleration of the gallop leads to the light gallop, with the formula $1-2-1-0-1-2-1-0$. Here the forefoot lands after the hind foot has lifted (Figure 28), so that there is a second stage of flight in the cycle. This is called the stage of extended flight, as the forelimbs are at this time stretched forward and the hind limbs backward. Such a gallop is frequently called the full gallop in horse breeding. Sprinting records have been set by horses running at a full gallop. Whereas for horses the light gallop indicates the extreme degree of tension, as it arises only during very fast running over a short distance, for small ungulates and most carnivores this is the usual high-speed gait (Figure 29). Further acceleration leads to the fast and very fast gallop. With the fast gallop, formula $1-2-1-0-1-0-1-0$, the second forefoot comes down after the first forefoot has lifted, and therefore there is a third stage of flight. With the very fast gallop, a stage of flight ensues after the thrust of each limb. The support formula assumes the aspect $1-0-1-0-1-0-1-0$. According to its formula, this gallop is identical with the very fast walk, a hypothetical symmetrical gait.

The very fast gallop was recorded only once, from tracks on loose solonchaks of the Badkhyzskii State Reservation. Very clear traces of Pamir argalis allowed us to determine the prints of each hoof. A little to the side

---

* Each of these forms may have a lateral and diagonal sequence of leg movement. And both these may in turn be direct or converse. The latter differ in the limb beginning (leading) the cycle. After five or six cycles the leading limb is generally replaced with another, and therefore there is no further need to discuss converse asymmetrical sequences.

of the group ran a young argali; the distance between its tracks increased gradually, and it went through several cycles of 230-cm jumps after the thrust of each limb. These jumps are clear evidence of the presence of a stage of flight after the lift of each leg.

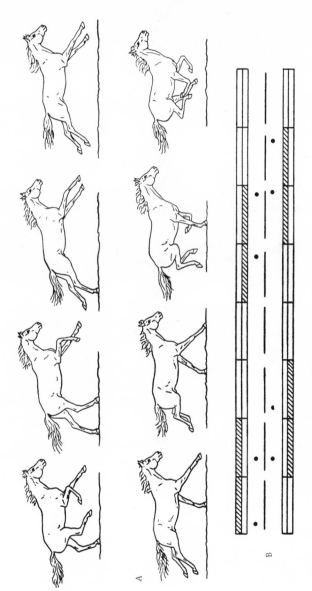

FIGURE 28. Light diagonal gallop of the horse (A) and its support graph (B)

FIGURE 29. Light lateral gallop of the cheetah (Acinonyx jubatus) (A) and its support graph (B) (after Hildebrand, 1959)

The half-paired gallop (half-bound, after Howell, 1944) (Figure 30), with the support formula 2—0—1—2—1—0, is characterized by the fact that after the simultaneous thrust of the hind limbs there is a stage of flight, ending in landing on the forefeet successively. A thrust of the forelimbs then makes for a second flight. Thus, the half-bound is identical with the light gallop, in which the hind limbs work in unison. With the paired gallop, (bound, after Howell, 1944), formula 2—0—2—0 (Figure 31), the fore and hind feet push off in unison. Howell considers that the bound arises in cases when mammals have to run under very difficult conditions — deep snow, viscous mud, etc. But in fact the bound is the usual fast gait of a number of small mammals (sousliks, weasels, pikas, etc.).

FIGURE 30. Half-bound of the hare (Lepus europaeus) (A) and its support graph (B)

FIGURE 31. Bound of the long-tailed Siberian souslik (Citellus undulatus) (A) and its support graph (B)

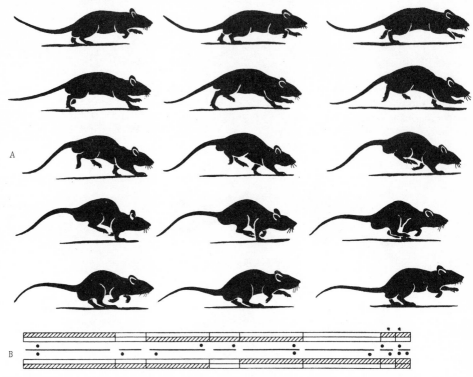

FIGURE 32.  Slowed-down primitive ricocheting jump of the Norway rat (Rattus norvegicus) (A) and its support graph (B)

Probable stages are marked by an asterisk on the graph.

Just as in the gallop, in primitive ricocheting saltation a number of different forms are distinguished in connection with the gradual acceleration of movement.  The slow primitive ricocheting jump (Figure 32) with the support formula 2—1—2—1—2—1—2—3 is characterized by the fact that the last forefoot leaves the ground after the first and even the second hind foot have landed.  After this three-point stage of support a fore and then also a hind foot take off, but the second hind foot does so only after a forefoot has touched down.  A certain acceleration of this movement leads to typical ricocheting saltation (Figure 33) with the support formula 1—2—1—0—1—2—1—2, which is similar to that of the heavy gallop.  However, with the heavy gallop the stage of flight in the air sets in following the push-off of the forefeet, whereas with primitive ricocheting saltation it ensues after the push-off of the hind feet.  After the stage of flight the animal lands on the forefeet consecutively, but the first hind foot manages to touch down before the second forefoot takes off.  Hence there is no second stage of flight after the forefeet leave the ground.  This gait is speeded up by an increased stage of flight after the push-off of the hind feet.  Since a forward deflection of the hind limbs in proportion to the length of the jump

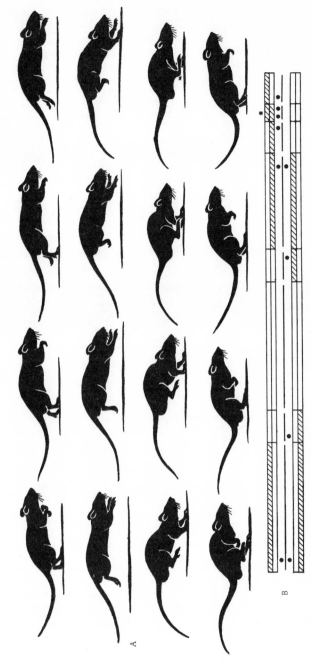

FIGURE 33.  Primitive ricocheting jump of the Norway rat (R a t t u s    n o r v e g i c u s) (A) and its support graph (B)

Probable stages are marked by an asterisk on the graph.

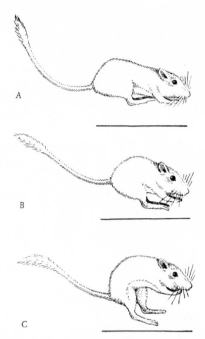

FIGURE 34. Increase in the degree of deflection of the hind limbs as a factor of the lengthening of the jump:

A) by 45 cm;  B) by 55 cm;  C) by 70 cm.

FIGURE 35. Primitive ricocheting saltation of the midday gerbil (Meriones meridianus) landing first on the hind and then on the forefeet

is typical for primitive ricocheting saltation, the hind and forelimbs gradually move closer together in the air (Figure 34), and at the next stage of
acceleration the hind feet land before the forefeet take off, so that a four-
point stage of support appears once more.   The support formula assumes
the following form: 2—0—1—2—4.   With a further increase in the size of
the jump, the hind limbs outstrip the forelimbs while in the air and landing
is accomplished on the hind feet a second time, after which the forefeet
follow (Figure 35).   When locomotion is speeded up still more and the size
of the jump increased, the hind limbs are drawn so far forward that they
wholly assume the function of support, since the forelimbs are no longer
touching the ground.   Running becomes bipedal (Figure 36).   Thus, the
appearance of bipedal running is associated with an increase of speed in
animals adapted to primitive ricocheting.   However, in forms specialized
for the bipedal ricochet, progression is achieved only on the two hind legs
even at low speeds of running.

FIGURE 36.  Bipedal ricochet of the mouselike hamster (C a l o m y s c u s   b a i l w a r d i)

The bipedal ricochet may be of three types: half-paired, paired or alternate.*  With the paired ricochet (Figure 37) the support formula is 2—0 (hop,
after Howell, 1944), and the hind limbs work in unison.   With the half-paired
ricochet the rhythm of the footfalls is somewhat disjunct, so that the formula
becomes 1—2—1—0 (Figures 38 and 39), while with the alternate form there
is a stage of flight after each foot leaves the ground (Figure 40).   The stage

* In giving the name running to all high-speed gaits which have a stage of flight, we unite under this term
  such forms of locomotion as the running of the kangaroo, jerboas, ungulates, carnivores, etc. It is cus-
  tomary when speaking about the kangaroo or jerboa to describe their movement as "running by jumps."
  But any type of running involves jumping. Running by jumps differs from the jump proper in that the stages
  of flight are more or less equal to each other in the successive cycles, while with jumping one stage of
  flight is much longer than the others. A jump may be from one spot, when it is begun from a position of
  rest, or on the run, when a number of preliminary jerks are performed prior to a major stage of flight.

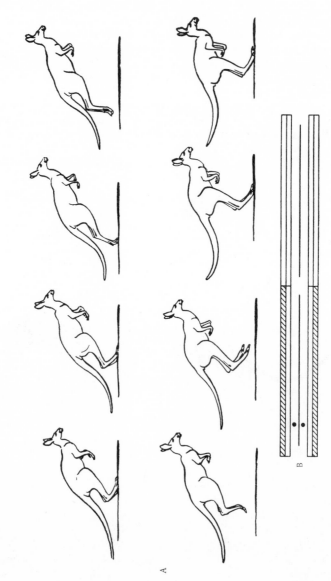

FIGURE 37. Paired ricochet of the kangaroo (Macropus rufus) (A) and its support graph (B)

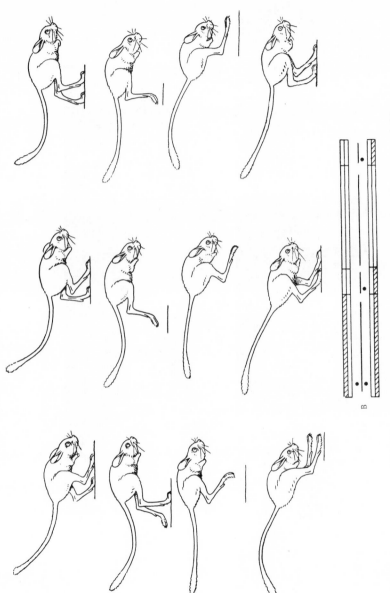

FIGURE 38.  Half-paired ricochet of Severtsov's jerboa (*Allactaga severtzovi*) (A) and its support graph (B)

FIGURE 39.  Different moments of the half-paired ricochet of the Northern three-toéd jerboa (Dipus sagitta).

Photo by I.M.Fokin.

FIGURE 40.  Alternate ricochet of the jerboa (Allactaga saltator) (A) and its support graph (B)

of flight is insignificant after the thrust of the first foot, but considerable after the thrust of the second. The support formula is expressed as 1—0—1—0.

The basic difference between the gallop and the ricochet manifests itself in the stage of flight, but these gaits can almost always be distinguished according to the support formula as well. If at the beginning of the stage of flight the degree of forward deflection of the hind limbs increases in proportion to the length of the jump, this is a primitive ricocheting jump. But if the degree of deflection decreases in proportion to the size of the jump, the gait is to be called a gallop.

In a case of similar gaits the features of the locomotion mechanics typical for the species must be studied, as otherwise it will be difficult to draw a line between the features which are associated with different gaits and those that are specific for the species. The scantiness of the starting data for comparative biomechanical investigations makes it most desirable to use miscellaneous material. We therefore give below a key to gaits (Gambaryan, 1967b) according to tracks and a procedure for making an approximate determination of a gait from individual photographs.

## Key to gaits according to tracks

1 (6).    Only tracks of hind feet present.

2 (3).    Footfalls distributed successively at an equal distance from each other . . . . . . . . . . . . . . symmetrical walk and run (Figure 41, *1*).

3 (2).    Footfalls not distributed at an equal distance from each other.

4 (5).    Footfalls distributed consecutively . . . . . . . . . . . . . . . . . . . . . . . . . . . . . . . . . . . half-paired and alternate ricochets (Figure 41, *2*).

5 (4).    Footfalls distributed in pairs . . . . . paired ricochet (Figure 41, *3*).

6 (1).    Tracks of both hind and forefeet present.

7 (8).    Footfalls alternate in the following order: hind—fore, hind—fore, or the hind footprint covers the fore footprint; the distance between the tracks on one side is always even . . . . . . . . . . . . . . . . . . . . . . . . . . . . . . . . . . . . . . . symmetrical gaits (Figure 41, *4—6*).

8 (7).    Footfalls alternate in the following order: hind—hind, fore—fore, or are distributed in pairs: hind feet—forefeet; if the tracks are distributed in another order, they are grouped in fours and each group is at a considerable distance from the preceding one.

9 (12).    Tracks of all feet distributed consecutively.

10 (11).    Footfalls of the right and left sides alternate . . . . . . . . . . . . . . . . . . . . . . . . . . . . . . . . . . . . . . . diagonal gallop (Figure 41, *7*).

11 (10).    Footfalls of the right and left sides do not alternate . . . . . . . . . . . . . . . . . . . . . . . . . . . . . . . . . . . lateral gallop (Figure 41, *8*).

12 (9).    The tracks of at least one pair of feet are set in pairs.

13 (16).    Footfalls of the hind limbs paired, of the forelimbs consecutive.

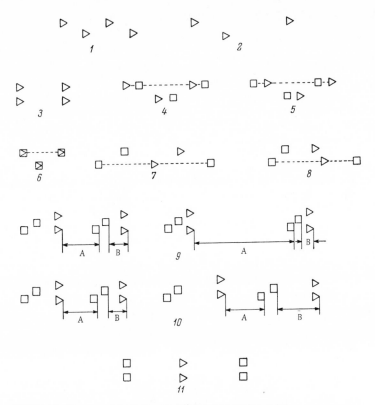

FIGURE 41. Diagram of footfalls with different mammalian gaits:

1) symmetrical walk and run; 2) half-paired and alternate ricochet; 3) paired ricochet; 4—6) symmetrical gaits; 7) diagonal gallop; 8) lateral gallop; 9) primitive ricochet at low and high running speeds; 10) half-bound at low and high running speeds; 11) bound. Triangles denote tracks of hind feet, squares tracks of forefeet. A) extended, B) crossed stage of flight; the dashed line shows the axis of the cycle.

14(15).    As a rule, the increase in the distance between the hind and the fore footfalls goes along with an increase in the distance between the tracks of the fore and hind feet . . . . . . . . . . . . . . . . . . . . . . . . . . . . . . . . . . . . . . . . . . . . . . . . . . . . . . half-bound (Figure 41,10).*

16(13).    Footfalls of both the hind and the forelimbs paired . . . . . . . . . . . . . . . . . . . . . . . . . . . . . . . . . . . . . . . . . . . . . . . . bound (Figure 41,11).

* Measurements should be carried out on obstacle-free areas, as an animal changes the nature of its jumping when clearing an obstacle, and the ratio of the distances between the tracks may prove atypical for the given gait.

## *Determining a gait from photographs*

We must have eight consecutive stages of support in order to determine a gait precisely. A single photograph usually allows us to judge three or four stages of support, but it still characterizes the gait quite accurately. We will give some examples. On the photograph in Figure 42 the lion is supported on its two right limbs, while the left hind foot has almost touched down. The right forelimb is approximately in the middle of the stage of support while the left is in the middle of the phase of transit and the right hind limb is approaching the end of the stage of support. These data are sufficient for a precise assessment of two stages of support and a very probable assumption of two more stages. The first, nonsupport stage is shown on the photograph (Figure 42, support graph 2), and the next definite stage is that of support on the two hind limbs and the right forelimb (3). We may now expect the right hind foot to leave the ground before the left forefoot lands (4). Finally, prior to the stage captured in the photograph, the lion was supported on the two forelimbs and the right hind limb (1). We proceeded from this to reconstruct four of the eight stages in the cycle and established that the lion was progressing by means of the normal walk. In the next photograph (Figure 43) we see a giraffe in the stage of crossed flight. It is seen that the left hind foot must be the first to touch down and the left forefoot will be the last to leave the ground. From this we infer that the giraffe was running at a lateral gallop. It is true that according to this photograph it is hard to judge whether the animal was snapped during a light or a heavy gallop.

FIGURE 42. Normal walk of the lion (F e l i s   l e o) (from Huxley, 1961).

The arrow shows the direction of progression. Other explanations in the text.

FIGURE 43. Lateral gallop of the giraffe (G i r a f f a  c a m e l o p a r d a l i s)
(from Huxley, 1964)

## REASONS FOR THE DIVERSITY OF GAITS IN MAMMALS

Undulating lateral flexures of the body, inherited from fish and synchronized with a symmetrical diagonal sequence of limb movement, became
fixed in the first terrestrial vertebrates by virtue of their advantageousness.  Lateral bending of the body is profitable because the length of each
pace is thereby increased with the same amplitudes of movement in the
limb joints (hip and shoulder joints) (Figure 44).  A diagonal symmetrical
sequence of limb activity is advantageous since all stages of three-point
support are in a state of stable equilibrium during the whole cycle, as in
this case a perpendicular dropped from the center of gravity does not extend beyond the limits of the triangle of support (Figure 45, A).  With a
lateral sequence of limb movement, on the other hand, stages of three-point
support prove unbalanced (Figure 45, B).

Emergence from the aquatic medium brought about the need to speed up
movement over land, and the primary form of locomotion (the very slow
diagonal walk) proved inadequate in this respect.  Terrestrial vertebrates
thus naturally had to find ways of moving more quickly.  As we have already
seen (p. 21), the direct method of speeding up the slow walk leads to the
appearance of the normal walk, in which there are two stages of lateral
support (Figure 10) which are maintained upon further direct acceleration

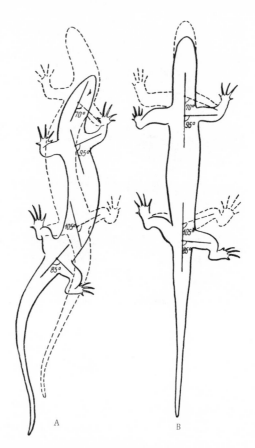

FIGURE 44. Increase of the length of a pace as a factor of body flexures at uniform pace angles in tetrapods:

A) pace with body flexures; B) pace without body flexures.

of movement. A stage of lateral support also occurs during racklike forms of the walk and during racks. These stages are particularly unstable when the limbs are straddled, the position characteristic for the lower tetrapods (amphibians, reptiles).* In these animals, therefore, acceleration of speed is possible only with a change in the rhythm of locomotion in the direction of trotlike gaits, in which a stage of lateral support is absent (Sukhanov, 1968). At the same time as a change in the rhythm of locomotion, a differentiation of limb functions may arise and develop in these animals during speeded-up movement over land. The hind locomotory limbs begin to assume the leading role, and the rhythm of their activity becomes somewhat

---

* The limbs are arranged in the segmental plane, when the femur and humerus are perpendicular to the median plane of symmetry.

different from that of the forelimbs.  The result are trotlike gaits and trots with predominance of the hind limbs.  It is on this basis that with further acceleration of movement the bipedal run of Rhynchocephalia arises (Sukhanov, 1963, 1967).  Bipedal running is typical for many recent lacertilians (Annandales, 1902;  Green, 1902;  Snyder, 1949, 1952, 1954).

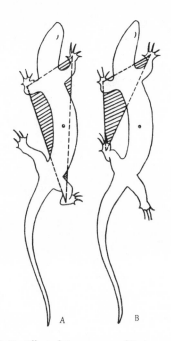

FIGURE 45.  Effect of the sequence of limb movement on equilibrium in the stage of three-point support in tetrapods:

A) symmetrically diagonal sequence — equilibrium is stable;  B) symmetrically lateral sequence — equilibrium is unstable.

Asymmetrical gaits are characteristic for high-speed locomotion among the Marsupialia and placental mammals.  Apart from this, unlike the lower tetrapods, all the mammals have the limbs arranged in the parasagittal plane.  Since both these aspects of structure and function of the limbs are inherent in all mammals,* we must assume that they appeared in their ancestors, which adapted to a mode of life for which an asymmetrical sequence and a parasagittal position of the limbs were unquestionably advantageous.

* Further, when talking about mammals, we will mean marsupials and placentals, not other groups (Monotremata, Multituberculata, etc.).

The mode of life of ancient mammals has interested all investigators studying the origin of this group of animals. Huxley (1880) believed that the predecessors of the Marsupialia were adapted to climbing trees. The same conclusion was reached by Dollo (1899a, 1899b), who made a detailed study of the structure of the hand in primitive marsupials. Bensley (1907a, 1907b) investigated the structure of the feet in recent marsupials and presents many arguments in favor of a primary arboreal mode of life for these animals. Matthew (1904, 1937) thinks that the climbing mode of life was the original one not only for the Marsupialia but also for all the placentals. A similar view is held by other authors (Gregory, 1913; Böker, 1926, 1935; Panzer, 1932; Simpson, 1937; Romer, 1945; Clark le Gros, 1949; and others).

The hypothesis that the climbing mode of life was primary for mammals stemmed from a study of the morphology of both fossil remains of ancient animals and the most primitive recent forms. It was found that in both of them the first digit (pollex and hallux) is situated opposite all the others and that in the forearm and tibia a well expressed adaptation for rotation is observed (rotatory mobility about the long axis of these bones). Both these properties are in fact a characteristic feature of recent highly specialized climbing animals.

Apart from this hypothesis, other points of view have been expressed. Greogry (1951) put forward the idea that the primitive mammals were eurybionts equally proficient in clambering, burrowing and jumping, and that only the direct ancestors of the Marsupialia had a climbing mode of life. Haines (1958) made a detailed comparison of the musculature and gross morphology of the hand of H e r p e s t e s  i c h n e u m o n with that of several other mammals and demonstrated that the development of the grasping function shows convergence in a number of groups of recent animals. In view of this, Haines concurred with Gregory that adaptation to climbing is secondary in the placental mammals. In our opinion, however, Haine's study confirms rather than contradicts the hypothesis on the climbing mode of life as being primary. As a matter of fact, the few remains that have been found of Jurassic mammals of the group Panthoteria, from which the recent marsupials and placentals probably descended, bear evidence rather that the first digits are placed opposite the others and that the tibia and forearm do rotate in these animals. In this case, convergent development of analogous characters in various groups of recent climbing mammals indicates that their ancestors also led a climbing life.

In order to accept the hypothesis that climbing was primary among the ancient mammals it is not enough to produce morphological facts; we must also prove that climbing must have led to an asymmetrical sequence of limb movement and a displacement of the limbs from the segmental to the parasagittal plane. We have already noted that all the recent mammals show these features of limb position and sequence of movement, and therefore these must have developed in their common ancestors.

It is obvious that climbing cannot bring about such changes in the position and sequence of movement of the limbs. The reason is as follows: we must distinguish between two forms of climbing trees — slow creeping from branch to branch and swift leaping from one branch to the other.

It is readily seen that slow creeping in an asymmetrical sequence does not provide any advantages over symmetrical movement, since the only outcome will be a shortening of the stride (Figure 46). Furthermore, a parasagittal position of the limbs does not make it easy to cling to branches. Fast leaping, on the other hand, is based on another type of movement, which apparently could not have been primary in the locomotion of mammals. This will be discussed in more detail (p. 58).

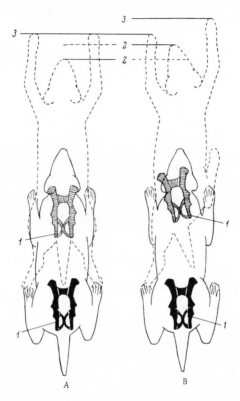

FIGURE 46. Effect of the sequence of limb movement on the length of a pace in tetrapods:

A) asymmetrical, B) symmetrical sequence; 1) pelvis; 2) tip of the snout; 3) hand.

The author has put forward another hypothesis (Gambaryan, 1967a) on the original semifossorial mode of life of mammals which, in contrast to the above, allows us to conceive of a process by which the parasagittal position of the limbs and the asymmetrical sequence of their movement arose. The morphological characters of fossil forms, which have generally served to prove that the climbing mode of life was primary, do not contradict this theory either.

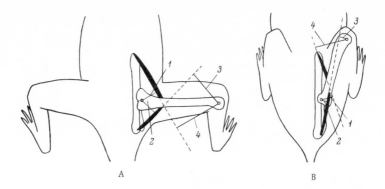

FIGURE 47. Effect of the position of the limbs on the muscular load during burrowing:

A) straddled position, B) parasagittal position of the limbs; 1, 2) arms of the levers
of the application of force in the hip joint; 3, 4) arms of the levers of the application
of force in the knee joint.

It is generally accepted that the ancestors of the mammals were small
animals the size of a rat or mouse, which fed mainly on invertebrates.
They probably lived in forests of horsetail, treelike ferns and primitive
Gymnospermae.  In seeking food they had to rummage among leaves, forest
litter and the upper layers of the soil, where invertebrates are most abundant.  Their hind limbs must have supported them, leaving the forelimbs
free to work at the same time or intermittently.  Support on the two hind
limbs destroys the primary symmetrical sequence of limb movement.
While raking through the leaves or the soil, the forelimbs move back and
forth or from the front to the side and then back.  In both cases the body
strains forward with an intensity proportional to the resistance of the foot
scratching on the substrate.  The tensing of the hind limbs counteracts the
forward movement of the body.  When the limbs are placed in reptile (straddled) fashion (Figure 47, A), all the resistance to motion derives from the
muscles attached to the shoulder, which is equivalent to a perpendicular
from the axis of motion in the knee joint to the direction of muscle pull.
With a parasagittal position of the limbs (Figure 47, B) the shoulder is
reduced in size and the bulk of the pressure passes through the bone-
ligament apparatus: the body becomes fixed with the minimum waste of
muscular energy.

Hence, both the phenomenon of an asymmetrical sequence of limb movement and the establishment of a parasagittal position of the limbs can be
satisfactorily explained by the fact that the ancestors of mammals had to
dig in order to obtain food.  The hypothesis we have proposed allows us to
acknowledge the possibility of diverse types of movement, although in the
author's opinion, the specific characters of mammals must have developed
in connection with digging activity, not directly with locomotion.

Since the original mammals lived surrounded by predatory reptiles,
their evolution must have taken a path whereby means of self defense were

worked out. Such means may have been swift fleeing from one refuge to another, perfection of fossorial ability, or adaptation to climbing. The parasagittal position of the limbs and the asymmetrical sequence of their movement were enough to provide for a change from the horizontal body flexures characteristic for reptiles to vertical flexures, giving the possibility for progression by jumps, during which flight through the air is achieved by pushing off with the hind limbs. Vertical jumps are not inherent in mammals alone. Amphibians (Figure 48), for example, jump in the vertical plane (Hirsch, 1931; Whiting, 1961), and the bipedal run of reptiles is also in fact a series of jumps in the vertical plane. But it is the mammals which have worked out the most economical and profitable method of running by means of vertical jumps. This is because in amphibians and reptiles the direction of movement of the thrusting end of the limb does not coincide with the line of flight owing to the straddled position of the limbs. Part of the thrusting force is uselessly damped by the opposite limb or is wasted on unproductive lateral bends of the trunk (Figure 49). Only in mammals, with the parasagittal limb position, does the trajectory of the vertical jump coincide with the direction of thrust of the proximal end of the femur, which makes for efficient utilization of the muscular force.

FIGURE 48. Frog jump (from Whiting, 1961)

With symmetrical gaits the limbs are carried forward immediately following the end of the phase of support, which makes for the necessary even rhythm of limb activity. The neuromuscular reflexes which make this steady movement possible may also be considered primitive. This primitive type of limb activity must originally have been retained in the asymmetrical gaits as well. Here, the hind limbs, providing the main thrust, manage to prepare for the following thrust while still in the air; this potentially promotes a separation of fore and hind limb functions. The hind limbs become propulsive and the forelimbs shock absorbers. Such a quadrupedal gait (Figure 33), when the hind limbs begin to progress forward immediately they have left the ground, is called by us "primitive ricocheting saltation" (see above, pp. 31—32 and 38—46).

When the speed of this gait is increased, each jump becomes longer, and the hind limbs are deflected farther and farther forward, which in the end results in a bipedal run. In the most specialized ricocheting animals support on the forelimbs may be deleted altogether from the cycle of locomotion, not only during swift running but also during slow movement by short jumps. This is one of the ways by which bipedalism arose, as is seen in the kangaroo jerboas and a number of other animals.

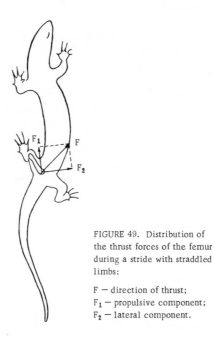

FIGURE 49. Distribution of the thrust forces of the femur during a stride with straddled limbs:

F — direction of thrust;
$F_1$ — propulsive component;
$F_2$ — lateral component.

The animals which mastered the primitive ricocheting jump must have become more mobile than their semifossorial ancestors, and therefore it is they which were able to begin a direct invasion of the most diverse biotopes. Some of them became better and better adapted to digging, others took to climbing trees, while still others emerged into the open, where they either adapted to swift running or specialized in digging refuge burrows. The forms which learned to climb must have known how to leap nimbly from branch to branch so as to escape from predators. Such leaps, however, called for doing away with the type of limb activity characteristic for primitive ricocheting saltation. The point is that during this kind of saltation the degree of forward deflection of the hind limbs in the air depends on the length of the jump. This is why landing must be performed on the hind legs when there is a large distance between branches, and on the forelegs when the distance is small. Such differences in the type of landing naturally make it difficult to coordinate a jump, to calculate it precisely and to prevent the body from falling. For this reason the climbing mode of life, involving leaping from branch to branch, led to a point where the animal began to inhibit the forward drift of the hind limbs while in the air (in the stage of free transit) and the hind limbs were drawn close to the forelimbs only after landing.

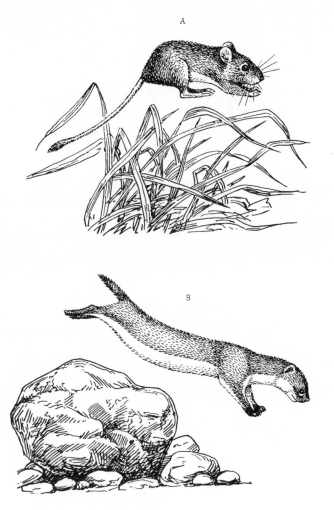

FIGURE 50. Jumps with different gaits:

A) primitive ricochet (the body is raised to the height of the obstacle plus the length of the limbs); B) gallop (the body is raised only to the height of the obstacle).

The secondary transition of climbing animals to the ground takes place with the new rhythm of activity of the neuromuscular apparatus preserved as being mechanically more profitable. The advantageousness of this new rhythm of limb work is manifested especially clearly when obstacles have to be overcome and stretched-back limbs (Figure 50, B) enable the animal to clear the obstacle without rising high and without having a steep flight trajectory. During the primitive ricocheting jump, the limbs are drawn down over the obstacle (Figure 50, A), which increases the steepness of the jump trajectory.

The new symmetrical gait arising in connection with secondary transition to the ground is called "gallop" (for the description of the gallop see pp. 31—37).* Here the activity of the limbs differs markedly from that during the primitive ricocheting jump. With the gallop the forelimbs not only cushion impacts but also effect a secondary thrust, with the result that a second stage of flight, crossed flight (Figures 24—31), appears in the cycle.

The incentive for the appearance of the gallop may have been not only adaptation to climbing but also, for example, the activity of leaping from stone to stone in animals which selected stony ground as their main habitat. An analogous change in limb activity may take place also in forest creatures, which come upon numerous obstacles while they are running. But whatever the reasons for the appearance of the gallop, it is difficult to go back from it to the primitive ricocheting jump, as then we would have to admit the possibility of a less advantageous gait appearing, and this contradicts the adaptive nature of phylogenesis.

In the light of the above we can readily understand why it is impossible to say that slow climbing and even jumping from branch to branch are primary forms of mammalian locomotion. Such means of locomotion inevitably lead to an inhibition of the forward deflection of the hind limbs right up to the point where landing is on the forelimbs, that is, bypassing the stage of the primitive ricocheting jump which is a property of many recent primitive groups of mammals (marsupials, insectivores and some rodents).

All these contradictions disappear if we recognize the terrestrial semi-fossorial mode of life of the ancestors of Marsupialia and Placentaria as primary. But in accepting this, we must find a new interpretation for the morphological characters that have been presented as proof of climbing being the primary mode of life. We have to find out why the primitive mammals had a higher mobility in the forearm and tibia and why their first digit is situated opposite the others. For it is just these characters which are typical for the recent climbing mammals and which were probably present in the ancestors of recent mammals.

There is no doubt that the mammals originated from the reptiles, with the straddled, not parasagittal, position of the limbs characteristic for this group. Under conditions of terrestrial locomotion the hand and foot of theriomorphic reptiles apparently became set parallel to the plane of symmetry of the trunk (Schaeffer, 1941). Animals with straddled limbs can maintain such a position of the hand and foot throughout the phase of support only if the forearm and tibia rotate. However, it seems that when movement is swift, the hand and foot turn outward at the end of the phase of support. The digit to touch the ground last is the first one, and as this is opposite the others it allows for the final thrust. It is interesting that in the recent

---

* The different forms of the gallop are described in many works, but the primitive ricocheting jump and gallop were strictly defined only in 1955 (Gambaryan, 1955).

Lacertilia, contraposition of the first digit is observed (Figure 51) even
in the presence of specialization for bipedal running (Snyder, 1954).*
The hand and foot can turn outward only when the forearm and tibia rotate.
Hence, both this rotation and the contraposition of the first digit must have
been present in the forerunners of the recent mammals, but these two features were preserved as a result of the wide-set positioning of the limbs
inherited from the reptiles, not as a consequence of adaptation to climbing.
With the primary semifossorial life led by the predecessors of mammals
both features were able to be retained for a long time, as rotation in the
forearm and a first digit opposite the others could be useful for digging.
This seems valid if we take into account that these two features are preserved even in highly specialized recent earth-diggers (Gambaryan, 1960).

FIGURE 51. Contraposition of the first digit (arrow) with
the bipedal run of Lacertilia (after Snyder, 1962)

    The fact that rotation in the forearm and the contraposition of the first
digit appear secondarily and independently in various groups of climbing
mammals (Haines, 1958) indicates that the predecessors of semifossorial
primitive animals went through a stage of living on the ground before they
took to climbing. Probably typical for these groups was adaptation to running by primitive ricocheting saltation, in which these structural features
of the stylo- and autopodia proved superfluous. Eventual transition to
climbing in many cases led to increased forearm mobility, but now on a
different basis and for other purposes.

* Studies of tracks of recent Lacertilia and Chelonia show that this turning-out of the limbs is expressed in
  the typical swirling of the imprints. A similar impression is obtained during a detailed analysis of motion
  pictures of caudate Amphibia.

A parasagittal position of the limbs, arising as a result of a primary semifossorial mode of life, brings about a situation where the stage of lateral support scarcely differs in its stability from the stage of lateral support. This is why a stage of lateral support may appear in the cycle of movement of mammals. Looking at the scheme of limb progression (Figure 52), we see that owing to the absence of a stage of lateral support, the hind leg is placed behind the foreleg in the slow trotlike walk (Figure 12) and slow trot (Figure 14). When the forelimb leaves the ground before the hind foot lands (stage of lateral support), the stride of the hind limb is increased (Figure 52). This aspect of locomotion becomes particularly important when long-legged forms are moving slowly. So as not to hinder the increase of the stride on account of the length of the legs, the forefoot must push off almost immediately after the hind feet have left the ground. As a result of this, the change in the rhythm of locomotion toward the trot, a change which is essential for the Rhynchocephalia, becomes unprofitable for mammals, and acceleration of their motion is not associated with a change in the rhythm of locomotion toward trotting, but in long-legged forms it even leads to a change of rhythm in the direction of racks. Thus, the slow racklike walk comes to be the normal gait of elephants and giraffes (Figure 13).

Interestingly, with asymmetrical gaits a lateral sequence of limb movement proves to be more usual than a diagonal sequence. Only heavy animals living in open spaces show adaptation to running at a diagonal gallop (large Bovidae and Equidae). However, even these animals, which normally run at a heavy diagonal gallop, return to the lateral gallop, according to Solomatin (1965), when jumping over obstacles; this was observed in an Asiatic wild ass leaping through the bush. Horses clearing obstacles present a similar picture (Muybridge, 1887, Plates 75, 76, 78).

The origin of bipedalism is one of the most important questions when dealing with phylogenetic changes in vertebrates. In the Rhynchocephalia, as we have noted above, the speed of motion over land is increased owing to a change in the rhythm of locomotion in the direction of the trot. The thrusting role of the hind limbs is enhanced at the same time, and trotlike walks and trots with the hind limbs dominant appear. With further acceleration this type of movement inevitably leads to the total detachment of the forelimbs from the ground and to the development of a symmetrical bipedal run (Sukhanov, 1968). In mammals bipedalism probably arose by various means.

Bipedalism in man is apparently connected with the primary function of the forelimbs, which is not associated with locomotion. Our rustic forefathers probably used their arms for gathering fruit and for building nests (like the recent anthropoid apes). These primary work habits freed the forelimbs of their locomotory task. With the transition to living on the ground the arms became specialized for various work processes and the legs became perfected in terrestrial locomotion (Nestrukh, 1957; Brull, 1962).

On the other hand, specialization for bipedal ricocheting saltation is the essential outcome of the accelerated primitive ricocheting jump. This acceleration leads to an increase in the size of jumps, which is possible

FIGURE 52. Increase in stride length due to the appearance of a stage of lateral support in mammals:

A) very slow walk; B) normal walk; $A_1$, $B_1$) tracks of animals with these gaits.

only if the hind limbs become more powerful and their deflection in the air increases with each jump up to the point where they begin to land before the forelimbs. More advanced specialization for bipedal ricocheting results in that running is accomplished only on the hind legs during small jumps as well.

It is puzzling why Howell (1944) considered that ricocheting appeared in connection with the need for small animals to cope with a large number of obstacles while running. Leaping over obstacles is difficult enough with the primitive ricochet and even more difficult with the bipedal ricochet. We should believe, rather, that the bipedal ricochet arises in animals which originally progressed by means of the quadrupedal primitive ricocheting jump because they could not camouflage themselves in the sparse grassy cover or because they became larger and were easily noticeable in the grass and hence had to have the possibility of escaping by swift running. In other words, adaptation to bipedal ricocheting arises both in dwellers of open spaces and in large animals. Small creatures develop a totally different type of escape from predators in the grass. Running with sharp turns, they avoid their pursuers by confusing them. In this situation any leaping about in the grass would expose them, which is obviously not desirable. Hence, the bipedal ricochet arose rather in connection with open habitats or with increased size of the animals.

*Chapter 3*

## BIOMECHANICS OF TERRESTRIAL LOCOMOTION

PHASES AND PERIODS OF LIMB ACTIVITY
DURING RUNNING AND JUMPING

The term "stage" (stage of three-point support, diagonal support, free flight, etc.) is used in describing gaits in order to define different moments of motion. There is no generally accepted terminology for defining the moments of motion of each limb. Free transit and support of the limbs are called sometimes "phases" (Kas'yanenko, 1947) and sometimes "periods" (Klimov, 1937; Donskoi, 1958; Zhukov et al., 1963; and others). In calling free transit and support of the fore and hind limbs phases,* we distinguish two periods in each phase. The phase of support has a preparatory period (of front support, or cushioning (Zhukov et al., 1963)) and a starting period (of hind support or thrusting), while the phase of transit comprises a drawing-up period (hind step) and an adjustment period (front step).** In the preparatory period of the phase of support, movements take place in the joints which provide the optimal conditions for the work of the locomotory organs in the next period. In the starting period the limbs impart an impulse to the body which brings about the appropriate acceleration of the center of gravity owing to which the animal moves. In the drawing-up period of the phase of free transit the limbs are drawn upward and forward, while in the period of adjustment they sink down again so that, placed on the ground, they allow for the next stage of support. Thus, in the phase of support the limbs act to propel the body forward, while in the phase of free transit they make ready for the next locomotory cycle.

Even though there are many different kinds of gaits and manifold types of mammals, the sequence of change of the flexor-extensor movements in the limb joints is very similar in the phases of support and free transit. The diversity of speeds and types of locomotion depends on the frequency of steps per unit time, the length of the limb segments, the amplitude of flexor-extensor movements in the joints and also the size of the pace angle.[†]

---

* For the forelimb the phase is of front support and transit and for the hind limb hind support and transit.
** The phase of free transit must not be confused with the stage of free flight of the whole animal, which occurs during some gaits when all four limbs leave the ground.
† The pace angle is formed by two straight lines from the point of support to the shoulder or hip joint at the beginning and end of the phases of front or hind support.

Flexor-extensor movements in the limb joints usually begin from the proximal joints and descend gradually in the distal direction. In the preparatory period of the phase of hind limb support, flexion takes place in the hip,* knee and talocrural joints and in the joints of the digits (Figure 53, *1* ). In the starting period of the phase of support the hip, knee and talocrural joints are extended and the toe joints flexed (*2*). In the drawing-up period of the phase of free transit the same movements which took place in the starting period continue in the limb joints at the first moment but they very soon change into flexion in the hip, knee and talocrural joints and also flexion in the toe joints (*3*). In the adjustment period of the phase of free transit the hip, knee and talocrural joints are extended and the toe joints flexed, in order that at the very end of the period all these movements will go in the opposite direction, continuing in the preparatory period of the phase of support (*4*).

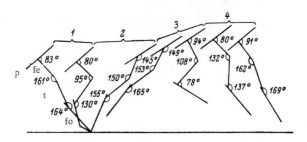

FIGURE 53. Phases and periods of hind limb activity:

1, 2) phase of support: 1) preparatory, 2) starting periods; 3, 4) phase of free transit: 3) drawing-up, 4) adjustment periods; p — pelvis; fe — femur; t — tibia; fo — foot.

In the preparatory period of the phase of forelimb support the shoulder and elbow joints are flexed and the wrist and finger joints are extended (Figure 54, *1* ). In the starting period the elbow and shoulder joints are extended and the wrist and finger joints flexed (*2*). In the drawing-up period of the phase of free transit the movement which took place in the starting period of the phase of forelimb support is at first continued, and then the shoulder and elbow joints begin to flex and the carpal and finger joints are extended (*3*). In the adjustment period of the phase of free transit the shoulder and elbow joints are extended and the carpal and finger joints flexed, but before the phase of support ensues the movement involved in its preparatory period begins in the joints (*4*).

There is thus a threefold change of flexor to extensor movements in the limb joints in the phase of fore and hind limb free transit, while there is only one change in the phase of support.

_____

* In the cycle the hip joint is generally flexed throughout the phase of transit and extended during the phase of support. At any rate, flexion of the hip joint is negligible in the phase of support.

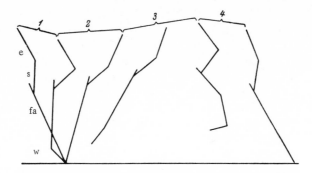

FIGURE 54. Phases and periods of forelimb activity:

e — elbow, s — shoulder; fa — forearm; w — wrist. Other notation
as in Figure 53.

This sequence of flexor-extensor movements in the joints during the
phases of free transit and support has undoubted advantages. The fact
that the flexor-extensor movements of the starting period of the phase of
support end in the drawing-up period means that the impulse lending acceleration to the body continues right up to the last moment of the phase
of support. At the end of the adjustment period the transition to the movements typical for the preparatory period of the phase of support causes
the movement of the distal elements to slow down, thus improving conditions for their landing.

Any animal may increase its speed either by increasing the angle of its
pace or by increasing the frequency of paces. A study of running in man
showed that with an increase of speed per unit time the number of paces
increases by 43.2%. But the length of each pace during this time increases
by only 8.6% (Zhukov et al., 1963). Each pace length increases mainly on
account of a prolonging of the stage of flight, from which it becomes obvious
that the angle of the human pace may even decrease upon an acceleration
of speed. When animals speed up their running, each stage of flight and the
frequency of strides also increase, and therefore a lessening of the angle of
each pace is typical when speed is accelerated. However, with very slow
speeds of locomotion the pace angle is again smaller than with average
speeds. For instance, with the slow trotlike walk of the horse the angle
of each pace is markedly smaller than with the animal's normal walk. It
is probably a general rule that during acceleration from the very slowest
forms of speed a speed increase is at first accomplished due to an increase
of the pace angle and of the frequency of paces, after which the pace frequency continues to increase without the pace of each angle increasing, so
that with a further rise in the speed of motion the frequency of paces would
increase and the pace angle diminish.

THEORY OF THE ECONOMY OF RUNNING AND JUMPING

In analyzing the conditions for improving and lessening the efficiency of running and jumping* in mammals, it is readily observed that there is a stage of free flight in both these movements.** Therefore, for a preliminary assessment of economy of movement we must characterize this basic stage. To make our further considerations more easily understood we will use the following assumptions: in all the diagrams analyzed in this section the center of gravity is strictly related to the hip joint, so that any movement of the center of gravity will correspond to movement of this joint. Only segments of the limbs can therefore be shown on the diagrams.

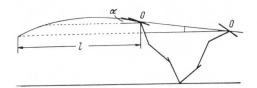

FIGURE 55. Stage of flight $l$ at angle of departure $\alpha$.

0 — hip joint (center of gravity).

In the phase of support the flexor-extensor movements of the joints impart an impulse to the center of gravity directed at angle $\alpha$ (angle of departure) to the horizontal plane (Figure 55).[†] Disregarding the air resistance, the path $l$ traversed by the center of gravity in the stage of free flight during time $(t)$ is determined by the formula:

$$l = v_{\text{horizont}} \ t_{\text{flight}} = v \cos \alpha \cdot 2 \frac{v \sin \alpha}{g} = \frac{v^2 \sin 2\alpha}{g}, \tag{1}$$

where $v$ is the velocity of departure and $g$ is the acceleration of gravity.

---

* As already pointed out, most types of running are running by jumps. Running proper will therefore be compared with jumping from one spot.

** The high-speed terrestrial gaits of all mammals (fast trots and racks, gallops and ricocheting jumps) have a stage of free flight. Only the elephants have secondarily lost the capacity for gaits with a stage of flight.

† All the formulas and assumptions presented below are to be found in more general form in various works on biophysics (Rashevsky, 1948, 1961; Bekker, 1955; and others).

As we proceed with the exposition it will be most convenient to examine two types of motion separately: the hop or jump, when the animal begins each movement from a state of rest, and running, when the horizontal component of the animal's speed may be considered constant.

With the jump, speed is gathered during the phase of support, and the most economical jump will be one which allows for the greatest distance to be covered with the least amount of work. The work* involved in jumping from a state of rest is determined by the formula:

$$A = \frac{mv^2}{2}. \tag{2}$$

The ratio between the length of the hop and the work performed during thrusting off equals:

$$\frac{l}{A} = \frac{v^2 \sin 2\alpha}{g} : \frac{mv^2}{2} = \frac{2 \sin 2\alpha}{mg}, \tag{3}$$

whereby the most economical angle of departure will be 45°, at which $\sin 2\alpha = 1$. The average force acting in the phase of support which allows the body to cover distance $l$ is found from the ratio between the work performed $A$ and the path $S$ traversed by the propelling end of the limb in the phase of support:

$$F_{av} = \frac{A}{S}. \tag{4}$$

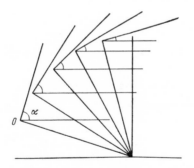

FIGURE 56. Changes of the direction of the thrust imparted to the body via the hip joint (O) as a factor of the position of the limbs. Explanation in the text.

* In the approximation in which all the work performed goes for imparting to the body a kinetic energy of $\frac{mv^2}{2}$. Here the part of the work expended in raising the center of gravity in the phase of support and in producing friction and the activity in the joints are not taken into account. With running by jumps and especially with one-time hops the raising of the center of gravity in the phase of support constitutes a negligible percentage in comparison with its raising in the stage of free flight. Our assumption should therefore hardly affect the calculations of the forces compared. In Figure 55 $l$ is shown 20—25 times smaller than the actual value.

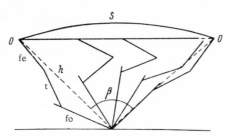

FIGURE 57. Path ($S$) of the hip joint ($O$) in the phase
of support:

fe — femur; t — tibia; fo — foot; h — length of limb;
β — pace angle.

If we picture the limb as a straight line the free end of which represents
the hip joint (and also the center of gravity, according to our conditions),
the angle of departure constantly changes from 90° to 0° when the limb
moves from the horizontal to the vertical position (Figure 56). This constant change of the direction of thrust is unprofitable, and therefore the
flexor-extensor movements in the limb joints promote as far as possible
linear movement of the propelling end of the limb (the capitulum of the
femur in the hip joint) more or less along a chord of the circle formed by
the limb of length $h$ with pace angle $β$ (Figure 57). The path $S$ covered
by the propelling end of the limb is

$$S = 2h \sin β/2, \tag{5}$$

and the average force

$$F_{av} = \frac{mv^2}{2} : 2h \sin β/2 = \frac{mv^2}{4h \sin β/2}. \tag{4'}$$

At the given length of the jump $l$ the velocity of departure is found from
the formula (1):

$$v^2 = \frac{gl}{\sin 2α}$$

and the force $F$ that the limb must develop depends on the angle of departure, according to the formula

$$F_{av} = \frac{mgl}{4h \sin β/2 \cdot \sin 2α}, \tag{6}$$

i. e., it is minimal at angle of departure $α = 45°$. Bernshtein (1940) showed
that as the angle of departure increases its relative efficiency is diminished.
Up to 30—32° with a uniform velocity ($v$) the jump becomes markedly longer
with an increase of the angle of departure. But a further increase of the
angle goes along with such an insignificant increase in the length of the jump
that it is practically of no importance.

Thus, from the point of view of the work involved and the forces developed in the phase of support, the most economical jump is that with an angle of departure around 45°.

Let us now discuss the conditions providing for economy of running.

What is significant here is not the length of jumps performed during running but the average speed attained.* Therefore, the result of each hop during running must also be assessed from the standpoint of the contribution it makes to the attainment of speed. In this assessment only the horizontal component of the velocity of departure ($v \cos \alpha$) proves useful, and the vertical component ($v \sin \alpha$), governing the rises and falls of the center of gravity inevitable in each cycle, may be considered lost. That part of the work performed during the thrust which is expended on these vertical oscillations of the center of gravity is equal to

$$A_{vert} = \frac{m \, (v \sin \alpha)^2}{2}. \tag{7}$$

We can determine the average force by considering the horizontal component of the speed to be constant during running. In this approximation the bulk of the forces developed in the starting period coincides with the vertical component of these forces. According to formula (7) the work of these forces equals

$$\frac{mv^2}{2} \sin^2 \alpha,$$

and to determine the forces themselves we must know the vertical component of the path traversed by the propelling end of the limb in the starting period. Let us say that the half of the path $S$ which is traversed by the propelling end of the limb in the phase of support corresponds to the starting period, since the second half of the path corresponds to the preparatory period, in which the shocks are absorbed. Consequently, the vertical component in which we are interested is

$$S_{vert} = \frac{S}{2} \sin \alpha$$

and, in accordance with formula (5),

$$S_{vert} = h \sin \beta/2 \cdot \sin \alpha. \tag{8}$$

Whence

$$F_{av} = \frac{A_{vert}}{S_{vert}} = \frac{mv^2}{2} \sin^2 \alpha : h \sin \beta/2 \sin \alpha = \frac{mv^2}{2h \sin \beta/2} \sin \alpha. \tag{9}$$

Thus, the useful component of the velocity of departure

$$v_{horizont} = v \cos \alpha$$

---

* Of course, some sort of general acceleration occurs at the beginning of running, but prolonged running may be considered uniform.

increases and the lost part of the thrusting work

$$A_{\text{vert}} = \frac{mv^2}{2}\sin^2\alpha$$

and the average forces developed in the starting period

$$F_{\text{av}} = \frac{mv^2}{2h\sin\beta/2}\sin\alpha$$

decrease as the angle of departure $\alpha$ decreases. If we add to this the fact that the damping of the thrusts is also related to a nullification of the vertical velocity component, the advantage of a small angle of departure for running becomes obvious.*

It follows from this analysis that the requirements for the angle of departure produced by adaptation to running and adaptation to single hops are directly opposite. The different angles of departure selected by different animals may be produced by small variations of the flexor-extensor movements of the limb joints. During jumps an increase in the angle of departure may be secured during the phase of support by active raising of the center of gravity by the extensors of the spine.


### INFLUENCE OF THE SIZE OF THE ANIMAL ON THE NATURE OF ADAPTATIONS TO RUNNING AND JUMPING

Cursorial and saltatorial specialization arises in animals of all sizes, with legs from a few centimeters to 3—3.5 m long and with a weight of the body from a few grams to several tons. This diversity naturally implies that there will be differences in the conditions of locomotion.

Let us take two examples to show the order of magnitude of the forces developed in the phase of support during a single hop of different-sized animals. First, the small jerboa, Allactaga williamsi, weight 200 g, leg length 12 cm, size of jump 3 m.** Second, the tur Capra caucasica, weight 50 kg, leg length 120 cm, size of jump 15 m. The pace angle is 80° for both species. These data may serve as a basis for calculating the average force developed during the phase of support. Taking the value of $\alpha = 45°$

---

* As mentioned above (p. 64), in the preparatory period of the phase of support movement in the limb joints is opposite to that in the starting period. In connection with this the inertia of the moving body tends to bend the limb joints, so that the shock is absorbed by active straining of the muscles (this is known as yielding strain of muscles). The shock-absorbing force: $F = \dfrac{mv\sin\alpha}{t}$, where $t$ is the duration of the preparatory period of the phase of support. Since the pace angle (the angle formed by the limb which is moving throughout the phase of support) changes negligibly with acceleration of motion, the shock-absorbing force increases owing to both the increase in the speed of motion and the decrease in the amount of time corresponding to the preparatory period of this phase.

** The jerboa does not usually make hops of more than 1 m while running. The single jump analyzed is of 3 m.

as the optimal angle of departure, the average force, according to formula (6), will be

$$F_{av} = \frac{mgl}{4\,h\,\sin\beta/2}\,,$$

and the relative force* developed during the jump

$$F_{rel} = \frac{F_{av}}{P} = \frac{l}{4h\,\sin\beta/2}\,.$$

We now have:
  for the jerboa

$$F_{rel} = \frac{3.0}{4\cdot 0.12\cdot\sin 40°} = 9.8$$

  and for the tur

$$F_{rel} = \frac{15}{4\cdot 1.2\cdot\sin 40°} = 4.9.$$

To sum up, in the jerboa the thrusting force is 9.8 times the weight of the body, and in the tur, despite the very large jump, only 4.9 times.

These two examples show that with saltatorial specialization the relative thrusting force decreases in proportion to the increase in the size of the body.

We will take three examples to characterize the orders of magnitude of the forces developed by different-sized animals during running, not during single hops:  1) the long-clawed ground squirrel, running at a speed of 9 m/sec, with a body weight of 0.5 kg and leg length 15 cm (size of jump 0.8 m);  2) the saiga, running at a speed of 20 m/sec, with a body weight of 50 kg and leg length 70 cm (size of jump 1.5 m);  3) the roe deer, running at a speed of 14 m/sec, with a body weight of 40 kg and leg length 1.2 m (size of jump 5 m).  The pace angle $\beta$ is generally smaller during running than during jumping.**  In the saiga it is 60° and in the ground squirrel and roe 70°.  To determine the average force developed in the starting period of the run we must know the angle of departure.  There is a theory (Bekker, 1955; Rashevsky, 1960) that during running the angle of departure of the center of gravity is 10—20° with respect to the horizontal plane.  However, knowing the approximate speeds and the size of consecutive springs during running, we can estimate the angle of departure more accurately.  Substituting the length of the jump and the terminal velocity of the phase of support in formula (1), we can calculate the angle of departure.  For this we rewrite formula (1) as

$$\sin 2\alpha = gl/v^2$$

---

*  The ratio between the force developed by the animal and the weight of its body is considered here as the relative force.

** The angle is usually smaller in large animals than in small ones.

and take $v$ as the average speed of running and jump length $l$ as the length of the spring.*

Then for our cases:

ground squirrel

$$\sin 2\alpha = \frac{9.8 \cdot 0.8}{81} = 0.0968, \quad 2\alpha = 5°33', \quad \alpha = 2°46', \quad \sin \alpha = 0.049,$$

saiga

$$\sin 2\alpha = \frac{9.8 \cdot 1.5}{400} = 0.03675, \quad 2\alpha = 2°9', \quad \alpha = 1°5', \quad \sin \alpha = 0.019,$$

roe

$$\sin 2\alpha = \frac{9.8 \cdot 5}{196} = 0.25, \quad 2\alpha = 14°30', \quad \alpha = 7°15', \quad \sin \alpha = 0.126,$$

i. e., in highly specialized runners, in accordance with the conditions for economy of running, the angle of departure is very small and increases only for the forest-dwelling roe deer.**

The relative forces developed by the ground squirrel, saiga and roe, calculated from formula (9), are:

ground squirrel

$$F_{rel} = \frac{F_{av}}{P} = \frac{(9 \text{ m/sec})^2 \sin \alpha}{2 \cdot 9.8 \text{ m/sec}^2 \cdot 0.15M \cdot \sin 35°} = 2.35,$$

saiga

$$F_{rel} = \frac{(20 \text{ m/sec})^2 \sin \alpha}{2 \cdot 9.8 \text{ m/sec}^2 \cdot 0.7 \cdot \sin 35°} = 1.11,$$

roe

$$F_{rel} = \frac{(14 \text{ m/sec})^2 \sin \alpha}{2 \cdot 9.8 \text{ m/sec}^2 \cdot 1.2 \text{ m} \cdot \sin 35°} = 2.47.$$

Thus, even with a twofold difference in running speed, the relative force developed by the ground squirrel is much greater (by a factor of 2:1)

---

* Since the average running speed is lower than the terminal velocity of departure ($v$), and the length of the spring is certainly greater than the shift of the body's center of gravity in the stage of flight ($l$), we obtain values of the angle of departure which are too high.

** The roe is always coming up against obstacles and therefore adapts to running with a very steep trajectory of each jump. The angle of departure markedly increases as a result, which leads to an appreciable enhancement of the relative forces developed in the phase of support; in this species these forces reach the values typical for small animals.

than in the saiga.   Consequently, adaptation to running, just like adaptation
to single hops, has the result that small animals possess much more
specific power than large ones.   This result is further compounded by the
fact that head drag is much more significant for small animals.  In practice
the drag force is defined as

$$Q = C_x S p v^2,$$

where $C_x$ is the coefficient of head drag, equal to 0.8—1.0;  $S$ is the middle
section, that is, the greatest cross sectional area of the animal perpendicular
to the direction of motion;  $p$ is the air density, equal to 0.125 kg/cm$^3$; $v$ is
the velocity of motion.

For the saiga the middle section equals 0.12 m$^2$ and for the ground
squirrel 0.01 m$^2$.   The relative force of head drag, i. e., the percent ratio
of head drag to body weight, is 12% for the saiga and 20% for the ground
squirrel, this despite the fact that the absolute running speed of the saiga
is more than double that of the ground squirrel.

It follows from all the above considerations that adaptation both to running and to single hops is manifested much more sharply in small animals
than in large ones.

INFLUENCE OF THE MECHANICS OF RUNNING AND JUMPING
AND THE SIZE OF THE ANIMAL ON THE LENGTH OF ALL
THE LIMBS AND ON THE SIZE OF INDIVIDUAL SEGMENTS

Mammalian limbs are multi-jointed levers which enable the body to
move and be supported.  If we depict a limb at the moment of landing
(Figure 58) as a straight line $a$b, and at the moment of its lifting as $a$b$_1$,
the size of the path from b to b$_1$ depends on the interval $a$b and on the
pace angle b$a$b$_1$. A steady increase of the pace angle with a constant limb
length $a$b causes an increase of the path traversed during the step by the
proximal end of the limb b at progressively decreasing intervals.   The
vertical oscillations of point b, increase at the same time, and this mainly
reflects vertical fluctuations of the body's center of gravity (Figure 58).
Flexor-extensor movements in the limb joints during the phase of support
may cause the proximal end of the limb to move in a straight line (Figure 57).
However, when the pace angle becomes larger, the flexor-extensor movements in the joints necessarily increase (Figure 59).

Let us imagine that the indexes of linear acceleration of the proximal
end of the limbs and of the body mass will be equal in two animals with
legs of different length $(h, h_1)$.  In this case the terminal velocity of the
phase of support $(v)$ proves to be higher in the form with longer legs
(Figure 60).   However, the moment of the force acting to cause breaking
of the limb is also enhanced in proportion to the length of the limb, and
this necessitates strengthening it.

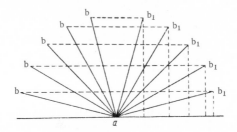

FIGURE 58. Progressive increase of vertical oscillations of the center of gravity and increase of the pace length with a steady increase of the pace angle ($bab_1$)

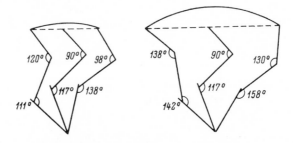

FIGURE 59. Growth of the amplitude of movement in the limb joints during horizontal movement and increase of the pace angle.

Explanation in the text.

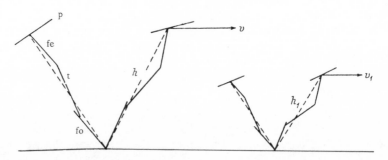

FIGURE 60. Increase in the velocity of departure ($v$) as a factor of the increase in the length of limbs $h$ and $h_1$ :

p — pelvis;  fe — femur;  t — tibia;  fo — foot.

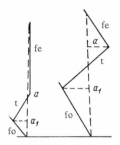

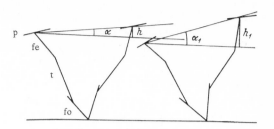

FIGURE 61.  Changes of moments $a$ and $a_1$ as a factor of the length ratios of the limb segments:

fe — femur;  t — tibia;  fo — foot.

FIGURE 62.  The angle of departure $\alpha$ as a factor of the difference in the position of the limbs at the beginning and end of the phase of support:

p — pelvis;  fe — femur;  t — tibia;  fo — foot; $h$ , $h_1$ — height to which the center of gravity is raised in the phase of support.

We shall examine the skeleton of the hind limb of mammals at rest.  If we accept that the force of gravity is directed vertically from the hip joint to the point of support of the leg, the moment of force on the knee and tarsal joints will be equal to the product of the body mass times the size of the perpendiculars from these joints on the vertical (Figure 61).  A shortening of the foot and tibia leads to a marked decrease of the force moments $(a, a_1)$ of the knee and tarsal (talocrural) joints.  Naturally, with very large increases of body mass or of the forces developed by the limb, one of the changes of the ratios in the limb levers must be a decrease in the relative size of the distal segments.

Proceeding from the above, we can formulate a number of hypotheses which make it easier to understand the conditions under which the length of the legs and the size ratios of their segments change in connection with cursorial and saltatorial adaptation.

1.  With equal velocities of departure $(v)$, angles of departure $(\alpha)$ and pace angles $(\beta)$ the relative force $(F_{rel})$ decreases in proportion to the increase of leg length (formulas (6) and (9)).

2.  With an equal length of the limb $(h)$ and equal velocities $(v)$ and angles of departure $(\alpha)$ the relative force $(F_{rel})$ decreases in proportion to the increase in the sine of half the pace angle $(\beta)$ (formulas (6) and (9)).

3.  With an equal value of the linear accelerations of the proximal end of the limbs the velocity of departure $(v)$ and the moment of the force acting to break the limb increase in proportion to the increase in leg length.

4.  An increase of the pace angle $(\beta)$, other conditions being equal, leads to a progressive increment of the vertical fluctuations of the center of gravity and to a smaller increment of the sections of the path traversed by the proximal end of the limbs, but if the center of gravity fluctuations are smoothed, the flexor-extensor movements in the limb joints increase.

A shortening of the tibia and especially of the foot leads to fewer oscillations of the moments of the forces acting on the knee and tarsal joints.

Adaptation to jumps begun from a state of rest is more economical (formula (6)) with larger angles of departure ($\alpha$) (around 45°). In this case the increase in the vertical fluctuations of the center of gravity in the phase of support is even profitable, as it provides for an increase in $\alpha$ (Figure 62). Adaptation to running, on the other hand, is more economical (formula (9)) with small angles of departure (approaching 0°). Here an increase in the vertical fluctuations of the center of gravity in the phase of support now has an adverse significance. The negative role of center of gravity fluctuations in the phase of support is particularly apparent upon an increase in the animal's body mass. It is natural to expect that the pace angle ($\beta$) will increase with saltatorial specialization and decrease with cursorial specialization. These differences cannot be particularly marked, however, because the advantage of increasing the pace angle is associated with an increase of the path traversed by the proximal end of the limb in the phase of support, but with a steady increase of the pace angle the path becomes progressively shorter (hypothesis 4). On the other hand, harmful increases of center of gravity fluctuations are leveled out starting from angles of 50—60°, especially for small animals. As a rule, only very large increases of body mass or leg length lead to pronounced decreases of the pace angle, while in other animals the angle in fact fluctuates between 60 and 80°.

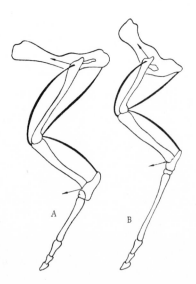

FIGURE 63. The limb as a lever (from Egorov, 1955):

A) built for strength; B) built for speed.

As shown above (the examples of the jerboa and tur on the one hand and of the ground squirrel, saiga and roe on the other), saltatorial adaptation requires that a much greater relative force be developed in the phase of support than with specialization for swift running both in open spaces and, to a lesser extent, on broken terrain.   Taking into account that the body mass increases in proportion to the cube of linear quantities while bone strength is proportional to the square, it becomes clear that even for medium-sized animals saltatorial specialization must have an effect on the reorganization of the limbs toward decreasing the moments of the forces affecting the talocrural and knee joints.   This reorganization of limb segments causes the bone-muscle system of animals adapted to salta-tion to gain in strength and lose in distance (Figure 63), while the opposite is the case in swiftly running animals (Egorov, 1955).

The numerous obstacles in shrub and forest habitats result in both a lessening of the absolute speed of running and an increased steepness of jumps.   The relative forces developed in the phase of support are in this case, too, much smaller than with one-time hops (see above with reference to the roe deer) and somewhat greater than during running in open spaces. So that the increase in the relative force might not appear excessive, we must expect animals adapted to living in forests to show both a greater leg length (hypothesis 1) and a larger pace angle (hypothesis 2) than animals adapted to living in open spaces.   At the same time these relative forces are not so great as to promote a reorganization of the limbs toward a short-ening of the distal elements (hypothesis 5).

Whereas the relative length of the body regions is constant in a number of animals, with an absolute increase of leg length the body mass increases. If the linear accelerations of the proximal end of the limbs and the size of the pace angle were preserved, there would be a rapid increase in running speed, but there would have to be a simultaneous enhancement of both the strength of the bones (hypothesis 3) and the relative forces developed in the phase of support (formulas (6) and (9)).   Therefore, with an increment of body mass the values of proximal limb end acceleration diminish, leading to a decrease of the relative force in large animals in comparison with small ones (for more details see the preceding section).

A further increase of body mass must be reflected also in changes of the length ratios of the limb segments.   This is because in comparison with the linear increase of the limb, the cubic increase of body mass leads to a very pronounced increase in the absolute forces in the phase of support. The need therefore arises to diminish the fluctuations of the force moments in the distal joints (hypothesis 5).

GROUPS OF MUSCLES AND THEIR LOADS IN DIFFERENT
PHASES OF LOCOMOTION

Limb activity, as we have noted above, consists of two phases:  support and free transit.   The legs naturally bear a heavier load in the phase of support as during this time they promote movement of the body and cushion shocks and vibrations.   In the phase of free transit, on the other hand, the entire work of the limbs consists in developing positive and negative accelerations of the distal components,* which are maximally lightened.

Before examining the activity of individual muscles we must analyze the common properties and fundamental differences in the work of the fore and hind limbs. At the moment the animal lands on the hind feet the point of hind limb support is situated in front of the perpendicular from the overall center of gravity (Figure 64, A). The overall center of gravity now moves forward and downward at an angle, and this shift leads to flexion of the hip, knee, and talocrural joints. The resistance to these flexing moments imparts an impulse to the center of gravity, contributing to a change in the direction of its motion (absorption of the shock). If the leg muscles were not actively working, this impulse from the limbs would have to be directed straight from the point of support to the center of gravity. But the action of the leg muscles sets up a force couple. One of the forces is applied to the proximal end of the limb and is directed forward (Figure 64, $P$) while the other is applied to the leg itself and is directed backward (Figure 64, $P_1$). Together they put pressure on the ground and cause the legs to create a force which is directed not to the center of gravity but in front of it (Figure 64, $F$). The general direction and the force of this impulse depend on the development and topography of the muscles, the dimensions of the legs themselves and on the relative position of the center of gravity and point of support. As long as the center of gravity is situated behind the point of support, the sum effect of the propulsive forces developed by the limb, the force of inertia and the counteraction of the limb may even inhibit movement and at the same time change the direction of motion of the center of gravity. The moment the center of gravity intersects the vertical from the point of support the propulsive activity of the legs finds especially favorable conditions. The reason for this is that a forward thrust is facilitated if the point of support is situated behind the center of gravity.

With landing on the forefeet, as on the hind feet, the perpendicular from the center of gravity passes behind the point of support. It thus seems that the action of these limbs is essentially the same. However, owing to the increasing distance between the center of gravity and the point of limb support (Figure 64, B), propulsion of the forelimbs encounters favorable conditions much later, as it is only at the very end of the phase of support that the perpendicular from the center of gravity intersects the point of support of the forelimbs. Because of this the forelegs inhibit movement for a longer time and have a greater cushioning and a lesser propulsive value than the hind legs.

It would therefore seem that in essence the differences between fore and hind limb activity are only quantitative. They have, however, far-reaching consequences. The hind limbs find conditions advantageous for propulsion sooner, and the forelimbs remain longer in conditions promoting shock absorption. The significance of these differences is more clearly pronounced

---

* According to Donskoi (1958), the great muscular strength of the proximal joints is due to the considerable dynamic moments of the terminal limb components. We, however, believe that this muscular strength is an adaptation to moving the body in the phase of support. The muscles developed for this purpose are used also to obtain large dynamic moments of the terminal components. Proof of this is the fact that there is a relative and absolute weakening of the proximal joint muscles in bipedal animals, in which the dynamic moments of the terminal components are particularly large.

when an animal is moving by high leaps and it becomes especially important
for impacts to be cushioned; if, on the other hand, the animal is moving with
a very smooth trajectory, the propulsive role of the fore and hind limbs becomes more and more equalized.   These differences in the part played by
the limbs, which seem insignificant at first glance, have additional consequences.   The hind limbs, which impart the impulse, are immovably fixed
to the spine, while the forelimbs, which in transmitting the impulse, mitigate
its force, are attached to the spine by elastic muscular connections.

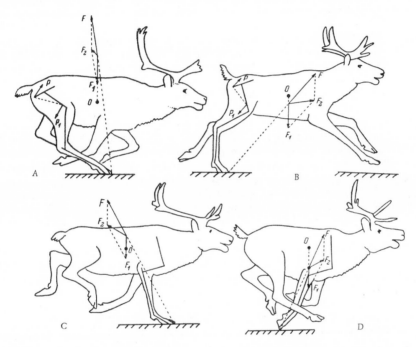

FIGURE 64. Interaction of the forces of limb propulsion and the animal's gravity in the phase of
hind (A, B) and fore (C, D) limb support:

$O$ — overall center of gravity; $F$ — impact force of limb; $F_1$ — force of gravity; $F_2$ — their resultant; $P, P_1$ — force couple of leg muscle activity.

A force couple (Figure 64, $P, P_1$) is applied to the limbs of mammals.
The nearer the second force (directed backward) is applied to the proximal
end, the more speed is gained and force is lost and at the same time the
greater will be the force tending to break the leg.   Therefore, as the body
mass increases the relative size of the distal segments diminishes or else
the points of application of the leg muscle forces are shifted distally.
  Owing to the parasagittal position of mammals' limbs, both propulsion
and shock absorption are realized almost entirely by muscles acting in the
plane parallel to the plane of symmetry of the body; in other words,  the

main working elements will be the extensors and flexors of the leg joints. In determining the relative load on individual muscles and groups of muscles, therefore, we first have to examine the parts played by flexor-extensor movements in the limb joints in the phase of support.

In the preparatory period of the phase of hind limb support, there is slow flexion of the hip, knee and talocrural joints, while with forelimb support the shoulder and elbow joints are flexed and the wrist and finger joints are extended. Shocks are cushioned in this period. In the starting period, extension of the hip, knee, shoulder, elbow and talocrural joints is stepped up, and the wrist and finger joints are flexed. Naturally, the greatest load in the phase of support is borne by the extensors of the hip, knee and talocrural joints, which provide for a pulling back in the preparatory period and active contraction in the starting period. Correspondingly, in the forelimb the greatest load is borne by the extensors of the shoulder and elbow joints and the flexors of the wrist joint. The extensors of the hip joint comprise three groups of muscles: the gluteal and long and short postfemoral, the long group terminating on the tibia and the short group terminating on the femur. The gluteal group (Figure 65,1) includes m. m. gluteus medius, minimus and piriformis; according to its position, the superficial gluteal muscle should probably be referred to the hip joint flexors. The group of short postfemoral muscles (Figure 65,2) includes m. quadratus femoris, m. m. adductores femoris and sometimes m. pectineus, if its origin extends behind the axis of the hip joint. In carnivores, some rodents and a number of ungulates, m. semimembranosus anterior and m. praesemimembranosus are also included in this group. In some rodents and ungulates the group has in addition m. semimembranosus posterior. The group of long postfemoral muscles (Figure 65,3) comprises m. biceps femoris, m. semitendinosus, m. m. gracilis anterior and posterior and sometimes m. tenuissimus. In some rodents both m. m. semimembranosi belong to this group, while in carnivores and some ungulates only m. semimembranosus posterior.

The extensors of the knee joint are three wide capita of m. quadriceps femoris (m. m. vastus lateralis, medialis and intermedius). Originating on the ilium, m. rectus femoris acts to flex the hip joint rather than to extend the knee joint in the phase of hind support.

The extensors of the talocrural joint are two gastrocnemial muscles (m. m. gastrocnemius lateralis and medialis) and m. soleus. Apart from extending the talocrural joint, the two gastrocnemial muscles simultaneously act to flex the knee joint. Serving to extend the talocrural joint and at the same time flex the digits are also m. m. fl. digitorum longus, fl. hallucis longus and tibialis posterior plus m. plantaris.

The principal and indirect functions of the above groups of muscles make it possible to distinguish their roles. Apart from extending the hip joint the gluteal and short postfemoral groups act to extend the knee joint. The principle of their activity is illustrated by the diagram in Figure 65, B. A two-unit mechanism is shown with a fixed axis of movement in unit a (knee joint) and a moving axis in unit b(hip joint). The latter, in addition to its own movements, moves backward and forward with the whole system when there are angular movements in unit a.  If we denote the action of the gluteal group as force $F$ and of the short postfemoral muscles as $F_1$ and $F_2$, then separating them into horizontal and vertical components, we may visualize the resultant effect.  Taking the product of the weight $W$ times the lever bf to be equal to the sum of the products of $P_8$ times lever bg and $P$ times lever $bf_1$, we can consider the position of the hip joint to correspond to equilibrium.  Components $P_0$, $P_6$ and $P_3$ act to compress the femur. Components $P_5$ and $P_4$ are parallel and in the sum equal to the oppositely directed force $P_7$.  The same can be said of $P_1$ and $P_2$.  The lever of the application of forces, however, is different for all of them.  For $P_5$ it is $aa_1$, for $P_4$ it is $aa_2$, for $P_2$ and $P_7$ it is ab and for $P_1$ it is ae; since the product of ae times $P_1$ is larger then the product of ab times $P_2$, the gluteal muscle causes an angular forward motion of the system in unit a.  The effect of $P_5$ and $P_4$ is also smaller than that of $P_7$ in connection with the different sizes of the above levers.  On the whole, the moment produced exerts forces which move the proximal end of the femur forward, thereby extending the knee joint.

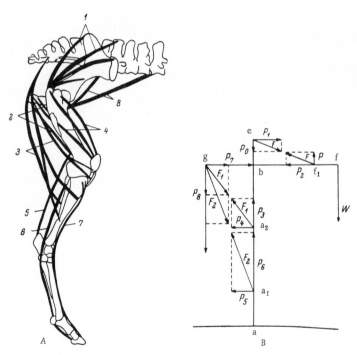

FIGURE 65.  Hind limb muscles (A) and the activity of some of them (B) in mammals:

1) gluteal; 2, 3) postfemoral: 2) short, 3) long; 4) knee joint extensors; 5) gastrocnemial; 6) long flexors of the digits; 7) long extensors of the digits; 8) hip joint flexors.  a—e is the femur; f—g is the pelvis; a and b are joints: a — knee, b — hip; $W$ — weight of body; $F$ — force of gluteal muscles; $F_1$ and $F_2$ — force of short postfemoral muscles; $P$, $P_0$, $P_1$, $P_2$ — components of gluteal muscles; $P_3$-$P_8$ — components of short postfemoral muscles; $a_1$, $a_2$, $f_1$ — points of application of forces.

The double-jointed position of the long postfemoral muscles makes it very complicated to distinguish their respective activity. It is therefore advisable to proceed from concrete schemes of limb skeleton movements, obtained by sketching the skeleton on successive frames of a moving animal. This type of analysis enabled us to pinpoint the causes of the reorganization of individual muscles (for more details see the part where muscle activity is discussed in different groups of animals).

The gastrocnemial muscles are also double-jointed. They can act to extend the talocrural joint in the phase of hind support only when the knee joint is flexed. We must therefore expect that the part of their contraction force which affects the flexion of the knee joint will be less considerable than the force of its extension, which depends on the total activity of all the knee joint extensors. It is this circumstance which gave rise to different types of origin of the gastrocnemial muscles from the femur. In a number of animals the muscles begin on the Vesalius sesamoid ossicles (Figure 66, A), and these ossicles are united with the lateral and medial epicondyles of the femur by ligaments directed almost parallel to the femoral axis. In this case when the knee joint is extended the Vesalius ossicles move backward along the epicondyle of the femur, and its activity scarcely affects the tension of the gastrocnemial muscle. Cursorial specialization in these animals naturally brings about a strengthening of the gastrocnemial muscles (Gambaryan, 1955). In other species (Figure 66,B) although Vesalius ossicles are present, some of the bundles of the gastrocnemial muscles originate directly on the femur. This lengthens the lever arm of the force of the gastrocnemial muscle in the knee joint. Therefore, for the talocrural joint to be extended, in this case the knee joint extensors must be stronger than the talocrural joint extensors. In a third instance (Figure 66, C) the gastrocnemial muscles originate directly on the femur almost at a right angle. Moreover, their origin extends proximally along the femur, enhancing the lever of the forces flexing the knee joint. Here the force ratios of the gastrocnemial muscles and the knee joint extensors must be similar to those of the second case. At the same time in the third example a clearly defined automatic connection is observed between the extension of the knee joint and of the talocrural joint (Gambaryan, 1960).

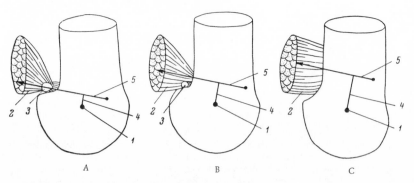

FIGURE 66. Types of gastrocnemial muscle insertion in mammals:

A) only on the Vesalius ossicle; B) on the Vesalius ossicle and on the ridge of the bone; C) only on the ridge of the femur. 1) center of movement in the knee joint; 2) gastrocnemial muscle; 3) Vesalius ossicle; 4) lever arm of force; 5) resultant of m. gastrocnemius.

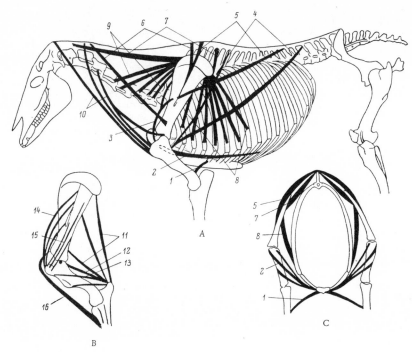

FIGURE 67. Muscles of the forelimbs and shoulder girdle of mammals:

A) lateral;  B) muscles of the free limb;  C) frontal section in the region of the shoulder girdle.
1) m. ectopectoralis;  2) m. endopectoralis;  3) m. sternoscapularis;  4) latissimus dorsi;
5) m. spinotrapezius;  6) m. acromiotrapezius;  7) m. rhomboideus;  8, 9) m. serratus ventralis: 8 — pars thoracalis, 9 — pars cervicalis;  10) m. brachiocephalicus + m. atlantoscapularis;  11–13) m. anconeus: 11) longus, 12) medialis, 13) lateralis;  14) m. supraspinatus;
15) m. infraspinatus;  16) m. biceps brachii.

For the forelimb, apart from the muscles examined above (p. 80), which
bear the greatest load during movements in the joints of the free limb, the
muscles confining and propelling the body between the limbs are also of
considerable importance.  These include all the pectoral muscles (m. m.
ectopectoralis, endopectoralis (Figure 67, 1 and 2);  pectoralis  abdomi-
nalis, sternoscapularis), m. serratus ventralis pars thoracalis (8),
m. spinotrapezius (5) and m. latissimus dorsi (4).

The muscles extending the shoulder joint include the shoulder joint
extensors proper:  m. supraspinatus (14) and frequently also m. infraspinatus (15),  and then the muscles of the shoulder girdle which influence
the extension of the shoulder joint:  m. acromiotrapezius (6), m. atlantoscapularis (10), m. rhomboideus (7), sometimes m. sternoscapularis (3),
which in addition acts to push the trunk forward between the legs, and
m. serratus ventralis pars cervicalis (9).

The muscles extending the elbow joint also fall into two groups.  There
are the elbow joint extensors proper — m. m. anconei (11–13), two capita

of which directly extend the joint while the long caput can do so only when the shoulder joint is extended simultaneously. This group also includes the tensor of the forearm fascia (m. dorsoepitrochlearis) and the lesser elbow muscles (m. epitrochleoanconeus medialis and, if present, m. epitrochleoanconeus lateralis). Among the muscles indirectly working to extend the elbow joint is the m. brachiocephalus complex (10), that is, m. clavodeltoideus, m. clavomastoideus, m. clavotrapezius and also m. atlantoscapularis, which is intergrown with them. On the other hand, m. latissimus dorsi (4), which raises the humerus, may serve to extend the elbow joint.

The flexors of the wrist and finger joints are m. palmaris longus, m. fl. carpi radialis, m. fl. carpi ulnaris, m. fl. digitorum sublimis and m. fl. digitorum profundus; in ungulates, m. ext. carpi ulnaris also belongs here.

The following conditions should be observed for these groups of muscles to work in the phase of fore support: the force of the shoulder joint extensors should prevail over that of its flexors; the force of the muscles drawing the body between the legs should be greater than that of the muscles inhibiting this movement. The complexity of muscle interactions in the shoulder girdle makes it possible for these conditions to be fulfilled in manifold combinations, and therefore a set of different cases will be presented below in the discussion on the changes taking place in the locomotory organs in each group of animals.

## SIGNIFICANCE OF STRUCTURAL REORGANIZATIONS OF MUSCLES

Muscles have quite a varied structure, and attempts to discover the patterns underlying their reorganization date a long way back. The relations between the structure of muscles and the speed and force of their contraction aroused especial interest. In classifying the forms of muscles according to the direction of the fibers most authors adhere essentially to the scheme proposed by Borellus (1710), who suggested dividing muscles into those having straight fibers (these are now called parallel-fibered), those with oblique fibers (unipinnate), those with diverging fibers (bi- and multipinnate) and those having curved fibers (concentric). Lesgaft (1905) presented a scheme of the activity of the pinnate muscle in which he stated the reasons for the partial loss of force in such a muscle. He believed, however, that the advantage underlying pinnation relates to a possible diversity of activity. Thus he writes: "Therefore, in such muscles (pinnate — P.G. ) some of the force is lost, but as much as they lose in force they gain in diverse function" (p. 242), and further, "... the physiological cross section of such muscles (parallel-fibered — P. G.) is relatively small, and yet the muscles can display quite considerable force with a low degree of tension" (p. 249). The diversity of pinnate muscles in terms of activity described by Lesgaft in most cases has no relevance where skeletal muscles are concerned. This is because they are attached to the bone by a tendon which is directed to the axis of the bone at a certain angle, so that no matter whether part of the muscle or the whole of it is tensed, the direction of its activity does not change.  Lesgaft's followers and students developed his theory and in so doing found a completely different interpretation for the significance of pinnate muscle structure.  For instance, in the manual compiled from records of Lesgaft's lectures we read: "Hence, the force of a muscle is proportional to the number of its fibers" (Krasuskaya et al., 1938, p. 17), and in the following manual a still more strictly defined statement: "If, for example, we compare a muscle with parallel fibers and a pinnate muscle, we see that given the same volume, the pinnate muscle will be stronger, since it will have a greater number of fibers" (Koveshnikova et al., 1954, p. 77).*  A similar opinion is held by a number of investigators (Schumacher, 1960, 1961;  Zhukov et al., 1963; and others).  On the other hand, completely different ideas regarding the advantages provided by pinnation have been put forward which come closer to Lesgaft's hypothesis.  Ukhtomskii (1952), for instance, considers that pinnation arises as an adaptation to the speed of motion, and this theory has a number of protagonists (Benninghoff and Rollhäuser, 1952; Donskoi, 1958; etc.).  It has also been confirmed experimentally (Hill, 1948, 1953).  In an attempt to examine all aspects of the structural reorganizations of muscles, Gans and Bock (1965) submitted their findings to comparative physiological and theoretical investigations and noted the following results when muscles assume a pinnate structure:  an increased number of fibers, leading to enhanced strength, although each fiber also loses some strength; a better distribution in confined places; diversity of activity according to the scheme suggested by Lesgaft; increased speed, etc.

Although our treatment of muscle pinnation does not contain anything fundamentally new, we nevertheless consider it useful to give as simple an explanation as possible of the significance of these structural rearrangements of muscles.  Some a priori assumptions are first necessary for general considerations.  The first of these is that the force and speed of contraction of the individual fiber of a pinnate and of a parallel-fibered muscle are considered to be equal; this has to be for if we want to assess the role of changes in muscle structure, we must accept that all other conditions of its activity will be equivalent.  It is only when we have worked out the effect of a pinnate structure, adopting this assumption, that we can specify the inequality of the work of the fibers themselves.

We can substitute the volume of a muscle graphically by an area and for a comparison of the number of fibers consider these to be distributed at equal intervals.  Drawing a rectangle (Figure 68, A) and two converging parallelograms (Figure 68, B) in which the length of the base is equal to the length of the base of the rectangle, while the sum of their areas is equal to the area of the rectangle, we can discern the following properties of these two muscles: on contraction of the parallel-fibered muscle the speed of contraction of the fibers and the speed of raising the load are equal to one another, whereas on contraction of the pinnate muscle the speed of raising the load is proportional to the product of the speed of fiber contraction times the secant of the angle of pinnation (this is the angle between the fiber and

* One author (Yakovleva, 1959) even calculates the coefficient of the static character of the muscle, i.e., the ratio of the physiological to the anatomical cross section. Unfortunately, however, she does not specify what she means by "static character." If we take it to mean the same as lesser mobility and lower speed, this definition diverges even further from Lesgaft's interpretation of pinnation (1905)

the central tendon of the muscle, that is, in the diagram between the base
and the side of the parallelogram). With this condition the muscle gains
in speed and loses in the force acting to raise the load. If we halve the
length of the base of the parallelogram (Figure 68, C) and leave the area
unchanged, in other words if we double the number of fibers of the pinnate
muscle in comparison with the parallel-fibered muscle, as long as the
pinnate muscle attains an angle of pinnation around 60° (sec = 60°=2),
its force will be greater than that of the parallel-fibered muscle, while
with a further increase of the angle of pinnation the force will again become lower than that of the parallel-fibered muscle.

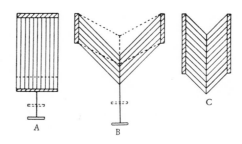

FIGURE 68. Diagram of the work of a parallel-fibered muscle (A)
and pinnate muscles (B, C)

Since the speed of fiber contraction is constant (as is assumed arbitrarily), shortening of the fiber leads to a lesser absolute scope of contraction
per unit time, i. e., load raising in the pinnate muscle with fibers that are
shorter than in the parallel-fibered muscle loses in speed and gains in force
Thus, until the pinnate muscle reaches a 60° angle of pinnation, its contraction (Figure 68, C) loses in speed in comparison with the parallel-fibered
muscle (Figure 68, A) and gains in force, while subsequently it gains in force
and loses in speed. At the same time fiber contraction in the pinnate
muscle diminishes also the possible absolute movements of the lever to
which it applies force. Hence, a comparison of long- and short-fibered
pinnate muscles (Figure 68, B, C) shows that movement of the terminal
tendon over equal distances accelerates the limit of the muscle's possible
contraction in proportion to the shortening of the fibers.

Other conditions being equal, the loss of force in the pinnate muscle
builds up rapidly as the angle of pinnation increases (i. e., during the work
of the muscle). This loss is twice as great at a 60° angle of pinnation,
three times as great at 70°, six times as great at 80°, and so on. To all
intents and purposes the fibers of a pinnate muscle can never contract
within the limits of the possible contractions of a single fiber. Otherwise,
the force of contraction would be uselessly wasted or else it would destroy
the intact structure of the actual muscle. As a result, the limit of useful
contraction in the pinnate muscle diminishes both on account of an increase
in the angle of pinnation and due to the shortening of the fibers. This is why

an increase in the physiological cross section with a uniform muscle volume is possible only in places where the movements of the terminal tendon are not considerable. In our opinion, the diversity of movements arising in connection with an asymmetrical sequence of nerve impulses transmitted to different parts of the pinnate muscle is of no value for the majority of skeletal muscles, as they end in a single tendon and therefore a pinnate structure could hardly appear in relation to a need to diversify muscle action. The diversity of movement is maintained by the separation of completely independent muscles, not by a reorganization of their inner structure.

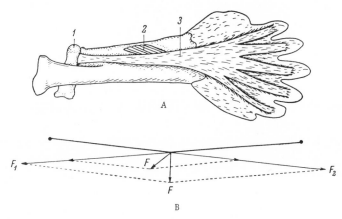

FIGURE 69. Work of m. fl. digitorum profundus of the mole:

A) structure of the muscle; B) sizes of the components as a factor of the angle at which the fibers are attached to the tendon. 1) origin of the muscle on the medial epicondyle of the humerus; 2) fibers of m. fl. digitorum profundus; 3) muscle ending; $F$ — force of muscle fibers; $F_1, F_2$ — force components.

One of the work variants of pinnate muscles is activity according to the principle of stretching (Gambaryan, 1957), when one tendon is inserted on two bones which form a joint or a system of joints, and the small venter of the muscle comes up to this tendon. In this case, when the tendon is strained by a load on one side of the joint, the component working to return the bone to its original position is automatically enhanced (Figure 69). In our example the force ($F$) is applied to the tendon (3). If it is directed perpendicular to the tendon, components $F_1$ and $F_2$ will be equal. But if it is directed obliquely, one of the components is enhanced more than the other. The muscle (2) therefore provides considerable pull for flexion of the digits and scarcely acts to detach the tendon from the medial epicondyle of the humerus at point 1.

BIOMECHANICS OF THE PRIMITIVE RICHOCHETING
JUMP AND GALLOP OF SMALL MAMMALS AND THE
RESPECTIVE ROLES OF THE LIMBS IN OTHER GAITS

We noted in Chapter 3 that asymmetrical saltatorial gaits, the primitive ricocheting jump and gallop, are typically adopted by mammals for land locomotion.  The differences between these two gaits are determined by the type of motion of the hind limbs in the stage of flight which follows the stage of hind support.  During the primitive ricocheting jump the hind limbs are deflected forward in proportion to the size of the jump.  Therefore, an increase in running speed, and with it an increase in the size of jumps, inevitably leads to the appearance of bipedal running with total detachment of the forelimbs.  With the gallop, on the other hand, forward movement of the hind limbs is inhibited in the stage of extended flight, and the animal always lands on the forefeet.  In view of .;at we have said, the mechanics of motion must also differ substantially in different species of animals using either of these two basic asymmetrical gaits.

An analysis of numerous motion pictures of animals adapted to primitive ricocheting saltation or to the gallop enabled us to formulate a number of theories concerning the mechanics of motion with these gaits which we wished to check by experiment.

These theories were based on a study of limb activity during both gaits. During the primitive ricocheting jump the hind limbs not only determine the size of the jump but already prepare themselves for the next impulse while in the air.  We may therefore expect that in animals employing this type of jump all locomotory function is imposed on the hind limbs, while the forelimbs merely cushion the impact up to the moment that the hind feet touch down.  With the gallop, however, the hind limbs begin to move forward only after the forefeet have touched the ground.  The forelimbs thus not only have to cushion the impact but also actively help pull the body up until the hind limbs move forward.  The result of this pulling action of the forelegs is that during the gallop the forelimbs frequently promote a second, crossed, stage of flight, which may exceed the extended stage of flight in terms of length.  Naturally, with the gallop, in contrast to the primitive ricocheting jump, the locomotory function is distributed between the fore and the hind limbs.  This is why we must expect that the differentiation of functions into propulsive hind limbs and shock-absorbing forelimbs to be more strongly expressed in animals adapted to primitive ricocheting saltation than in those adapted to galloping.

The force of shock absorption is calculated from the formula

$$F_{\text{sh. abs}} = \frac{mV \sin \alpha}{t}$$

(for more details see the second section of this chapter), in other words, it is inversely proportional to the time during which shock absorption is effected.  But in order to achieve the maximum length and speed of jumping, the propulsive thrust must be produced with the highest values of linear accelerations, that is, it is most effective with the lowest thrust times.

The above implies that during the primitive ricocheting jump the force upon pushing away will be greater than that during landing. For the forelimbs to allow for both shock absorption and propulsive thrust during the gallop, the index of the thrust force of the forelimbs will have to be relatively larger. Apart from this, we must expect that with an increase of running speed the quoted difference of limb activity will be especially pronounced.*

To prove these theories experimentally we chose animals which are typical gallopers and ricocheters. We also set up experiments with a souslik; these will be discussed below in the section on cursorial specialization in the Sciuridae. For greater homogeneity of the initial data we selected species of more or less the same size. As animals adapted to primitive ricocheting saltation we took three species of gerbils: Meriones persicus, M. blackleri and M. meridianus, and as gallopers the weasel Mustela nivalis (Carnivora) and the pika Ochotona pricei (Lagomorpha).

We could have demonstrated the correctness of our premises if we had succeeded in training the animals to jump a certain distance. Moreover, these jumps should have been similar to those known from observations in nature which are related to a change in the speed of running.

A study of the tracks left by the weasels and gerbils and scrutiny of the motion pictures of the pikas gave us an idea of the variations in the size of their jumps at different running speeds. With running at a moderate pace the jumps were of 25—30 cm, while with fast running they increased by a factor of 2—2.5. Still larger jumps were often observed. In our experiments we studied normal jumps and jumps double the size as those most closely matching the conditions for saltatorial variations with accelerated running in nature (Gambaryan and Oganesyan, 1970).

We used devices (see the description of the procedure in Chapter 1) to determine the force of the thrust during pushing away and of the impact upon landing in all the above animals with different sizes of jumps (Table 3).**
At least 10 experiments were set up for each specimen with all sizes of jumps (just in the pika the thrust and impact forces were calculated six times under conditions of a short jump).

The averaged data given in Table 3 prove that in animals adapted to the primitive ricocheting jump and the gallop the thrust and impact forces during pushing away and landing are distributed in different directions. Expressing the landing impact force as a percentage of the thrusting off force, we obtain the following results: in M. blackleri No. 2 it is 139% with a small jump and 146% with a large one, while in No. 1 the corresponding values are 149% and 206%. In M. persicus No. 19 the figures are 210 and 240%. The opposite is observed in the galloping animals: in the pika,

---

* The impulse of the force, that is, the index of the force multiplied by the time of its action, should be approximately the same during pushing away and landing with all types of cursorial specialization. However, in the one case for a large force the impulse corresponds to a shorter time, and in the other case the opposite is true. Here maximum force indexes are considered rather than impulses.

** The "dynamic force" of the thrust or impact of an animal and the "static force" of an applied weight are not necessarily the same. However, the required ratio of the thrust and impact forces can be correctly characterized thus.

84% with a small jump and 80% with a large one, and in the weasel 68% and 37%, respectively. We thus see that not only are the force ratios of propulsive thrusts and landing impacts opposite in galloping and ricocheting animals, but with an increase in the size of the jumps these differences become even more pronounced.*

TABLE 3. Relative force of thrusts and impacts upon pushing away and landing in galloping and ricocheting animals during different-sized jumps

| Species | Average size of jump, cm | Ratio of thrust or impact force to weight of body | | Average size of jump, cm | Ratio of thrust or impact force to weight of body | |
|---|---|---|---|---|---|---|
| | | pushing away | landing | | pushing away | landing |
| Meriones blackleri { | 22 | 2.01 | 1.35 | 43 | 3.30 | 1.6 |
| | 25 | 2.52 | 1.82 | 44 | 4.52 | 3.1 |
| M. meridianus ...... | – | – | – | 53 | 5.61 | 1.71 |
| M. persicus ....... { | – | – | – | 56 | 2.42 | – |
| | 29 | 3.1 | 1.48 | 51 | 2.96 | 1.23 |
| Ochotona pricei .... | 20 | 1.21 | 1.45 | 40 | 2.94 | 3.66 |
| Mustela nivalis ..... | 33 | 1.68 | 4.48 | 60 | 2.24 | 6.06 |

We thus succeeded in experimentally confirming our notions on the work of the fore and hind limbs being dependent on the asymmetrical gait typical for the animal.

A study of limb activity with different mammalian gaits led Goubaux and Barrier (1884) to the conclusion that the speed at which the distal end of the leg moves changes considerably in relation to the speed of the body's motion. By this they intended partly to explain economy of movement with various gaits. For instance, with the trot the leg moves only twice as fast as the body, but with the walk three times as fast, etc. The concept on the rhythm of the legs makes it possible in all cases to make an accurate comparison of the speed at which the body and the distal end of the limbs move. The greater the time of support relative to the time of free transit of the limbs, the greater the speed of movement of the hand and foot with respect to the speed of the body's motion. For example, with the very slowest walk of the horse, the speed of hoof movement is seven times as great, and to this we must add the speed of the body's motion. Hence, the hoof moves 8 times faster than the animal itself. With the slow rack and trotlike walk the speed of hoof movement is only three times as great, plus the body's speed, and so four times as fast. With the slow trot and the rack, and also with the normal walk, hoof movement is 2.66 times as fast as movement of the body, with the fast walk 1.6 times as fast, and so on.

* In all our experiments one jump was performed, and the subsequent and preceding jumps proved to be smaller than the experimental. In nature the transition to moving at high speed goes along with a change in the size of a series of jumps. We must therefore stress that there cannot, of course, be complete identification of the mechanics of running under experimental conditions and during acceleration of running in nature. Still, it may be considered most likely that the data obtained reflect the differences in the mechanics of running during the primitive ricochet and the gallop. This is all the more probable that in some cases the jump was carried out on the run, and the nature of the indexes did not alter.

Therefore, obtaining a support graph for any gait by an analysis of motion pictures and knowing the animal's average speed of motion, we can always calculate the speed ratio for the movement of the animal and of the distal limb segments.   These calculations are very important also for working out the acceleration of the distal segments which, together with other factors, is quite probably responsible for the maintenance of multi-jointed limbs.   The point is that with large angular accelerations of any body the body must break down at one of its particular centers of gravity. The studies performed by Brovar (1960) showed that the limb joints are distributed at individual centers of gravity.   He made sawed sections on a dismembered limb in the zone of its center of gravity.   The sawings were at the elbow and knee joints.   The centers of gravity of the distal parts thus obtained fell at first on the wrist and talocrural joints and in subsequent sawings on the metatarsodigital, carpodigital and digital joints.   As a result, the distal segments having the greatest acceleration strive primarily to break the limb in the region of the digital joints, but instead of a fracture, flexion or extension of these joints occurs.   Then, when the maximum tension is attained in the limits of this joint, the limb becomes rigid as far as the wrist or talocrural joint, in which the individual center of gravity of the whole distal part of the limb is situated.   With the attainment of maximum tension now in these joints as well, the limb should break at the knee or elbow joint, where the next individual center of gravity lies.   Therefore, instead of the limb's breaking, flexor-extensor movements appropriate to the loads take place in the joints, which not only prepare the limbs for the next phase of support but enable the necessary strength of the whole structure to be achieved with the minimal expenditure of building material.

The above-mentioned relationship between the speed of motion of the distal limb components and that of the body is additionally characterized by the well-known fact (Howell, 1944;  Sukhanov, 1968) that with accelerated motion the phase of support is progressively shortened while the phase of transit either is not shortened or decreases negligibly.   The reason for this is that in the phase of transit relatively weak muscles have to pull the powerful muscle groups working in the phase of support.

*Chapter 4*

## ADAPTATION TO RUNNING IN THE UNGULATA

### MECHANICS OF RUNNING

The phylogeny of the Perissodactyla and Artiodactyla* is one of the most studied pages in the chronicle on the class Mammalia. And yet no definite conclusions have been drawn concerning the position of these orders in the system, on the relationships with other orders and on the direct ancestors of these groups. Nor is much known about the original mode of life of these animals. It is probable, however, that unguligrade motion developed independently in the two groups as an adaptation to swift running, in which the limbs are narrowly specialized for locomotory function. This specialization led to a decrease in the diversity of possible movements that were characteristic for the more ancient mammals and to the formation of an additional segment which made the leg longer (isolation of the hand and foot from the ground and their lengthening).

The predecessors of the ungulates were probably primarily forest animals (Kovalevskii, 1874, 1875; Simpson, 1951; Gabuniya, 1956; Trofimov, 1956; etc.). Thickets of shrubs and piled up fragments of trees and branches created all kinds of obstacles in the way of these animals. For the ancestors of ungulates, still little adapted for running, and also for the primary mammals, the high-speed gait was possibly the primitive ricocheting jump, during which the hind limbs were deflected forward in proportion to the size of the jump. With this gait, at the moment of clearing an obstacle the hind limbs were situated under the body (Figure 50), which made jumping difficult. Cursorial adaptation in forest dwellers therefore led to an inhibition of the forward deflection of the hind limbs right up to the point where landing came to be carried out on the forefeet (Figure 50); a new asymmetrical gait thereby arose, the gallop, which in fact has become the high speed asymmetrical gait (lateral, diagonal, heavy, light, etc.) of all the recent ungulates.

Subsequent perfection of the hooves with reduction of the lateral digits and a decrease in the area of support probably appeared in connection with adaptation to running in open spaces (steppe, desert, etc.), where specialization for fast running became vital. This specialization came about not only due to a simplification of the limbs but also as a result of progress in the entire organization of the Ungulata. Now having an extremely reduced hand

---

* Owing to the close similarity of the paths of cursorial specialization taken by the Perissodactyla and Artiodactyla, it is justified to study the mechanics of their motion and the adaptive features in the structure of the bone-muscle system in one chapter.

and foot (one or two of the five digits preserved), the animals were able
to enter various habitats.   Since multiple assimilation in various habitats
took place in each group in its own ways, a study of the mechanics of running and of the structural features of the skeleton and muscles may serve
as guidelines for interpreting the paths of adaptation to different forms of
running in these groups of mammals.

The material at our disposal was very heterogeneous, and therefore we
do not claim to give a complete picture of the pathways of cursorial adaptation.   Still, we believe that we have covered the main trends of specialization
for land locomotion in these animals.

## Perissodactyla

We studied high-speed motion pictures of three species of the family
Equidae: the horse (E q u u s   c a b a l l u s), the zebra (E. q u a g g a   c h a p -
m a n i) and the Asiatic wild ass (E. h e m i o n u s   o n a g e r) and from the
family Tapiridae T a p i r u s   a m e r i c a n u s.

An analysis of the trot and gallop of the tapir (Figures 70 and 71), the
trot of the ass (Figure 72) and the trot and gallop of the horse (Figures 73
and 74) makes it possible to judge, as a first approximation, the phylogenetic
trend of changes in the mechanics of running of the Perissodactyla.   The
reason for this is that the tapirs are relict inhabitants of forest overgrowth
and reed thickets and probably run in the same way as the forest-dwelling
ancestors of horses did.

Examining the work of the limbs in two phases, support and transit, we
see that during a fast trot the angle of the talocrural joint changes in the
phase of support from 87° to 158°* in the tapir, from 106° to 123° in the ass
and from 125° to 147° in the horse, that is, if we compare the amplitude of
movement in this joint in the tapir and in the Equidae, it is clear that the
overall amplitude in the phase of support is considerably reduced in the
Equidae: from 70° in the tapir to 20—25° in the horse.   Furthermore, a
comparison of the minimal indexes of talocrural joint angle indicates that
there is a progressive straightening of the joint in this series of perissodactyls (87° in the tapir, 106° in the ass and 125° in the horse).   In the
phase of free transit of the limbs flexor-extensor movements in the talocrural joint attain 110° in the tapir but not more than 70° in the horse and
ass.   It is much harder to measure the amplitude of movement in the knee
joint, and especially in the hip joint, according to motion pictures because
the position of the femur, which constitutes one of the sides of the angle,
is determined only approximately.   So is the position of the pelvis and hip
joint, making it even more difficult to make accurate measurements of the
angles.   Doubtful differences in the amplitude of movement in the knee joint
during the phase of support are therefore discarded.   In the phase of free
transit of the limbs flexor-extensor movements in the knee joint are around
80—85° in the tapir and 30—60° in the horse and ass.   The amplitude of

* For each phase the minimum and maximum indexes of the angles in the joints are given rather than the
   indexes at the moment of landing and the moment of pushing away.

flexor-extensor movements in the hip joint in the phase of support is about 60° in the tapir and only 10–15° in the horse and ass, while in the phase of free transit it constitutes 65–70° in the tapir and 20–25° in the Equidae. Comparing the gallop of the tapir and of the horse (Figures 71 and 74), we may note some smoothing out of the differences in the amplitude of movement of the hind limb joints, since in the horse there is an increase in comparison with the trot while in the tapir there is a decrease. Nevertheless, as with the trot, the overall nature of the differences is more or less preserved, in other words, a much greater amplitude of flexor-extensor movements in the hind limb joints is observed in the tapir than in the horse and ass. It is somewhat more difficult to determine the difference in the amplitude of movement in the joints of the forelimbs. However, a comparison of the drawings showing the trot and gallop of these animals (Figures 70–74) shows that the forelimbs also display the same differences, that is, there is a lesser amplitude of movement in the forelimb joints in the Equidae than in the tapir.

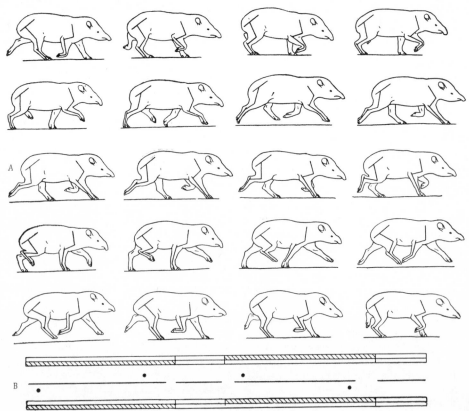

FIGURE 70. Movement of the appendicular skeleton of the tapir (T a p i r u s  a m e r i c a n u s) during a trot (A) and its support graph (B)

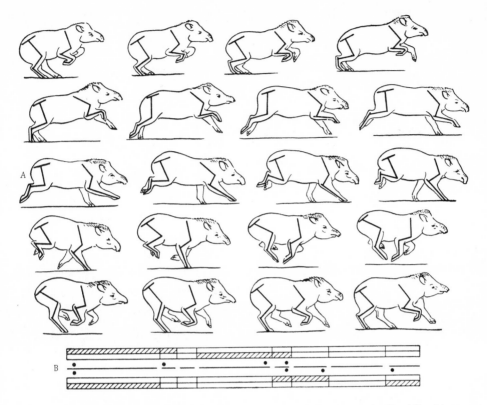

FIGURE 71. Movement of the appendicular skeleton of the tapir during a light lateral gallop (A) and its support graph (B)

Apart from these differences between the Tapiridae and Equidae in the flexor-extensor movements in the limb joints, considerably more clearly defined vertical fluctuations of the center of gravity are observed during the gallop. During an experimental run of the tapir and a pony* around a circular arena in the Erevan Zoological Garden it turned out that in the pony the vertical movements of the withers relative to the ground did not attain 6% of the animal's height in the withers, whereas in the tapir they were at least 14%.

Comparing the trot and gallop of the tapir, we see that the vertical fluctuations of its center of gravity are markedly larger with the gallop than with the trot. The opposite is true of the Equidae. This is demonstrated by an analysis of both motion pictures and tracks. The latter showed that during the fastest trot of the ass, free flight attains 150 cm and even 175 cm,

* We chose a pony for the experiments as it is closest in height to a tapir (in the tapir the height in the withers is 120 cm and in the pony 115 cm). However, the pony reached a speed of 45 km/hr and the tapir only 35 km/hr.

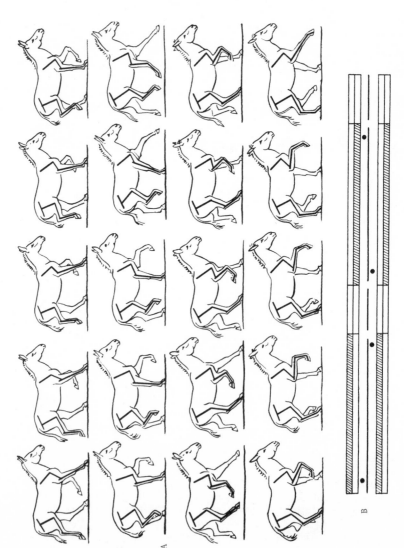

FIGURE 72. Movement of the appendicular skeleton of the Asiatic wild ass (Equus hemionus) during a fast trot (A) and its support graph (B)

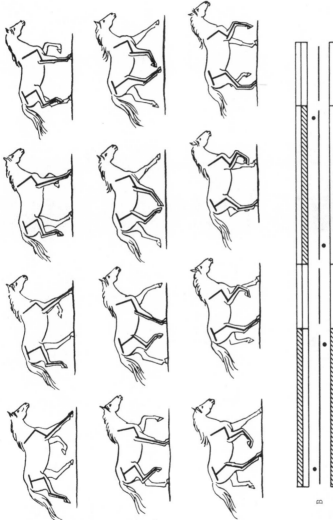

FIGURE 73. Movement of the appendicular skeleton of the horse during a fast trot (A) and its support graph (B)

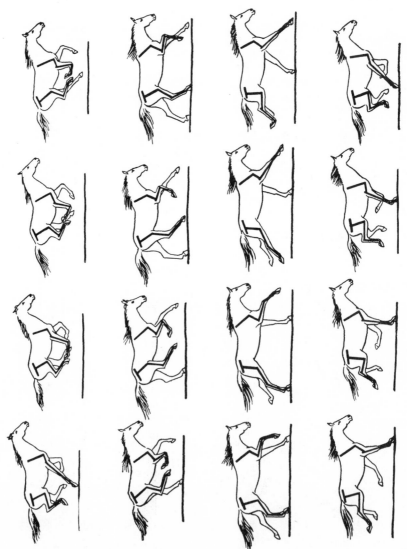

FIGURE 74.  Movement of the appendicular skeleton of the horse during a heavy diagonal gallop

and during the gallop 150—165 cm, i. e., in fact no more than during the trot.  Calculations of running speed from an analysis of motion pictures showed that with the fastest trot the speed of the ass does not exceed 45 km/hr while with the gallop it reaches even 80 km/hr.  Computing the angle of departure from formula (1), we can determine also the approximate vertical movements of the center of gravity in the stage of flight during the trot and the gallop according to the formula

$$Y = \frac{V^2 \sin^2 \alpha}{2g} ,$$

where $y$ is the height attained by the center of gravity.  We find that in the ass the absolute vertical movements of the center of gravity exceed 2.3 cm during the trot but do not attain 0.7 cm during the gallop, in other words, the difference is more than triple.  Apart from being better adapted for swift running, the Equidae also show greater stamina than the Tapiridae.  Unfortunately, no direct experiments on studying endurance in tapirs are available. But where the Equidae are concerned we can judge their endurance from records of running times and distances in domestic horses and from observations of Asiatic wild asses in nature.  For instance, a month-old ass was captured and turned loose after nine dummy horses, and the animal was able to be corraled only at the 120th kilometer (Kolmakov and Vasil'ev, 1936). Observations of a herd of asses in spring 1962 also demonstrated the exceptionally great stamina of these animals.  One of the observations was as follows.  On 6 May 1962 the herd's stallion drove off intruding males eight times in the course of four hours.  In order to chase away the rival, each time it had to run around the entire herd, which consisted of 147—149 females with young.  On two occasions this tour around the herd lasted more than ten minutes.  Running speed during pursuit was close to 80 km/hr (calculated from motion pictures).  Apart from this, within the same period the stallion chased a mare six times and mounted her four times.  Once it managed to catch up with the rival male and scuffled with him for 3—4 minutes. During these four hours the stallion did not spend more than 7 or 8 minutes resting or grazing.  Hence, the stallion of a herd hardly gets a minute's rest.  He keeps up this hectic schedule for several days, and then usually part of the herd is taken over from him or else a less tired rival takes his place (Solomatin, 1964, 1965).

Considering the findings on the mechanics of running of tapirs and representatives of the Equidae employing different gaits, we may note the following.  Dwelling in the forest, with its manifold obstacles in the form of twigs, fallen tree trunks, shrubs, etc., led to the result that with its slow trotting movement the tapir became adapted to pulling its legs up with minimal displacements of the center of gravity.  During a gallop tapirs clear obstacles by raising the center of gravity high and not drawing the limbs up so much. Hence, we observe greater amplitudes of movement in the limb joints and smaller movements of the center of gravity in a trotting tapir than in a galloping animal.  Horses, on the other hand, encounter obstacles very rarely while running, and therefore the amplitude of movement in the limb joints is increased in order to accelerate motion, i. e., during the gallop.  At the same time, high speeds of motion call for maximum economy, and this is achieved by sharply reducing the vertical movements of the center of gravity during galloping. With the trot, however, vertical movements of the center of gravity make it easier for the limbs to cope with uneven ground. The lower speeds attained during trotting do not place such strict demands on lessening the vertical movements of the center of gravity, and these become markedly larger both absolutely and relatively.

The differences in types of running discussed above characterize two main trends of specialization of the Perissodactyla for land locomotion. The running typical for the tapir is called the "battering-ram" form, while the form adopted by the Equidae is known as the "cursorial" form. Both these forms appear in the Artiodactyla as well. For instance, the battering-ram form has been developed in swine.* Habitat in dense thickets has left a mark not only on the type of flexor-extensor movements in the limb joints of tapirs and swine, but also on the whole appearance of the animal, which has to part a thick shrub with its head while running. Characteristic for such animals are: deep-set eyes, covered by a pecten, ears sunken in a fold, and a "compressed" type of exterior.

### Artiodactyla

Unlike the Perissodactyla, the number of species and genera of which is very limited in the recent world fauna, the diversity of recent Artiodactyla is very great. Cursorial specialization is also naturally much more varied in this order. The phylogenetic changes of artiodactyls, like those of perissodactyls, are related to a progressive enhancement of the locomotory function of the limbs. At all stages of cursorial specialization groups appeared which were associated with different habitats: forests, rocky areas and open spaces. In open spaces running was the only means of escaping from enemies, and therefore in such habitats the swift and enduring type of running (cursorial type) was developed, similar to the analogous type in the Perissodactyla. In a forest and over rocky ground precisely calculated leaps and sharp turns are often of greater value than speed. Hence, settlement in rock, forest and shrub habitats primarily, and probably also secondarily, led to a decrease in the speed of running.

Strict confinement to any of the above habitats presents certain requirements regarding the mechanics of running. However, the paths of adaptation even to identical conditions may be completely dissimilar. Moreover, open spaces, forests and crags can be of very different types. For instance, certain types of tundra, steppe, semidesert and desert are all open spaces. The mechanics of motion in tundras, which are covered with snow for more than eight months in the year, will of course differ from that in steppes. We must also remember that there are very many different kinds of tundras. It is hard to compare conditions for running in deserts too, as for example running

---

* Swine are also specialized for digging (Gambaryan, 1960), and this makes it hard to explain the connection between the mechanics of their motion during running and the type of change of the bone-muscle mechanisms. This is one of the reasons that these animals are not included in this work.

in sand and running on takyrs, or in stony deserts, etc.   But all these
different varieties of open spaces make one basic demand on the mechanics
of running — adaptation to speed and endurance as the sole means of fleeing
from predators.

Forest habitats are also varied.   A high-stemmed sparse forest enables
animals to muster quite considerable speeds, while dense undergrowth and
shrubs slow them down.   The presence of many fallen tree trunks and
branches turns running into an obstacle course.   Thus the demands made
on the mechanics of running of ungulates confined to forest habitats are
most diverse.   There are much better possibilities for hiding from an
enemy in a forest than in an exposed area.   Racing from one densely overgrown part to another, an animal keeps out of sight.   Therefore, while adaptation to speed is still of value where forest habitats are concerned, endurance comes to rank next in importance.

There are likewise many different types of rocky habitats.   Mountain
spurs crisscrossed with gulches and with rock outcrops force animals to
leap great distances from crag to crag.   One such successful leap may
render further pursuit of an ungulate pointless for the predator.   In rocky
habitats, high-speed running may prove simply dangerous, and each jump
is performed from a spot, the distance precisely calculated.   Adaptation to
life among rocks therefore brings about a decrease of speed.   Much less
stamina is also needed, and in place of it the animal acquires the ability to
perform larger, more accurately measured jumps.

But it is not only choice of habitat which affects the mechanics of running.
Variations of body mass (from 1 kg to several tons) and mode of feeding may
have a very significant influence on the mechanics of running.   As the most
graphic example of the influence of one of these factors let us present the
mode of feeding of the giraffe.   Plucking leaves from tree crowns has led
to the need for much longer legs, and lengthening of the legs in turn brought
about a change in the conditions of locomotory mechanics if only because
with uniform pace angles longer legs lead to an increased range of vertical
movements of the center of gravity in the phase of support (Figure 75).

Probably also of some importance is the previous history of the group,
as a result of which the paths whereby the locomotory mechanics changes
may differ even in the presence of identical trends of specialization.

It is particularly difficult to analyze different trends of locomotory
mechanics in view of the lack of special experiments and because of the
heterogeneity of the initial material.   Still, the differences between the extreme variants are so great that they are differentiated clearly enough.
To establish criteria for different types of mechanics of running we must
first of all examine endurance and the vertical fluctuations of the center of
gravity.   Differences in endurance can be gauged from a series of observations on animals adapted to one of the habitats described above (rocks,
forests, open spaces).   The artiodactyls adapted to open spaces which we
chose to study were the saiga Saiga tatarica and the reindeer Rangifer tarandus.   When pursued by a motorcycle traveling at 40 km/hr, a
group of saigas ran for several hours without pausing to rest and without
any visible signs of fatigue.   When a three-year-old male was chased out
of the group and pursued at a speed of 75—80 km/hr, he kept up the chase

for 15—16 minutes, after which he collapsed in total exhaustion. According to Sokolov (1951), a saiga ran at 80 km/hr for 15 minutes.

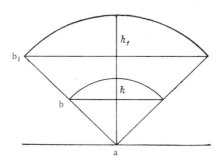

FIGURE 75. Absolute vertical movements of the center of gravity ($h$ and $h_1$) as a factor of leg length (ab and $ab_1$) with uniform pace angle

Reindeer (seven males aged from three to eight years) frightened by a shot were chased by a husky for 35—40 minutes, during all which time they maintained a gallop with a speed of at least 60 km/hr.

When we followed two 6—7-year-old wild goats (Capra hircus aegagrus) from a helicopter we found that they develop a maximum speed of 45—46 km/hr and become utterly exhausted after keeping up such a speed for 5—6 minutes. The goats were running along a well-beaten track with a slight downward slope (around 5—6°).

During a helicopter census of ungulates in the Khosrovskii Reservation one male axis deer (Cervus nippon), 5—6 years old, strayed from the herd and raced down a slope, doing about six kilometers in six minutes. The animal sped through several areas of the forest and a number of meadows bordered by shrub thickets. It easily sprang across a bush even 1.8 m high. At the end of the path the deer hid among the bushes, sank down on its belly and, lowering its head, breathed heavily. These observations show that the champions in speed and endurance are the inhabitants of open spaces, the saiga and the reindeer. The forest-dwelling axis deer can more or less compete where speed is concerned, but has nowhere near as much stamina (15 min at a speed of 80 km/hr and 6 min at a speed of 60 km/hr are indexes of very different endurance). Similar data on running speeds were obtained for the roe deer. This inhabitant of rocky areas yields to both the above groups in both speed and endurance (5—6 min at a speed of 45—46 km/hr). Observations of a goitered gazelle showed that this animal has roughly the same stamina as a saiga. Its maximum speeds of running are scarcely above 65 km/hr (60 km/hr according to Sokolov) and at such a speed the length of each running cycle is greater than in the saiga (Table 4).

TABLE 4. Types of footfalls and angles of departure in some ungulates during the gallop and the trot

| Species | Average length of part of pace, % of pace length* | | | | Absolute length of pace during the gallop, m | Galloping speed, km/hr | Angle of departure during the gallop | Absolute length of pace during the trot, m** | Trotting speed, km/hr | Angle of departure during the trot |
|---|---|---|---|---|---|---|---|---|---|---|
| | hind step | extended flight | front step | crossed flight | | | | | | |
| Cervus nippon ......... | 8 | 65 | 13 | 13 | 6.0 | 55 | 5° | 3.5 | 35 | 3°30' |
| Capreolus capreolus .... | 5 | 74 | 11 | 10 | 6.5 | 55 | 6° | 3.5 | 35 | 3°30' |
| Rangifer tarandus ....... | 16 | 40 | 24 | 20 | 4.3 | 65 | 1°30' | 5.0 | 40 | 6° |
| Saiga tatarica .......... | 15 | 37 | 16 | 31 | 4.0 | 80 | 0°50' | 4.5 | 40 | 5°8' |
| Gazella subgutturosa ... | 18 | 49 | 14 | 18 | 5.0 | 65 | 2°10' | 4.0 | 40 | 4°30' |

* The hind step is equal to the distance between the hind footfalls, extended flight equals the distance from hind to front footfall, crossed flight is equal to the distance from front to hind footfall, and the front step equals the distance between the front footfalls (this is measured along the direction of the animal's path).

** During a fast trot free flight is more or less equal to half a full pace.

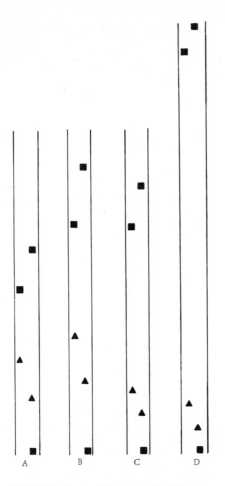

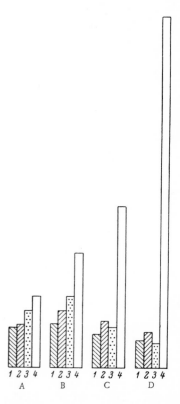

FIGURE 76. Tracks of Ungulata with different
running speeds:

A) similar tracks of the saiga, gazelle and
mouflon with a low running speed (30–40 km/hr);
B) tracks of Saiga tatarica, 70 km/hr;
C) tracks of Gazella subgutturosa, 65 km/hr;
D) tracks of Ovis orientalis, 60–65 km/hr.
Triangles denote the hind feet, squares the forefeet.

FIGURE 77. Diagram showing the changes
in the size of different parts of a pace depending on the speed of running:

A) slow gallop; B, C, D) running at maximum speeds: B) saiga (70 km/hr); C) gazelle (65 km/hr); D) mouflon (60 km/hr).
1) hind step; 2) front step; 3) crossed
flight; 4) extended flight.

It is interesting that the differences in the mechanics of movement, and
therefore also in the tracks, are brought out most distinctly at maximum
running speeds. For instance, it was found that during a lateral gallop at
a speed of 45 km/hr it was hard to distinguish among the tracks of the saiga
(high-speed running), mouflon (typical saltatorial-cursorial running) and

gazelle (saltatorial-cursorial running tending toward cursorial running) (Figures 76, A; 77, A). At this speed the front and hind step of all of them was small and extended flight was hardly any more considerable than crossed flight. In the saiga running at a speed of 70 km/hr (Figures 76, B; 77, B) all the elements of the full pace increase almost to an equal extent (hind step by 13%, front step by 25%, extended flight by 56% and crossed flight by 29%). In the gazelle running at a speed of 65 km/hr extended flight increases markedly (by 130%) while the other elements increase negligibly (by 5—6%, or they even decrease by 15—20%). In the mouflon, on the other hand, at a speed of 60—65 km/hr (Figures 76, D; 77, D) extended flight is immensely increased (by 400%) while the hind and front steps and crossed flight decrease appreciably (by 20—55%).

Tracks indirectly give us an idea of the vertical fluctuations of the center of gravity. Knowing the approximate running speeds and measuring the magnitude of free flight according to tracks, we can calculate the angle of departure, which may serve as an approximate index of these vertical fluctuations (Table 4).

We see from Table 4 that in artiodactyls adapted to living in forests, the axis and roe deer, extended flight will constitute the largest part of a full pace, and this with an overall increase of the full pace and a small decrease of running speed leads to a marked enlargement of the angle of departure in comparison with that in artiodactyls which live in open spaces (reindeer, saiga, and to a lesser extent the gazelle). Apart from this it is to be noted that in a forest-dwelling artiodactyl the length of a full pace (cycle) is smaller during trotting than during galloping, whereas in inhabitants of open spaces, on the contrary, it is larger. This shows that in the latter animals the vertical fluctuations of the center of gravity are larger with the trot than with the gallop, while in forest artiodactyls the opposite is true.*

Specialization for one-time large leaps comes about through adaptation to life among rocks. The following observation can give us an idea of possible sizes of jumps. Five male wild goats aged between two and six years nimbly leaped across an abyss 13.5 m wide. Before making the jump, each of them came to a halt on a small ledge, bent all their legs under and leaped. They landed on a bench on the opposite side of the abyss which was situated at an angle of 100° to the line of the animals' flight. Landing on their forefeet, the goats threw the hind part of their body sideways and upward, so that the hind feet also touched down on the bench, which was not more than 20 cm wide. The bench was at a slightly higher level than the ledge, so that a rope stretched from the ledge to the bench formed an angle of 1—2° with the horizontal plane.

* This statement contradicts the view held by Sokolov, Klebanova and Sokolov (1964), who write (p. 346): "When an animal is traveling at a rack or trot the body moves primarily in the horizontal plane and undergoes very little vertical displacement." And further on: " ... progression by the gallop or saltation involves considerable displacements of the body in the vertical plane as well." In fact, in all artiodactyls and perissodactyls adapted to life in open spaces, and at any rate in the saiga and gazelle (which are discussed in the work cited), both the relative and the absolute movements of the center of gravity in the vertical plane are probably always larger with the trot than with the gallop. It is not for nothing that in describing the gallop of steppe animals, the expression "the animal seems to creep along the ground" is often used. This different degree of displacement of the center of gravity during the trot and gallop is well known by all riders, who receive severe jolts when a horse is trotting.

FIGURE 78.  Heavy diagonal gallop of the reindeer (Rangifer tarandus) (A) and its support graph (B)

FIGURE 79.   Trot of the reindeer

FIGURE 80.   Group of reindeer running at a heavy lateral gallop.   Photo by the author.

FIGURE 81.  Movement of the reindeer's appendicular skeleton during a heavy lateral gallop (according to the photograph in Figure 80)

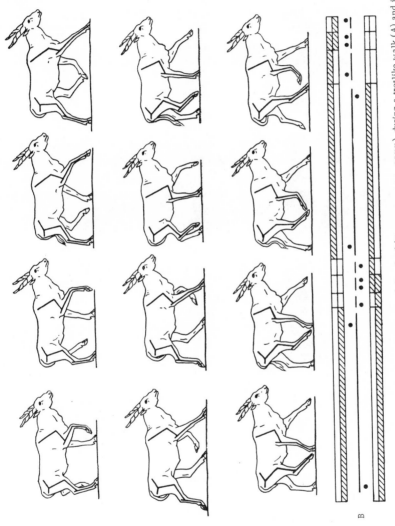

FIGURE 82. Movement of the appendicular skeleton of the eland (Taurotragus oryx) during a trotlike walk (A) and its support graph (B) (after Muybridge, 1887)

FIGURE 83. Movement of the appendicular skeleton of the gnu (Connochaetes gnou) during a heavy diagonal gallop (after Muybridge, 1887)

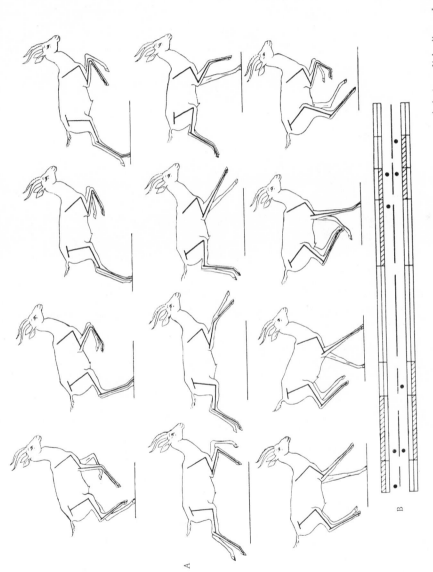

FIGURE 84. Movement of the appendicular skeleton of the goitered gazelle (Gazella subgutturosa) during a light diagonal gallop (A) and its support graph (B)

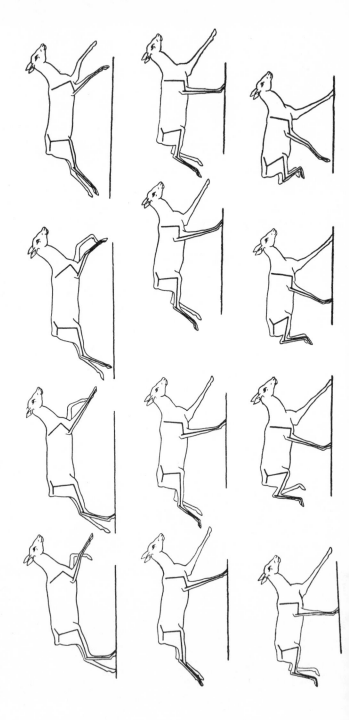

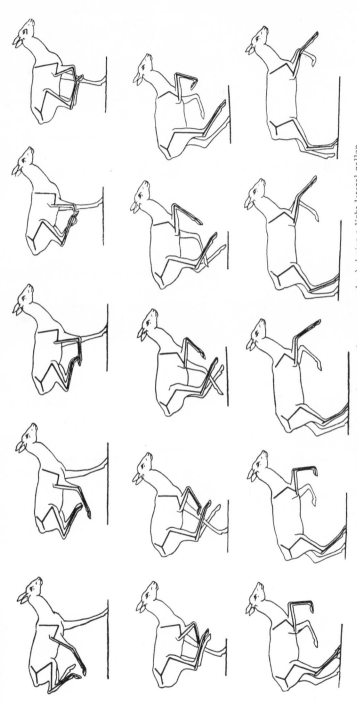

FIGURE 85.  Movement of the appendicular skeleton of the roe deer (Capreolus capreolus) during a light lateral gallop

In comparing the amplitudes of movement in the limb joints during the trot and gallop of the reindeer (Figures 78—81), the trot of the eland (Figure 82) and the gallop of the gnu (Figure 83), gazelle (Figure 84) and roe deer (Figure 85), no distinct differences can be detected in these animals which are adapted to different habitats. The probable reason for this is that the various species were filmed under quite different conditions. Whereas the reindeer were photographed while running at top speed, the gallop of the gazelle and roe deer was filmed in the presence of very low speed indexes.

Furthermore, even animals which are adapted to similar habitats sometimes show a different amplitude of movement in the limb joints, that is, the same effect may be achieved in different ways. We will return to this problem several times when discussing the causes of morphofunctional changes of the locomotory organs. At this point we will take a specific example to analyze various biomechanical forms of adaptations to similar habitat conditions and mode of life.

A behavior study of the chamois and tur, carried out in a reservation in the Caucasus, showed that these animals make for rocks to hide when danger looms. Leaping from crag to crag is their favorite way of shaking off pursuers. Observations of their running speeds revealed that neither species can probably muster a speed above 40 km/hr. The usual running speed indexes when the animals were heading for nearby rocks where they would be inaccessible were at most 20—28 km/hr. We never noted a speed of above 30 km/hr. These two species must therefore be placed in the group of artiodactyls which are adapted for one-time jumps with characteristic features in their mechanics of motion: adaptation for large leaps, considerable vertical fluctuations of the center of gravity, low running speeds and poor endurance. The similarity of their habits, choice of habitat, typical speeds of running and stamina seem to indicate that the whole mechanics of motion would be similar. There are, however, marked differences between the two forms in the movement of the femur. If we construct a vertical plane from the hip joint, in the chamois the distal end of the femur does not extend back beyond this in any of its movements (Figures 86 and 87), while in the tur it readily does so even during the walk (Figure 88), and especially during the gallop (Figure 89).

Interestingly, this feature of the femur's motion is typical for the whole tribe Caprini (defined according to Simpson, 1945) but not for other representatives of the families Bovidae and Cervidae. The mere fact that both during jumps and in various poses where we might expect the femur to be strongly deflected backward it is in fact so deflected only in members of the tribe Caprini is enough to show that this restriction in the movement of the femur is fixed. For instance, comparing the leap of Ovis musimon (Figure 90) and Cervus dama (Figure 91), or the vertical stance of the Siberian ibex (Figure 92) and of the gerenuk (Figure 91), we see that there is much more of a backward deflection of the femur in the Caprini than in the other Bovidae. These examples bear evidence that even similar demands can be fulfilled by different biomechanical systems.

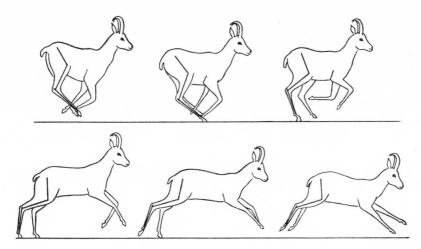

FIGURE 86. Movement of the hind limb skeleton of the chamois (Rupicapra rupicapra) during a light lateral gallop

FIGURE 87. Position of the hind limb skeleton of the chamois standing on its back legs

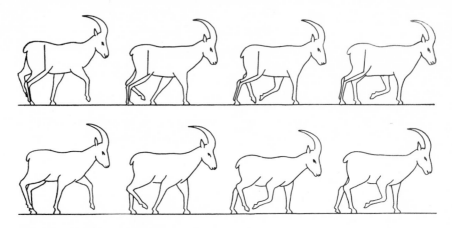

FIGURE 88. Normal walk of the tur (Capra caucasica)

The phylogeny of the Artiodactyla and Perissodactyla was frequently marked by the appearance of branches in which forms with markedly increased body dimensions were observed.* A greater body mass makes special demands on economy in running. One way to raise the efficiency of running will be to decrease the vertical fluctuations of the center of gravity, and this in turn can be done by decreasing the pace angle and the amplitude of flexor-extensor movements in the limb joints, especially the central elements (elbow and knee joints).

A comparison of the gallop of the bison (Figure 93) with the gallop, trot and even the walk of other artiodactyls and perissodactyls (Figures 70—74, 78—86, 88 and 89) shows clearly that in the bison 1) the pace angle decreases and 2) there is a lesser amplitude of flexor-extensor movements in the knee and elbow joints. The angle of the bison's hind step is 38° while in the other animals shown in Figures 70—75 and 78—89 the angle ranges between 47° and 65°. The front step angle is 53° in the bison and between 60° and 95° in the other animals. In the knee joint the amplitude of flexor-extensor movements in the phase of support is 24° in the bison and more than 30° in the others. In the elbow joint the respective figures are 23° and at least 35°.

In the giraffe, specialization for feeding on leaves from the top of tree crowns led to a marked elongation of the legs. Along with increasing in length, the limbs became straighter, making the animal still taller. For complete straightening of the leg the sum of the angles in the talocrural and knee joints must equal 360°. In the giraffe the sum of the angles of these joints in the phase of support is not less than 250° (Figure 94) and at least 270° during a slow racklike walk (Figure 95). In the other ungulates, only with an extreme degree of limb straightening may the sum of

---

* An increased body mass, like other structural features of animals, is the result of an adaptive phylogenesis. There are probably various reasons that the body increases in size, but at any rate the result is that new conditions for the mechanics of motion arise, and the animal adjusts to them.

FIGURE 89.  Gallop of the tur

the angles in these joints approach the value in the giraffe, while in the phase of support the sum is often below 200°.  The lower limit is 220° only in the bison.

FIGURE 90.  Leap of O v i s   m u s i m o n  (A)   (after Hájek, 1954) and leap of  C a p r a  a e g a g r u s  across an abyss (B) (drawn by the author)

Gregory (1912) thought that phylogenetic changes of the ungulate skeleton depend on the degree of adaptation to speed and endurance and that they reflect the influence of the weight of the body.  In view of this he wanted to give a clearer picture of the evolution of these animals and classified their running into four forms: 1) subcursorial, which is analogous to the battering-ram type we have described for the tapir; 2) cursorial, which characterizes

the ungulates living in open spaces;  3) mediportal, typical for large artiodactyls and perissodactyls;  4) graviportal, which developed earlier in the
group of Proboscidea (see p. 168).   Gregory was thus the first to put some
order into the classification of running in the Ungulata, which contributed
greatly to a correct interpretation of the structure of the skeleton in these
animals.

The mechanical features of running and saltation in different groups of
artiodactyls that we have discussed above enable us now to define the following forms of running.

**Cursorial form,** analogous to the cursorial form in the Perissodactyla.
Characterized by adaptations to maximum speeds and endurance.   High endurance is promoted by the very insignificant vertical fluctuations of the
center of gravity, which are conditioned by small angles of departure in
each cycle.   The vertical fluctuations of the center of gravity are particularly small in ungulates which are adapted to cursorial running during their
fastest gait, the gallop.   With slower gaits (the trot and probably the rack)
the angles of departure are larger and the absolute fluctuations of the center
of gravity greater.   The cursorial form of running has been developed in
dwellers of open spaces, as for these animals swift and prolonged running
is the only means of escape from enemies.

A                                              B

FIGURE 91. Leap of the fallow deer (C e r v u s   d a m a ) (A) (after Muybridge,  1887)
and the gerenuk (L i t o c r a n i u s   w a l l e r i) standing up on its hind legs (B) (after
a photo by Grzimek taken from Schomber, 1962).

**Saltatorial-cursorial form.** Characterized by less stamina plus increased vertical fluctuations of the center of gravity during the gallop and decreased fluctuations during the trot. The reason for enhanced vertical fluctuations of the center of gravity is that there is a marked increase in the angle of departure in each cycle, which leads to a lengthening both of the stage of free flight and of the full pace as a whole. This type of running has developed in forest animals. Narrow specialization for individual types of forest may lead to a considerable variation in the mechanics of running. Both endurance and speed are diminished in dense thickets, whereas in high-stemmed, sparse forests speed may show a negligible decrease in comparison with that of animals adapted to the cursorial form of running. This is especially typical for animals living in sparse shrub habitats (the gazelle) whose running speed approaches the cursorial.

**Saltatorial form.** Characterized by adaptation to one-time jumps, the economy of which is raised in proportion to the increase in the angle of departure of the center of gravity. Therefore, whereas for the first two forms of running cursorial adaptation is accompanied by a decrease in the angle of departure in each cycle of motion, with the saltatorial form development proceeds in the opposite direction. At the same time, adaptation to saltatorial running leads to a reduction in speed and endurance. The mechanics of motion of individual forms with saltatorial adaptation may in turn differ according to the type of movement of the femur in the phase of support. In a number of forms (the genera Capra and Ovis) the femur is deflected backward more strongly than in other ungulates. And in other forms (the genus Rupicapra) the norm of femoral deflection remains characteristic for all the Ungulata.

**Mediportal form.** Characterized by a decrease in the pace angle and in the amplitude of flexor-extensor movements in the elbow and knee joints in the phase of support, this causing an additional decrease in the vertical fluctuations of the center of gravity. This form of running develops in ungulates in connection with an increase of the body mass.

**Stilt form.** Characterized by a straightening of the elbow, wrist, knee and talocrural joints in the phase of support. It develops in giraffes as a special adaptation to taking food from tree crowns, and leads to very long, straight legs.

Despite the marked differences between these forms of running in different groups and species of ungulates, the one feature they have in common is that the hind and forelimbs function exclusively as locomotory organs. Apart from this, during the whole running cycle the vertebral column of all Ungulata remains rigid, owing to which the spine cannot promote an increase of the size of a full pace. For this reason we have coined a general term for the running of ungulates — "dilocomotory dorsostable" mode of running.

### STRUCTURAL FEATURES OF THE SPINE

The rigidity of the vertebral column typical for all ungulates during running is promoted by a system of spinous processes and supraspinous ligament and also by a system of transverse processes of the lumbar vertebrae with the intertransversal ligament and the hinges of the articular processes of the lumbar vertebrae. All these formations have much in common in all representatives of the Ungulata. The spinous process of the first thoracic vertebra is usually small and directed vertically (Figure 96). Spinous processes are either absent or weakly developed in the cervical vertebrae and have a gradually increasing forward inclination. The spinous processes of thoracic vertebrae II and III and sometimes up to V become increasingly longer and then progressively shorter again to the last one. At the same time each successive spinous process is sloped more and more backward as far as thoracic vertebra X. Further on the processes again become increasingly vertical, and the 11th, 12th or 13th processes, known as "anticlinal" (Dombrovskii, 1935), are vertical. The forward-inclined spinous processes of the lumbar vertebrae come up to the anticlinal process. The processes of the lumbar vertebrae are wide bony plates distributed in the sagittal plane; their apexes are widened anteroposteriorly (Figure 97), so that the interspinal distance becomes minimal.*

FIGURE 92. Siberian ibex (Capra sibirica) plucking leaves from a shrub

---

* An exception are the smallest Tragulidae, in which a widening of the apexes is scarcely perceptible.

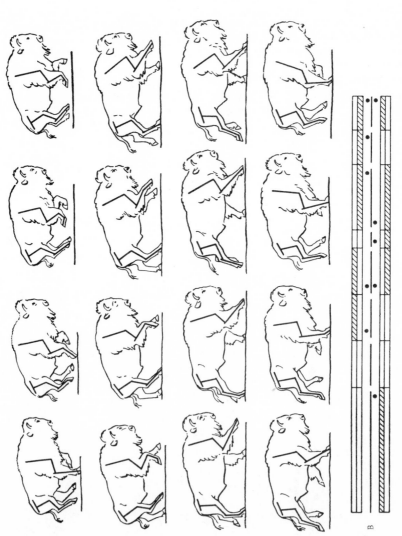

FIGURE 93. Movement of the appendicular skeleton of the bison (B i s o n  a m e r i c a n u s) during a heavy diagonal gallop (A) and its support graph (B)

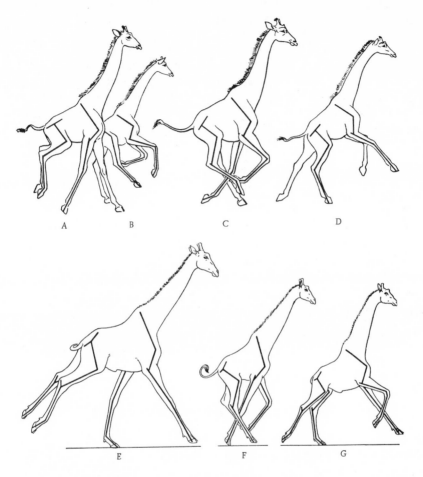

FIGURE 94. Movement of the appendicular skeleton of the giraffe during a heavy lateral gallop:

A, B, D) after photos by Grzimek taken from Bentley, 1961; C) from a photo by Grzimek taken from Das Tier, 1970, 12:43; E, F, G) after a photo by Okapia taken from Das Tier, 1966, 9.

The supraspinous ligament is stretched between all the spinous processes. It is strongest in the anterior part of the thoracic region, and it becomes denser and thicker again in the lumbar region. Brovar (1935, 1940) compared the spinous processes and the supraspinous ligament to a girder of rigidity (the term is used in the study of the strength of materials). Here the processes will be beams of compression and the ligament beams of expansion (Figure 98). This girder acts to restrict the possibilities for the spine to straighten. Lengthening of the spinous processes

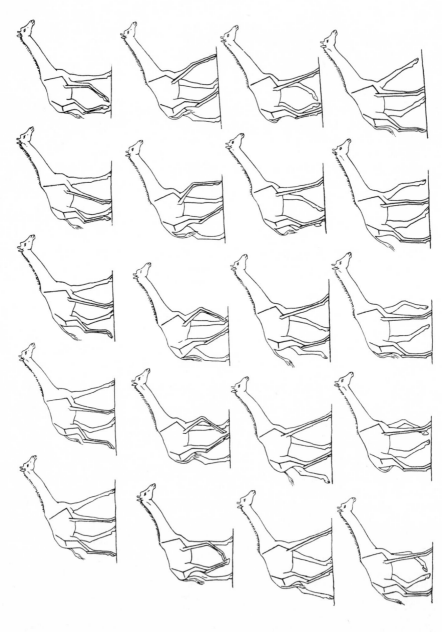

FIGURE 95. Movement of the appendicular skeleton of the giraffe during a slow racklike walk (every third frame is shown, filming speed 64 frames/sec)

thereby reflects an increase of extensor tension in various regions of the
vertebral column.   The greatest resistance to extension of the spine must
naturally be concentrated in the region where the forelimbs are attached
to the body (the region of the withers on the spine).   The tension of this
region at extension was graphically illustrated by Kummer (1959a, 1959b)
(Figure 99), who drew the spine as a beam resting on two supports.   The
part of the beam extending beyond the front support (region of the neck and
head) creates marked stress tending to straighten the spine in the region
of the front support (i. e., the region of the withers).   A specially marked
extensor effect on the spine in this region arises at the moment the animal
lands on the forefeet (Figure 100).

Expansion of the spinous processes with an increasingly pronounced
slope backward is useful from another point of view as well.   This is because the interspinous parts of the supraspinous ligament are thereby
lengthened, and the ligament, despite its fibrous, little-elastic structure,
still allows for limited changes of the parameters necessary for mitigating
the impacts imparted to the body via the forelimbs at the end of the stage
of extended flight.

The anteroposterior expansion of the apexes of the spinous processes
shortens the intervertebral parts of the supraspinous ligament (Figure 96),
and this enhances the rigidity of the lumbar region.   It is interesting that
the apexes of the processes are widened particularly strongly in ungulates
which are adapted to large leaps (Capra aegagrus, C. sibirica,
Moschus moschiferus, etc.).   The abrupt changes of body position
during jumps apparently increase the need for keeping the lumbar region
rigid.   On the other hand, the nimbly leaping small Tragulidae do not show
any widening of the apexes of the spinous processes in the lumbar vertebrae,
as all the other ungulates do.*  Because of the small size of these animals,
straightening moments which would necessitate special rigid arrangements
of the spine cannot arise in the lumbar region.

Apart from a reduction of dorsal flexion in the spine, there are diminished
possibilities for lateral flexures in the lumbar region, which are promoted
by expansion of the transverse processes and also by the formation of hinges
in the articular processes of the lumbar vertebrae.   In the Ungulata the
transverse processes of the lumbar vertebrae are situated in the horizontal
plane, are strongly developed and become markedly wider toward their ends,
to which the distinctly expressed intertransversal ligament is attached
(Figure 101).   The distinctive hinges formed by the articular processes
of the lumbar vertebrae are supplementary rigid devices.   Their nature
is expressed in the expansion of the cranial articular processes which en-
compass the caudal processes not only from the side but also from above.
They thus both restrict lateral flexures and create additional obstacles for
vertical flexures of the spine (Figure 97).   All these structures together
make up a sufficiently rigid formation which impedes lateral and vertical
bends of the spine.

The sacrolumbar region is most subject to flexor-extensor influences.
The heavy loads to be borne in this region cause mobility between the sacral

* Lack of material unfortunately prevented us from studying this family of artiodactyls in more detail.

and last lumbar vertebrae to be much greater than in the loin itself.
Increased mobility in this region is reflected in the growth of the spinous
processes, which are sloped forward on the last lumbar vertebra and backward on the first sacral vertebra.   This sloping of the processes leads to
a lengthening of the supraspinous ligament of this region. An interesting
point is that in the genus C a p r a  the ligament is lengthened owing to a
shortening of the spinous process of the first sacral vertebra, so that the
supraspinous ligament passes from the last lumbar directly onto the second
sacral vertebra (Figure 102) (Gasparyan, 1969).

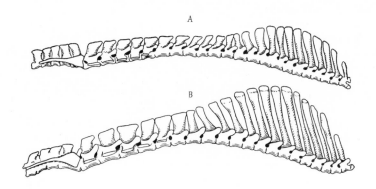

FIGURE 96.  Structural of the vertebral column in the Ungulata (lateral):

A) E q u u s  h e m i o n u s;  B) B i s o n  a m e r i c a n u s.

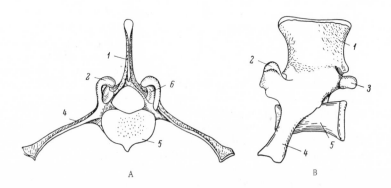

FIGURE 97.  Third lumbar vertebra of C a p r a  c a u c a s i c a:

A) anterior view;  B) lateral.  1) spinous process;  2, 3) articular processes:  2) cranial,
3) caudal;  4) transverse process;  5) body of the vertebra;  6) hinge of articular process.

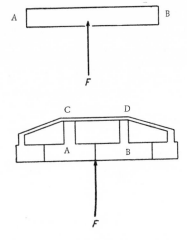

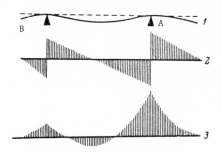

FIGURE 98.  Action of force $F$ on side AB and girder of rigidity ABCD:

AC and BD are beams of compression; CD is the beam of expansion.

FIGURE 99.  Curves of the distribution of forces on the spine (after Kummer, 1959a):

A and B  are triangles of support.  1) bending of beam;  2) breaking forces at points of support;  3) bending moments.

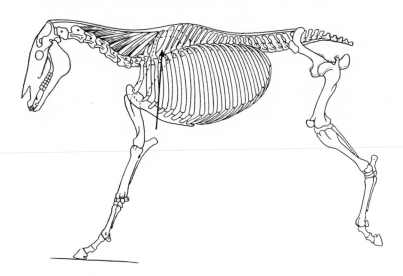

FIGURE 100.  Effect of the reactive force (arrow) on the vertebral column of the horse at the moment of touching down

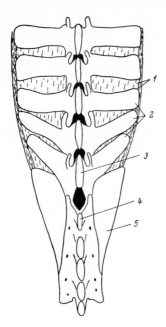

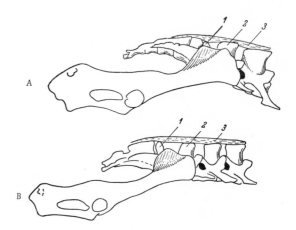

FIGURE 101. Structure of the lumbar region in E q u u s  h e m i o n u s :

1) intertransversal ligament; 2) transverse process; 3, 4) spiny processes of the last lumbar (3) and first sacral (4) vertebra; 5) wing of the sacrum.

FIGURE 102. Sacrolumbar region of the spine in C a p r a a e g a g r u s (A) and R u p i c a p r a  r u p i c a p r a (B), lateral:

1, 2) spiny processes of the first sacral (1) and last lumbar (2) vertebra; 3) supraspinous ligament.

It would seem that it would be easiest to maintain rigidity of the spine throughout the cycle of motion if the vertebrae were fully intergrown, as is the case in birds, for example. However, concrescence of the spine into a single unit greatly diminishes its shock-absorbing role, which is vital with high-speed gaits having a stage of free flight (all forms of the gallop).

## DISTRIBUTION OF LOADS ON THE LIMBS

A study of the loads on the individual limbs and of the various amplitudes of movement in the joints is necessary if we are to explain the causes of their changes upon specialization for different forms of running in the Ungulata. The high-speed gaits of animals adapted to battering-ram, saltatorial, saltatorial-cursorial and sometimes the cursorial form of running will be the light lateral gallop and, more rarely, the diagonal gallop; in other words, in all these animals the end of the phase of hind support is followed by a stage of free flight in the air. As a rule, the phase of support of one hind limb does not come to an end before the second limb enters into action.* The motion and role of the two limbs in the cycle are somewhat different. The hind limb to come down first is usually placed more vertically than the second, that is, the angle between the horizontal plane and the direction of the landing foot is 5—10° larger in the first limb than in the second (Figure 103). When the phase of support has ended, it is, on the contrary, the limb which landed first that pushes away at a greater angle to the horizontal plane than the second. This pattern of landing and pushing away probably comes about due to the following conditions of limb activity. We noted on pp. 64—65 that in each cycle of movement the limb joints change flexor movements into extensor movements four times and then vice versa, while in the phase of free transit three such switches of movements are made, and in the phase of support one switch. Before the leg comes down, the talocrural, knee and hip joints begin to be flexed while the digital joints begin to be extended, preparing for shock-absorbing movements. The moment the first limb touches down, the digital joints are greatly extended to begin to cushion the impact. Extension of the digital joints during flexion of the talocrural, knee and hip joints helps lower the body a little. Therefore, for trouble-free landing of the second limb in front of the first, it has to be more bent in the knee and talocrural joints before it makes contact with the ground (Figure 103). On the other hand, at the end of the starting period of the phase of hind support, extension of the talocrural, knee and hip joints and flexion of the digital joints of the first limb do not promote a stage of free flight in the air, but in the case of the second limb they do. In connection with this, extension and flexion of the joints are almost completed in the phase of support before the first limb pushes away, whereas in the case of the second limb these movements are completed in the first moments of the drawing-up period of the phase of free transit. We may naturally assume that the first limb will bear a greater load in the phase of support, since it takes the main weight of the body and brings to an end practically all movement in the joints. Studies of the load dynamics on the leg of a galloping horse showed that the greatest load actually falls on the first leg to land (Marei, 1875). This is probably why the leading limb is usually changed after a number of cycles.

With the heavy gallop, which is typical for the high-speed gait of the stilt and mediportal forms of running, and also in some animals adapted to the cursorial form of running, up to the end of the phase of support the load is distributed on the forelimbs as well.

CAUSES OF CHANGES IN THE LENGTH RATIOS
OF THE HIND LIMB SEGMENTS

All the species investigated in this work, adapted to different forms of the dilocomotory dorsostable mode of running, are animals that are highly

* One case of a very fast lateral gallop of a young Pamir argali (discovered from tracks), in which there was a stage of flight after each limb pushed away, may be considered an exception.

specialized for running. It is thus quite probable that the differences in the ratios of the limb segments characterize the form of running rather than the extent of specialization for it in the group analyzed.

The general theories on the changes in length of the segment which we examined in the preceding chapter allow us to proceed to analyze them in individual representatives. It is, of course, more correct and more convenient to conduct such an analysis within generic groups, so that we can then make even more general comparisons.

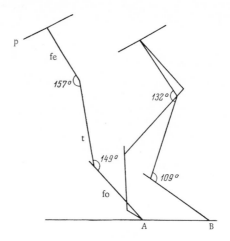

FIGURE 103. Sizes of angles in the hind limb joints as a function of the time of their landing:

A) leading leg;  B) second leg;  p — pelvis;  fe — femur; t — tibia;  fo — foot.

It may be considered that the few perissodactyls discussed in this book to a certain extent reflect the pathways of change in the length ratios of the hind limb segments in their phylogeny. Kovalevskii (1873) with good reason considered the tapirs, in which the ratio of the segments is similar to that in the ancestors of the horses, to be living fossils among the Perissodactyla. In the wild ass and zebra the foot and tibia are markedly longer and the femur slightly shorter than in the tapir (Table 5). The overall length of the limb segments is in the ass 24% and in the zebra 14% greater than in the tapir. Accepting that the ratios of the segments of the appendicular skeleton characteristic for the tapirs are original for the Perissodactyla, we have to assume that in the phylogeny of these animals a lengthening of the limbs took place, especially of the foot and tibia. However, due to the increase in the size of the perissodactyls, the relative size of the hind limb segments secondarily decreased; we can already see this in comparing the ass with the zebra, for in the latter the total length of the limbs is relatively reduced.

TABLE 5.  Changes in the size of the hind limb segments in connection with adaptation of ungulates to various forms of running

| Species | Relative size | | | | Absolute length of limb, mm | Weight of body, kg | Maximum running speed, km/hr | Form of running |
|---|---|---|---|---|---|---|---|---|
| | femur (a) | tibia (b) | foot (c) | total (a+b+c) | | | | |
| Tapirus americanus ......... | 37.3 | 30.0 | 38.6 | 105.9 | | 250 | 40 | Battering-ram |
| Equus quagga chapmani ..... | 32.8 | 34.6 | 53.5 | 120.9 | 1,085 | 400 | 70 | Cursorial |
| E. hemionus onager ......... | 36.0 | 38.2 | 57.6 | 131.8 | | 250 | 80 | Cursorial |
| Connochaetes gnou ......... | 32.6 | 39.3 | 51.7 | 123.6 | 980 | 300 | 90 | Cursorial |
| Taurotragus oryx ........... | 35.3 | 36.4 | 50.5 | 122.2 | 1,540 | 400 | 60 | Cursorial |
| Saiga tatarica ............. | 33.0 | 38.6 | 54.5 | 126.1 | 665 | 35 | 80 | Cursorial |
| Gazella gutturosa ........... | 31.3 | 37.9 | 55.0 | 124.2 | 725 | 30 | 80 | Cursorial |
| G. subgutturosa ............. | 36.7 | 45.8 | 65.7 | 148.2 | 710 | 30 | 75 | Cursorial with saltatorial-cursorial elements |
| Rupicapra rupicapra ......... | 43.8 | 55.0 | 68.0 | 166.8 | 865 | 50 | 40 | Saltatorial |
| Ovis ammon polii ........... | 39.7 | 49.0 | 59.2 | 147.0 | 964 | 80 | 60 | Saltatorial-cursorial |
| O. am. cycloceros .......... | 38.6 | 47.8 | 60.0 | 146.4 | 813 | 50 | 60 | Saltatorial-cursorial |
| Capra aegagrus ............. | 38.0 | 45.0 | 49.0 | 132.0 | 770 | 30 | 45 | Saltatorial |
| C. caucasica .............. | 38.9 | 45.6 | 49.5 | 134.0 | 685 | 70 | 45 | Saltatorial |
| Bison bonasus ............. | 35.8 | 37.0 | 46.6 | 119.4 | 1,410 | 1,500 | 55 | Mediportal |
| B. americanus ............. | 34.5 | 36.0 | 43.3 | 113.8 | 1,400 | 1,500 | 80 | Mediportal |
| Cervus elaphus sibiricus .... | 36.3 | 42.8 | 60.0 | 139.0 | 1,230 | 350 | 60 | Saltatorial-cursorial |
| C. el. elaphus ............. | 38.8 | 46.3 | 66.0 | 151.1 | 1,170 | 180 | 60 | Saltatorial-cursorial |
| C. dama .................. | 37.9 | 44.5 | 62.0 | 144.4 | 970 | 80 | 55 | Saltatorial-cursorial |
| Capreolus capreolus ....... | 41.7 | 51.7 | 69.9 | 163.3 | 830 | 50 | 60 | Saltatorial-cursorial |
| Rangifer tarandus ......... | 36.2 | 42.3 | 60.2 | 138.7 | 1,000 | 120 | 80 | Cursorial |
| Giraffa camelopardalis...... | 48.5 | 55.5 | 106.0 | 210.0 | 2,080 | 800 | 60 | Stilt |

Note.  The length ratio of the segments is given as a percentage of the sum of lengths of the lumbar and thoracic regions of the spine.  The forms of running have been described above.

With a very marked increase in the size of the body, the relative dimensions of the limb segments begin to decrease secondarily (Gregory, 1912).
We can ascertain this from the length ratio of the third metatarsal bone
and the femur (Table 6).*

TABLE 6. Changes in the ratios of the third metatarsal bone (Mtt III) in the phylogeny of the Perissodactyla
(after Gregory, 1912)

| Group, species | Length of femur, mm | Ratio of Mtt III and length of femur, % |
|---|---|---|
| Perissodactyla, Tapiroidea | | |
| Heptodon calciculus ...................... | 175 | 43 |
| Tapirus americanus........................ | 262 | 41 |
| T. indicus ................................ | 320 | 37 |
| | | |
| Perissodactyla, Hippoidea | | |
| Eohippus sp. ............................. | 162 | 50 |
| Mesohippus sp. ........................... | 178 | 68 |
| Hypohippus osborni ....................... | 278 | 78 |
| Neohipparion whitneyi .................... | 249 | 101 |
| Equus kiang............................... | 313 | 88 |
| E. scotti ................................. | 370 | 71 |
| E. caballus (racehorse) ................... | 392 | 73 |
| Hippidion neogeum ....................... | 340 | 62 |

The size of animals can be judged as a first approximation according
to the absolute length of the femur. As is seen in Table 6, the evolution
of the Perissodactyla is marked by a progressive lengthening of the foot,
this being expressed in an increase in the relative size of the third metatarsal bone. However, when the femur exceeds a length of 300 mm, the
relative size of the third metatarsal bone undergoes a secondary decrease,
so that according to the indexes it comes close to that in the ancestors of
the whole group of Eohippus sp. and Mesohippus sp.

In the recent perissodactyls studied here, the ratios of the limb segments
and their relative and absolute length agree well with their type of running.
Low running speeds in the presence of the constant need for jumping over
obstacles lead to the appearance of frequent one-time large loads (jumping
from one spot), for which legs constructed according to the principle of a
lever of force (long femur and shortened foot) are more advantageous.
Acceleration of running speed leads to a lengthening of the legs. The absolute length of the legs is markedly greater in the zebra than in the wild ass,
so that to keep pace with the ass the zebra can maintain the same number
of paces per unit time with lesser amplitudes of flexor-extensor movements

* The length ratio of the foot and femur is about the same in the zebra and ass.

in the joints, or, preserving the same amplitude of these movements, it may take fewer steps per unit time.  In all probability, the zebra's type of running changes along the lines of both fewer cycles of motion and diminished flexor-extensor movements in the joints.  This no doubt explains the lesser absolute speeds characteristic for the zebra and also the fact that it more rarely switches to the light gallop.  While the light gallop is the normal gait for the wild ass (observations in the Badkhyzskii Reservation), it is uncharacteristic of the zebra, judging from photographs of a herd fleeing in panic from an aircraft (Bentley, 1961).

Among the Cervidae the cursorial form of running is typical only for the reindeer; all the others show the saltatorial-cursorial form (Table 5). The light lateral gallop (Figure 85) is the basic high-speed gait characteristic for all deer; however, the reindeer more often employs the heavy lateral or the diagonal gallop (Figures 78 and 80).

As mentioned above, for the roe deer the extended stage of the gallop at a running speed of 60 km/hr is 6.0 m, while for the reindeer running at a speed of 65 km/hr it is only 1.8 m.  If we take into account that the reindeer has a greater absolute limb length (100 cm) than the roe (83 cm), it is clear that the roe needs to develop much greater relative forces in each cycle than the reindeer (more details in the previous chapter).  Lengthening of the legs promotes some lessening of the relative force (formula (9)).  Moreover, during each jump, relatively longer legs come closer to the vertical line dropped from the center of gravity, thereby facilitating an increased angle of departure with each jump.  The saltatorial-cursorial form of running naturally must lead to the development of relatively longer limbs than the cursorial form.  And, in fact, looking at Table 5, we note that of all the Cervidae, the reindeer shows the least relative leg length (138.7% of the length of the lumbar and thoracic regions of the spine, 144—163% in the other deer).  Only in the maral, which weighs 1.5—2 times as much as the reindeer, does this index approach that in the reindeer, being equal to 139.1%. It is interesting to note that it is precisely the foot, which has the highest indexes in the saltatorial-cursorial Cervidae, that becomes relatively shorter in the maral than in the tundra subspecies of the reindeer (we drew an analogy above for the Perissodactyla, in which an increase in body weight led to a relative shortening of the third metatarsal bone).

A considerable lengthening of the hind limb segments is observed in the giraffe, in which the total length of the segments is 2.1 times the length of the lumbar and thoracic regions of the spine.  It would seem as if the giraffe should be able to make particularly large leaps.  However, lengthening of the legs is in this case associated with the special mode of obtaining food.  The giraffe has therefore developed a specific form of running, known as the "stilt" form, in which there is a marked reduction of flexor-extensor movements in the limb joints and in the pace angle, the result being that the animal runs at a heavy gallop which does not have a stage of extended flight (Figure 94).  The giraffe's foot is particularly long, both absolutely and relatively (Table 5); yet during the phase of support it changes its position relative to the horizontal plane by at most 25°, which, naturally, does not contribute to large jumps.

The representatives of the Bovidae studied display a number of basic types of length ratios in the hind limb segments.  The first two types are characteristic of cursorial animals, in which the relative length of all the hind limb segments is not substantial, constituting 122.2—127.1% of the length of the lumbar and thoracic regions of the spine.  The longest of the three segments is the foot (50.5—55.0% of the length of the truncal region of the spine).  In the lighter species, the saiga and Mongolian gazelle (G a z e l l a  g u t t u r o s a), the foot indexes are 54.5 and 55.0%, while in the heavier eland and gnu they are 50.5 and 51.7%.  Thus, also in the Bovidae, an increase in body weight causes a secondary reduction in the size of the foot while the femur begins to increase in size.

The goitered gazelle (G. s u b g u t t u r o s a) approaches the saltatorial-cursorial animals according to its form of running and also according to the indexes of the relative length of the limbs and size of the foot (Table 5).

Whereas in the cursorial animals the reduction in the relative size of the foot and the enlargement of the femur find their explanation in an increased body mass (comparing the gnu and eland with the saiga and Mongolian gazelle), when we come to the two genera of Bovidae which are specialized jumpers (C a p r a and R u p i c a p r a), we see that the differences in the ratios of the segments cannot be explained by their size. Certain species of C a p r a are known to be either larger or smaller than the chamois. The difference cannot be attributed to the degree of saltatorial specialization. Take, for instance, the above-described 13.5-m leap of the wild goats.  In a Caucasian reservation we measured a chamois jump at 11.3 m.  During the leap the animal raised its level slightly: the spot on which it landed was 1.3 m higher than the horizontal plane constructed from the place of the jump.  In fact, if we translate these two jumps into horizontal ones, the difference in their length is nullified.  Proceeding from these data, we may assume that the differences in the length ratios of the segments in these two forms are brought about by differences of mechanics in the phase of support prior to the jump, and not by the extent or depth of their specialization for jumping.

We have already pointed out the different nature of movement of the femur in the tur and chamois.  A more detailed analysis of jumps performed by the tur and other species of the genus C a p r a on the one hand and by the chamois on the other showed the following points of similarity and dissimilarity in the saltatorial mechanics of these animals.  Before leaping, both the chamois and the goats squat on all fours.  They now tilt themselves backward so that the center of gravity often passes beyond the line of support of the foot.  As a result, either the animal lands full on the foot or it even adopts a sitting pose for a moment, looking like a sitting dog.  Both the chamois and the tur raise themselves from this position primarily due to extension of the elbow and shoulder joints, which raise and draw the trunk forward.  This is followed by extension of the hip, knee and talocrural joints, which promote a further acceleration of the forward motion of the body, right up to the point that the animal leaves the ground.  Then synchronized extension of the knee and talocrural joints begins in the chamois, while in

the tur extension of the talocrural joint is slightly delayed.* This difference also comes out in an analysis of the dynamics of change in the angles
of the hind limb joints in the chamois and tur in the course of the phase of
support during a lateral gallop.  In the chamois, therefore, the thrust of the
hind limbs, involving extension of the knee joint and synchronized talocrural
joint, maintains the favorable direction forward and upward up to the point
where the head of the femur and the talocrural joint intersect the vertical
planes, the first of which is drawn from the knee joint and the second from
the point of support of the limb (Figure 104).  Further extension of the knee
joint already impedes the necessary direction of motion of the head of the
femur even with continued extension of the talocrural joint.  This is because,
passing beyond the vertical erected from the point of hind limb support, the
talocrural joint itself begins to descend (Figure 104).  In the tur, only after
the head of the femur has intersected the vertical plane erected from the
knee joint does extension of the talocrural joint set in (Figure 104) which
gives the necessary direction when the head of the femur imparts the thrust
to the trunk in conditions of continuing extension of the hip and knee joints.
As a result, in the tur we find a much greater amplitude of hip joint extension, a somewhat greater amplitude of knee joint extension and lesser talocrural joint extension in the phase of support of the gallop than in the case
of the chamois.  Although the final size of the angle of the talocrural joint
is smaller in the tur than in the chamois, the intensity of its extension at the
end of the phase of support is greater in the tur.

The different sequences in which the joints go into action in the chamois
and tur and the different degree of extension of the joints at the end of the
phase of support indicate that there is an increase of loads on the talocrural
joint at the final moment of the phase of support in the goats, in which this
joint sets the direction of movement of the whole body.

In Chapter 3 we mentioned that the relative forces developed in the
phase of support are much greater with single jumps than during a gallop
(pp. 70—72).  Moreover, the increased loads on the talocrural joint lead to
such a considerable intensification of the forces acting to break the foot that
even if the body mass of the animal is relatively small, conditions provoking
shortening of the foot begin to have an effect (as was shown also for cursorial animals).  But in the chamois, synchronized extension of the knee
and talocrural joints does not bring about excessive loads on the foot, and
consequently the foot does not have to be shortened.

Shortening of the foot in connection with an increased body mass was
demonstrated in a comparison we drew between the saiga and Mongolian
gazelle and the heavier cursorial animals, the gnu and eland, and when comparing different species of deer, the heaviest of which (maral) displayed the
shortest foot. A further decrease in the relative length of the foot and of the
limb as a whole (Table 5) is observed in the European and American bisons,
which are adapted to the mediportal form of running.

---

* During the gallop this difference in the work of the hind limb joints is expressed in that the posterior part
of the body is thrown upward with each cycle of running in the chamois, while the goat's body moves
smoothly during the gallop. This distinctive difference between the two forms hits one in the eye the first
time one conducts observations in nature. We analyzed the mechanics of motion of the chamois and tur
in collaboration with K.M.Gasparyan.

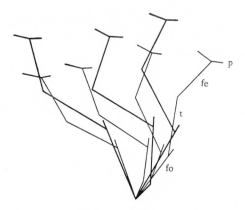

FIGURE 104. Diagram showing the movement of the
hind limb skeleton of the chamois (R u p i c a p r a
r u p i c a p r a) (bold lines) and the tur (C a p r a   c a u -
c a s i c a) (fine lines) during the phase of support in
the gallop:

p — pelvis; fe — femur; t — tibia; fo — foot.

An analysis of the length ratios of the hind limb segments in the Ungulata
brings us to some moot points concerning the systematics and evolution of
these animals.  Sokolov, Klebanova and Sokolov (1963, p. 345) write: " . . . as
is known (Sokolov, 1953), the general trend of evolution of the subfamily
Caprinae is adaptation to life in the mountains, to moving over very broken
terrain.  One of the morphological indexes of this trend is progressive
shortening of the limbs, especially of their distal elements.  The ancestors
of the saiga apparently did not escape this either," and further (p. 347):
"While in the goitered gazelle this process took the course usual for many
other groups of mammals — lengthening of the limbs and development of the
gallop, for the saiga, which has a common origin with the group of wild goats
and sheep, this path proved to be closed:  the level of specialization for mountain life attained by the predecessors of the saiga (shortening of the limbs)
at the moment of their divergence from the common root of the Caprinae
was already so high that a backtrack of evolution in the direction of secondary lengthening of the limbs was found to be impossible, and adaptation to
running in this line of development proceeded along a path of working out
other morphophysiological adaptations."  The presence of certain differences
in the ratio of the limb segments between the goitered gazelle and saiga does
not seem to us to justify concluding that the saiga originates from the mountain forms.  The point is that the saltatorial-cursorial forms have relatively
longer limb segments than the cursorial forms.  This is seen both when comparing the relative size of the limb segments in the reindeer  and  in the

forest-dwelling genera Capreolus and Cervus and when comparing
the Mongolian gazelle and saiga with the goitered gazelle. For the cursorially adapted Mongolian gazelle, saiga and reindeer, the relative size
of the limb segments is markedly smaller than in the saltatorial-cursorial
goitered gazelle and forest deers.

Sokolov (1953) includes the chamois in the tribe Caprini together with
the genera Capra, Ovis and Hemitragus. We believe that the different mechanism of jumping in Capra and Rupicapra is an indication
that these two genera went over to life among the rocks independently of
each other. It is thus very unlikely that they are close systematically, and
so cannot be placed in the same tribe.

GROUPS OF HIND LIMB MUSCLES AND THEIR REORGANIZATION
IN CONNECTION WITH ADAPTATION TO DIFFERENT
FORMS OF RUNNING

We noted above (pp. 77, 80) that the main load is borne by the hind limb
muscles in the phase of support, during which the most burdened muscles
are in turn the extensors of the hip, knee and talocrural joints and the flexors
of the digits. Whereas for single-jointed muscles a change of their length
is directly related to flexor-extensor movements in the corresponding joint,
double-jointed muscles change in length in dependence not only on the flexor-
extensor movements in both joints but also on the topographical position of
the origin and end of each muscle. In order to characterize the work of one
of the most fundamental groups of hip joint extensors, the long postfemoral
group, we therefore give a diagram showing the movement of the hind limb
skeleton in five species of ungulates (Figure 105). In the figure the length
of the segments is taken to be proportional to the data in Table 5.

We unfortunately cannot give analogous diagrams for all ungulates because the relevant biomechanical data are not available. The diagrams
given must thus be considered only as models making it easier to figure
out how double-jointed muscles work during different forms of running.
In these models the topography of the actual position of each muscle is not
given, but just three points are shown, at equal distances from the knee
(Figure 105, d, e, f) and hip (a, b, c) joints. If we join up the points distributed
on the ischium with the points on the tibia in all nine possible combinations
in three moments of the phase of hind support, we can ascertain the course
of change of the "nominal long postfemoral muscles" in this phase. The
following moments were chosen on the diagram: 1) moment of landing (beginning of the phase of hind support, beginning of the preparatory period);
2) middle of the phase of support (end of the preparatory period and beginning of the starting period); 3) moment of pushing away (end of the phase
of support and end of the starting period). Expressing the change in the
length of the muscles in percent of their length in the middle position (2)
of the appendicular skeleton (Table 7), we can conjecture the most favorable
conditions for the work of the long postfemoral muscles in the phase of support in the animals studied (Figure 105). The general premises for profitable

muscle work are as follows: the length of the muscle must decrease in the starting period. If it does not, and especially if it increases, it will interfere with rather than promote extension of the hip joint in the starting period of the phase of support.* If the muscle becomes shorter in the preparatory period, it does so very insignificantly, as otherwise the energy of contraction would be to a large extent utilized ineffectually for the propulsive thrust, which is the main function of the phase of support.

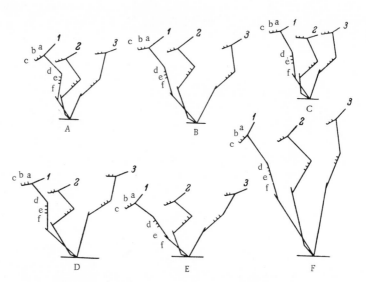

FIGURE 105. Diagram showing the movement of the hind limb skeleton of ungulates during the phase of support:

A) tapir (Tapirus americanus); B) Asiatic wild ass (Equus hemionus); C) bison (Bison americanus); D) roe deer (Capreolus capreolus); E) wild goat (Capra aegagrus); F) giraffe (Giraffa camelopardalis). 1) beginning, 2) middle, 3) end of the phase of support; a, b, c) reference points on the ischium; d, e, f) on the tibia.

In ungulates the long postfemoral group of muscles includes m. biceps femoris, m. gracilis and m. semitendinosus. The first muscle is divided into m. biceps femoris anterior and m. biceps femoris posterior. The former is again divided into two parts, proximal and distal. The bundles of the proximal part depart fanwise from the spinous processes of the sacral vertebrae and the sacroischial ligament and converge toward the proximal end of the tendon passing to the tibia along the anterior margin of this muscle (Figure 106, 1). The distal part is the rectangular venter whose fibers pass from the anterior margin of m. biceps femoris posterior

_____

* Extension of the hip joint is the most important function promoting propulsion in the starting period, as it counteracts the weight of the body.

to the anterior margin of m. biceps femoris anterior obliquely forward
and downward (Figure 106,2).   M. biceps femoris posterior originates
on the tuber ischiadicum and the posterior margin of m. biceps femoris
anterior and diverges in a fan, terminating on the tuberculum majus and
fascia of the tibia.   A strand of connective tissue passes from the posterior
margin of m. biceps femoris posterior to the Achilles' tendon (Figure 106,5).
M. m. biceps femoris anterior and posterior are so closely intergrown
that only the direction of the fibers serves to draw the boundary between
them.   The functions of the two portions of m. biceps femoris anterior
and of m. biceps femoris posterior are different.

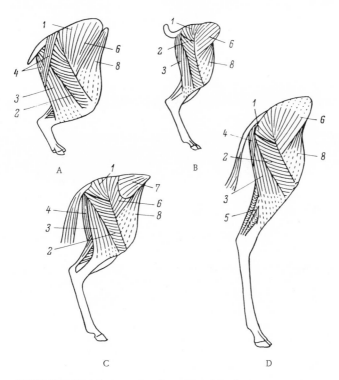

FIGURE 106.  Hind limb muscles of ungulates, lateral:

A) tapir (T a p i r u s   a m e r i c a n u s); B) wild goat (C a p r a   a e g a g r u s);
C) Asiatic wild ass (E q u u s   h e m i o n u s); D) giraffe (G i r a f f u s   c a m e l o -
p a r d a l i s).  1—3) m. biceps femoris: 1) anterior pars proximalis, 2) anterior
pars distalis, 3) posterior; 4) m. semitendinosus; 5) tendinous band to
m. gastrocnemius; 6—7) m. gluteus: 6) superficialis, 7) medius; 8) m. tensor
fasciae latae.

TABLE 7. Length of "nominal postfemoral muscles" at various moments of the phase of support in percent of their length in the middle position (2) according to Figure 105

| Species | Position of skeleton | Combination of reference points | | | | | | | | |
|---|---|---|---|---|---|---|---|---|---|---|
| | | a | | | b | | | c | | |
| | | d | e | f | d | e | f | d | e | f |
| Capreolus capreolus ... | 1 | 134 | 123 | 112 | 135 | 126 | 113 | 132 | 123 | 113 |
| | 3 | 101 | 92 | 88 | 110 | 105 | 92 | 118 | 112 | 102 |
| Bison americanus ...... | 1 | 115 | 108 | 103 | 114 | 110 | 103 | 113 | 110 | 102 |
| | 3 | 91 | 88 | 85 | 97 | 95 | 89 | 101 | 98 | 95 |
| Capra aegagrus ........ | 1 | 142 | 128 | 112 | 145 | 131 | 113 | 145 | 130 | 117 |
| | 3 | 103 | 88 | 74 | 117 | 101 | 87 | 133 | 116 | 103 |
| Equus hemionus onager | 1 | 129 | 121 | 112 | 129 | 123 | 112 | 128 | 123 | 114 |
| | 3 | 99 | 93 | 88 | 103 | 99 | 94 | 112 | 107 | 102 |
| Tapirus americanus ... | 1 | 117 | 107 | 102 | 113 | 109 | 101 | 115 | 112 | 103 |
| | 3 | 106 | 97 | 88 | 112 | 104 | 94 | 106 | 113 | 103 |
| Giraffa camelopardalis | 1 | 108 | 103 | 96 | 112 | 107 | 99 | 113 | 108 | 103 |
| | 3 | 101 | 97 | 89 | 108 | 101 | 96 | 110 | 106 | 99 |

The anterior tendon on which the proximal part of m. biceps femoris anterior ends occupies a position similar to that denoted as c—f in Table 7. In all the mammals presented in Table 7 the value c—f changes minimally. In the wild goat, roe deer, wild ass and tapir c—f in the starting period even becomes longer, while in the bison and giraffe it is slightly shortened. Therefore, slight strains of the proximal part of m. biceps femoris anterior are sufficient to keep the anterior tendon of the muscle permanently tensed. The posterior tendon of this muscle, on which the muscle fibers also of m. biceps femoris posterior originate, occupies position b—f in Table 7. In the starting period of the phase of support b—f in the phase of support is negligible: in the giraffe 4%, in the tapir 7%, in the bison 14%, in the ass 16%, and only in the wild goat 26% of the length of b—f at the moment the animal lands. The muscle fibers of the distal part of m. biceps femoris anterior in C. aegagrus, C. falconeri (markhor) and C. caucasica (Gasparyan, 1967) are directed toward the caudal tendon, on which the fibers of m. biceps femoris posterior also originate, at an angle of 50—55°; this angle is 70—75° in the giraffe, while in the other ungulates it is larger than in the Capridae and smaller than in the giraffe (Figure 106).

In the preparatory period of the phase of hind support, the caudal tendon of m. biceps femoris anterior is in all ungulates except the giraffe contracted by 1—13% of its original length, while in the giraffe it is elongated by 1% (Table 7, b—f). This muscle continues to be shortened, and during the whole course of the starting period of the phase of support it becomes 4—13% shorter than at the end of the preparatory period. Whereas the

cranial and caudal tendons of m. biceps femoris anterior are tensed
throughout the phase of support, so that the distance between them does
not alter, contraction of the fibers of the distal part of this muscle pulls
the ischium at a rate proportional to the rate of contraction of the fibers
and the secant of the angle at which these fibers are inserted on the tendon.
In the phase of support this angle increases in size as the fibers of the
distal part of m. biceps femoris anterior contract, whereby not only does
the speed of movement of the ischium increase but the component tending
to deflect the cranial and caudal tendons of this muscle is enhanced.  Deflection of these tendons deprives the whole construction of meaning, and so
m. biceps femoris posterior, and also the proximal part of the anterior
muscle, must help preserve the tension of the cranial and caudal tendons
of m. biceps femoris anterior throughout the phase of support. M. biceps
femoris posterior occupies the zone from b—d to b—e (Table 7).  In three
of the species discussed b—d undergoes a more than 25% reduction of its
original length in the preparatory period of the phase of support (35% in
Capreolus capreolus, 45% in Capra aegagrus and 29% in
Equus hemionus).  In these forms, however, the fibers of the "nominal
muscles" become 3—17% longer in the starting period than at the end of the
preparatory period, that is, although the distal part of m. biceps femoris
anterior actually becomes gradually more influential in deflecting the tendon,
the tension of the posterior muscle, which is working under a yielding regime,
also gradually increases, especially in the crucial starting period.   The
tonus of contraction of m. biceps femoris posterior is probably maintained
throughout the phase of support, but in the starting period its fibers are
lengthened by virtue of their position, so that there is particularly powerful
strain on the tendon.  In Bison americanus b—d contracts during the
entire phase of support, and yet the total amplitude of contraction is small,
permitting quite considerable tension to be maintained in the tendon.   In
Tapirus americanus and Giraffa camelopardalis the muscle
contracts in the preparatory period almost to the same extent that it is extended in the starting period (Table 7). As for the proximal part of m. biceps
femoris anterior, as we said above, its minimal changes of length during the
phase of support are fully able to strain the cranial tendon of the muscle,
on which, in addition, m. tensor fasciae latae terminates.

Hence, the distal part of m. biceps femoris anterior constitutes the
working part of the muscle, while the other two parts together with m. tensor
fasciae play an auxiliary role, promoting optimum working conditions for
the main part.  We may naturally expect that in ungulates this part will also
be the most powerful.  And, in fact, the distal part makes up 70—80% of the
entire mass of this muscle.  A different function of the parts and subdivisions
of m. biceps femoris is possible in ungulates only when they are intergrown
into a single whole.  This phenomenon is all the more interesting that differentiation of a muscle into parts generally arises simultaneously with the
appearance of different functions of its parts, which in the final analysis
leads to a subdivision of the muscle into independent portions, or capita.

Most ungulates have two gracilis muscles.  M. gracilis anterior originates on the fascia of m. psoas minor and m. psoas major, its second

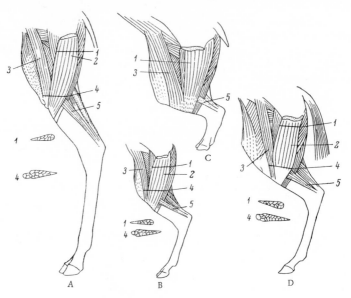

FIGURE 107. Hind limb muscles of ungulates, medial:

A) G. camelopardalis; B) C. aegagrus; C) T. americanus;
D) E. hemionus. 1—4) m. gracilis: 1) section at the proximal end, 2) posterior,
3) anterior, 4) section at the distal end; 5) tendinous band to m. gastrocnemius.

stem sometimes beginning on the wing of the ilium.* This stem is attached
medial to m. iliacus.   M. gracilis posterior originates on the pelvic
symphysis and the tendinous attachment of the adductors.   Both muscles
end on the proximal end of the tibia (Figure 107).   From the posterior
margin of the terminal tendon a connective-tissue strand extends to the
Achilles' tendon (Figure 107, 5).   In its position, m. gracilis posterior
is similar to b—e and b—f in Table 7, in other words, its distal bundles are
elongated in the starting period of hind support in some animals, while the
proximal bundles continue to contract.   The position of the proximal
bundles is obviously more advantageous, and therefore a tendency is observed in ungulates for m. gracilis posterior to move proximally along
the tibia.   A cross section of this muscle shows that at its origin the most

---

* M. gracilis anterior is usually called m. sartorius (Ellenberger and Baum, 1893, and all subsequent publi-
cations — Klimov, 1927, 1937; Yanushevich, 1931; Gindtse, 1937; Sadovskii, 1953; Akaevskii, 1961; etc.).
However, as shown by Gasparyan (1967), characteristic for the true m. sartorius is a close connection with
m. tensor fasciae latae and an origin on the ilium lateral to m. iliacus. The simultaneous presence of
m. gracilis anterior and m. sartorius in several mammals (Gambaryan, 1960) especially clearly points to
the principles according to which we may homologize them, and therefore the author considers it more
correct to name this muscle m. gracilis anterior rather than m. sartorius in view of its origin, which is
typical for the former but not for the latter.

thickened part will be the posterior portion of the venter (Figure 107, *1*),
while at its end, on the contrary, the anterior portion will be thickest
(Figure 107, *4*).  M. gracilis anterior (Figure 107, *3*) adheres to the anterior margin of m. gracilis posterior at its end.  Beginning in front of the
hip joint, it is extended in all ungulates in the starting period of hind propulsion, straining m. gracilis posterior.  This muscle has an active effect
not only on account of the contraction of its fibers but also due to the tension imparted to it by m. gracilis anterior.  Tension of the latter may be
transmitted even if there is tendinous degeneration of m. gracilis anterior.
In ungulates this muscle is heavily streaked with tendinous fibers and is
virtually in a stage of tendinous degeneration.

In most ungulates, m. semitendinosus (Figure 106, *4*) is similar to a—d
in Table 7, but in the giraffe to a—e.  As we see from the table, a—d in the
starting period of the phase of support continues to contract only in the
ass and bison, and a—e in the giraffe also contracts in this period.  Interestingly, in the bison, wild ass, zebra (which moves similarly to the ass),
and giraffe, this is stronger than in other ungulates, accounting for 6.0—8.3%
of the weight of all the hind limb muscles (Table 8), whereas in the others
it constitutes only 2.6—5.9%.

The group of short postfemoral muscles may be divided into two subgroups, proximal and distal.  In the Ungulata the proximal subgroup comprises the adductors (there is usually only one adductor, the product of the
fusion of m. m. adductor brevis, adductor maximus and adductor longus;
m. adductor longus is only very rarely separated from the other part).
The proximal subgroup also includes m. quadratus femoris and m. pectineus.  The distal subgroup comprises m. m. semimembranosus anterior
and posterior, which in ungulates are generally intergrown into a single
muscle.  The proximal subgroup differs from the distal in the position of
its origin on the ischium: the distal subgroup originates from the caudal
part of the ischium, the proximal subgroup much closer to the hip joint.
The two subgroups differ in their ending, the distal muscles being inserted
on the femur almost parallel to the axis of the bone, the proximal muscles
at a large angle to it.

Contraction of the short postfemoral muscles in the phase of support
brings about extension of the hip joint all the more intensively the further
away from the center of the joint their origins are situated and the more
the angle between them and the pelvis approaches 90°.  The proximal
muscles usually begin at a smaller angle to the pelvic axis and the distal
muscles almost at a right angle.  The absolute amplitude of contraction
and elongation of the fibers of the short postfemoral muscles during flexor-
extensor movements in the hip joint increases in proportion to the shift of
their endings distally along the femur.  The muscles to be strengthened in
this group will, of course, be those that originate further caudally on the
ischium and end further distally on the femur.  In the Ungulata these are
m. m. semimembranosus (Table 8).

In ungulates the adductors originate on the pelvic symphysis and the
tendinous attachment which extends distally from the symphysis. Muscular
bundles are inserted to the left and right on the attachment; the bundles of
the right side often penetrate to the left side by means of teeth, and those
of the left side penetrate to the right, thereby liquidating the attachment

TABLE 8. Relative weight of the hind limb muscles in the Ungulata ( % of total weight of all

| Muscles | Tapirus americanus | Equus quagga | Equus hemionus | Connochaetes gnou | Strepsiceros cudu | Taurotragus oryx | Saiga tatarica | Gazella sub-guttutosa | Rupicapra rupicapra | Ovis ammon | Ovis ammon polii |
|---|---|---|---|---|---|---|---|---|---|---|---|
| m. gluteus medius | 13.1 | 19.1 | 22.6 | 11.0 | 7.6 | 8.2 | }9.5 | 7.5 | }7.5 | 10.1 | 7.7 |
| m. gluteus minimus * | 1.2 | 2.2 | 0.9 | 2.1 | 1.5 | 2.4 | | 0.9 | | 1.7 | 2.0 |
| Gluteal | 14.3 | 21.3 | 23.5 | 13.1 | 9.1 | 10.6 | 9.5 | 8.4 | 7.5 | 11.8 | 9.7 |
| m. pectineus | 1.4 | 1.0 | 0.8 | 1.0 | 1.5 | 1.6 | 1.2 | 1.7 | 1.4 | 1.5 | 1.4 |
| m. adductor | 4.0 | 3.8 | 3.7 | 5.3 | 4.4 | }18.7 | 4.6 | 4.5 | 7.1 | 7.4 | 6.8 |
| m. semimembranosus | 13.0 | 7.5 | 8.6 | 14.9 | 12.7 | | 21.5 | 21.4 | 12.5 | 13.8 | 12.9 |
| m. quadratus femoris | 0.2 | — | 0.1 | 0.1 | 0.2 | 0.2 | 0.2 | — | 0.1 | 0.1 | 0.3 |
| Short postfemoral | 17.2 | 11.3 | 12.5 | 20.3 | 17.3 | 18.9 | 26.3 | 25.9 | 19.7 | 21.3 | 20.0 |
| m. biceps anticus | 10.2 | 15.3 | 14.3 | }19.9 | 16.8 | }16.9 | }15.4 | 14.8 | 12.1 | 13.5 | }15.8 |
| m. biceps posticus | 2.8 | 3.1 | 3.3 | | 2.1 | | | 1.1 | 1.9 | 1.6 | |
| m. semitendinosus | 5.5 | 7.4 | 6.2 | 5.3 | 5.9 | 5.1 | 4.9 | 6.0 | 4.8 | 4.1 | 5.0 |
| m. gracilis anticus | 1.0 | 0.8 | 1.1 | 1.5 | 1.1 | }4.4 | 0.6 | 1.0 | 0.6 | 0.5 | 0.6 |
| m. gracilis posticus | 3.1 | 3.4 | 3.1 | 2.8 | 2.7 | | 2.2 | 2.4 | 2.7 | 2.3 | 2.7 |
| Long postfemoral | 22.6 | 30.0 | 28.0 | 29.5 | 28.6 | 26.4 | 23.1 | 25.3 | 22.1 | 22.0 | 24.1 |
| Hip joint extensors | 54.1 | 62.6 | 64.0 | 62.9 | 55.0 | 55.9 | 58.9 | 59.6 | 49.3 | 55.1 | 53.8 |
| m. vastus lateralis | 7.0 | 4.5 | 5.1 | 6.6 | 7.9 | 6.7 | 7.6 | 7.7 | 8.4 | 7.7 | 8.0 |
| m. vastus medialis ** | 4.6 | 3.8 | 3.8 | 2.9 | 3.7 | 3.9 | 5.6 | 4.1 | 5.7 | 5.0 | 5.2 |
| Knee joint extensors | 11.6 | 8.3 | 8.9 | 9.5 | 11.6 | 10.6 | 13.2 | 11.8 | 14.1 | 12.7 | 13.2 |
| m. gastrocnemius † | 3.3 | 2.2 | 3.2 | 3.5 | 4.6 | 5.0 | 4.1 | 3.6 | 6.3 | 5.7 | 4.8 |
| m. psoas minor | 0.6 | 1.0 | 0.9 | 0.9 | 0.8 | 0.6 | 1.2 | 0.9 | 1.4 | 1.3 | 1.3 |
| m. iliopsoas | 9.2 | 7.3 | 6.6 | 5.2 | 7.7 | 7.2 | 5.6 | 5.5 | 5.9 | 6.2 | 5.5 |
| m. gluteus superficialis | }6.3 | 1.9 | 1.7 | }2.8 | 3.7 | 3.0 | 2.5 | 2.7 | 3.5 | 3.7 | 3.3 |
| m. tensor fascia latae | | 2.8 | 3.1 | | | | | | | | |
| m. obturator †† | 1.6 | 1.5 | 1.0 | 1.5 | 0.7 | 1.7 | 1.0 | 0.6 | 1.2 | 1.0 | 1.4 |
| m. popliteus | 0.8 | 0.8 | 0.6 | 0.7 | 0.7 | 1.0 | 0.7 | 0.6 | 0.9 | 0.6 | 0.9 |
| m. rectus femoris | 5.9 | 6.1 | 5.4 | 4.6 | 6.6 | 6.3 | 5.1 | 8.3 | 7.2 | 6.4 | 6.5 |
| m. plantaris | 0.2 | 0.3 | 0.3 | 0.9 | 1.3 | 1.1 | 1.4 | 1.1 | 2.3 | 1.7 | 1.7 |
| m. tibialis anterior | 1.5 | 0.6 | 0.6 | 0.3 | 0.5 | 0.5 | 1.2 | 0.1 | 1.8 | 1.1 | 0.4 |
| m. ext. digitorum longus | 1.3 | 1.1 | 0.8 | 1.6 | 1.4 | 1.5 | 0.4 | 0.4 | 0.5 | 0.6 | 1.5 |
| m. peroneus | 0.2 | 0.4 | 0.3 | 1.2 | 0.8 | 1.0 | 1.3 | 1.0 | 1.2 | 1.1 | 1.9 |
| m. m. fl. digitorum et tibialis posterior | 2.0 | 2.6 | 1.8 | 3.2 | 2.7 | 2.7 | 2.8 | 1.9 | 2.8 | 1.4 | 2.5 |

\* m. gluteus minimus + m. piriformis.
\*\* m. vastus medialis + m. vastus intermedius.
† m. m. gastrocnemius lateralis + medialis.
†† m. m. obturator externus + internus.

hind limb muscles)

| Capra caucasica | Capra aegagrus | Capra falconeri | Capra sibirica | Bos indicus | Phoëphagus mutus | Bison bonasus | Bison americanus | Capreolus capreolus | Cervus elaphus | Cervus nippon | Cervus dama | Rangifer tarandus | Giraffa camelo-pardalis |
|---|---|---|---|---|---|---|---|---|---|---|---|---|---|
| 8.5 | 9.2 | }11.5 | 8.2 | 9.4 | 8.3 | 8.8 | 9.2 | 5.8 | 8.4 | 6.3 | 8.3 | 8.0 | 5.8 |
| 1.8 | 2.6 |  | 1.5 | 1.9 | 2.7 | 1.7 | 1.5 | 1.4 | 1.1 | 1.5 | 2.4 | 1.7 | 1.7 |
| 10.3 | 11.8 | 11.5 | 9.7 | 11.3 | 11.0 | 10.5 | 10.7 | 7.2 | 9.5 | 7.8 | 10.7 | 9.7 | 7.5 |
| 1.7 | 1.6 | 1.4 | 1.5 | 1.4 | 1.6 | 1.6 | 1.6 | 0.9 | 1.3 | 1.4 | 1.6 | 1.4 | 1.6 |
| 6.2 | 5.7 | 7.9 | 6.4 | 7.8 | }14.6 | 5.6 | 5.2 | 6.5 | 3.9 | 4.2 | 5.4 | 5.1 | 6.3 |
| 10.0 | 11.1 | 8.9 | 10.8 | 13.6 |  | 11.5 | 13.0 | 16.9 | 13.0 | 13.8 | 12.4 | 17.0 | 7.8 |
| 0.3 | 0.3 | 0.2 | 0.3 | 0.2 | 0.5 | 0.2 | 0.2 | 0.1 | 0.1 | 0.2 | 0.3 | 0.2 | 0.1 |
| 16.5 | 17.1 | 17.0 | 17.5 | 21.6 | 15.1 | 17.3 | 18.4 | 23.5 | 17.0 | 18.2 | 18.1 | 22.3 | 14.2 |
| 12.3 | 13.0 | }15.9 | }15.2 | }18.7 | }22.7 | 14.2 | 14.2 | }12.5 | }20.3 | }19.4 | }17.5 | 12.7 | 15.7 |
| 1.8 | 1.9 |  |  |  |  | 2.8 | 3.8 |  |  |  |  | 1.9 | 3.3 |
| 4.6 | 4.5 | 3.9 | 4.2 | 8.3 | 5.8 | 5.5 | 6.5 | 5.6 | 5.2 | 5.5 | 4.0 | 4.9 | 4.8 |
| 0.6 | 0.5 | }2.7 | 0.6 | 0.8 | 0.8 | 1.1 | 1.1 | 1.0 | 0.7 | 1.2 | }4.5 | 0.8 | 1.9 |
| 2.6 | 2.6 |  | 3.2 | 3.9 | 3.3 | 3.3 | 3.4 | 2.2 | 2.5 | 2.2 |  | 2.7 | 4.1 |
| 21.9 | 22.5 | 22.5 | 23.2 | 31.7 | 32.6 | 26.9 | 29.0 | 21.3 | 28.7 | 28.3 | 26.0 | 23.0 | 29.8 |
| 48.7 | 51.4 | 51.0 | 50.4 | 64.6 | 58.7 | 54.7 | 58.1 | 52.0 | 55.2 | 54.3 | 54.8 | 55.0 | 51.5 |
| 7.2 | 7.4 | 6.9 | 7.5 | 6.7 | 5.7 | 8.6 | 7.2 | 8.5 | 7.9 | 9.0 | 7.3 | 7.1 | 8.0 |
| 5.2 | 5.0 | 4.6 | 5.2 | 2.8 | 3.2 | 3.9 | 3.0 | 5.4 | 6.0 | 3.7 | 5.4 | 5.2 | 3.1 |
| 12.4 | 12.4 | 11.5 | 12.7 | 9.5 | 8.9 | 12.5 | 10.2 | 13.9 | 13.9 | 13.7 | 12.7 | 12.3 | 11.1 |
| 5.9 | 5.2 | — | 6.5 | 4.6 | 4.9 | 6.0 | 5.3 | 5.6 | — | 5.0 | 5.0 | 5.9 | 8.4 |
| 1.2 | 1.1 | 1.3 | 1.2 | 0.8 | 0.9 | 0.9 | 1.2 | 1.2 | 1.1 | 0.8 | 1.4 | 1.1 | 0.5 |
| 7.0 | 7.2 | 6.9 | 6.3 | 5.9 | 7.1 | 6.0 | 6.4 | 6.6 | 5.0 | 5.9 | 6.7 | 5.9 | 6.9 |
| }3.6 | 3.5 | 3.7 | 3.3 | 3.7 | 2.8 | 2.7 | 3.4 | 3.1 | 3.4 | 3.3 | 3.1 | 2.4 | 4.4 |
| 1.7 | 1.5 | 1.4 | 1.2 | 0.9 | 1.4 | 1.1 | 1.1 | 0.7 | 0.6 | 1.0 | 1.8 | 1.5 | 1.8 |
| 0.9 | 0.8 | 0.9 | 0.8 | 0.7 | 0.9 | 1.0 | 1.0 | 0.6 | 0.7 | 0.8 | 1.0 | 0.8 | 1.3 |
| 7.9 | 7.3 | 6.8 | 7.1 | 5.1 | 5.2 | 5.4 | 5.3 | 6.7 | 6.9 | 7.2 | 6.5 | 6.0 | 5.4 |
| 2.1 | 2.1 | — | 2.6 | 0.8 | 1.1 | 1.5 | 1.1 | 1.7 | — | 1.2 | 1.7 | 2.1 | 1.7 |
| 1.6 | 1.3 | }2.2 | 1.4 | 1.0 | 1.3 | 1.5 | 1.3 | 0.7 | 1.6 | 0.8 | 1.5 | 0.1 | 0.5 |
| 0.9 | 0.3 |  | 0.8 | 0.8 | 0.9 | 0.8 | 0.6 | 1.7 | 0.6 | 1.6 | 0.7 | 0.4 | 2.2 |
| 1.2 | 1.3 | 1.0 | 1.2 | 0.7 | 1.4 | 0.9 | 0.9 | 1.2 | 2.2 | 0.8 | 0.7 | 0.9 | 1.4 |
| 2.7 | 2.4 | 2.7 | 2.8 | 2.0 | 1.9 | 2.8 | 2.3 | 2.9 | 2.7 | 2.7 | 3.7 | 2.5 | 2.2 |

and forming a sort of intergrowth of the right and left adductors. In the preparatory period of the phase of support they are greatly elongated, so that in the starting period they will contract sharply. When the preparatory period of the second limb sets in, the leading limb is usually already entering the starting period. Extension of the adductors in the preparatory period of the second limb per se increases the tension of the contracting adductors of the leading limb. The starting period of the second limb sets in when the leading limb has entered the drawing-up period of the phase of free transit, during which time the hip joint is flexed, thereby increasing the tension of the contracting adductors of the second limb. Hence, intergrowth of the origin of the right and left adductors and of m. m. gracilis is a peculiar adaptation to enhancing their tension in the crucial periods of running.

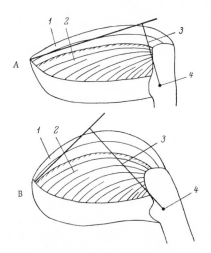

FIGURE 108. Changes in the angle of force application for m. gluteus medius in relation to tension of m. gluteus minimus:

A) relaxation of the muscle; B) tension of the muscle. 1—2) m. gluteus: 1) medius, 2) minimus; 3) lever arm; 4) center of movement in the hip joint.

The group of gluteal muscles (m. gluteus medius, m. gluteus minimus and m. piriformis) acts to extend the hip joint with a force which is intensified the more the angle between the ilium and the direction of the origin of the muscle bundles approaches 90°. M. m. gluteus minimus and piriformis are situated under m. gluteus medius, and therefore when m. gluteus minimus contracts, its convexity is increased, as a result of which the force lever arm of the bundles of m. gluteus medius from the ilium is enhanced (Figure 108) and the muscle becomes more efficient.

The indirect influence of the development of the gluteal or postfemoral muscles on extension of the knee joint (see the principle of action on p.80) may have the result that with specialization for running a strengthening of

the gluteal or postfemoral muscles causes a weakening of the knee joint
extensors proper.  This relationship is observed very often in ungulates.
The apparent reason for it is that the resultant of the gluteal muscles is
almost perpendicular to the axis of the femur, while that of the short postfemoral muscles is almost parallel to it.  Naturally, under these conditions
the gluteal muscles have a greater indirect influence on extension of the
knee joint.  Such a relationship is present in the Artiodactyla, but it is
particularly well expressed in the Perissodactyla.  For instance, in the
gnu the weight of the gluteal group equals 13.1% of the weight of the hind
limb muscles, while in the other artiodactyls it ranges from 7.5% to 11.8%.
On the other hand, the extensors of the knee joint make up 9.5% of the
weight of the hind limb muscles in the gnu and in the other artiodactyls
from 10.6 to 13.9%.* In the tapir the weight of the gluteal muscles accounts
for 14.3% of that of the hind limb muscles whereas in the artiodactyls highly
specialized for running — the zebra, ass and domestic horse — their weight,
according to Udovin and Yanshin (1951), constitutes 21.3—23.5%.  As for the
knee joint extensors proper, their weight equals 11.6% of that of the hind limb
muscles in the tapir but 8.3—8.9% in the zebra and ass.

The group of hip joint extensors proper consists of three femoral capita
of m. quadriceps femoris: m. m. vastus lateralis, vastus medialis and
vastus intermedius.  Of these three capita the longest lever arm for extension of the knee joint belongs to m. vastus lateralis, which begins in ungulates at the anterior margin of the proximal end of the trochanter major of
the femur (only in the tapir do a few bundles of the lateral caput originate
also on the lateral ridge of the femur (Gambaryan, 1964)).  In addition, the
lateral caput of m. quadriceps femoris is inserted on the femur at an angle
of 40—50°, while the other capita are inserted at an angle of 10—20°.  The
lateral caput is naturally the most strongly developed of the knee joint extensors proper (Table 8).

The extensors of the talocrural joint fall into two groups: the gastrocnemial group and the talocrural joint extensors proper.  The gastrocnemial
muscles, m. gastrocnemius lateralis, m. gastrocnemius medialis and
m. soleus, with ventral support should act to flex the knee joint, but the
force of extension of the knee joint is much greater than the force of the
gastrocnemial muscles, and therefore in the stage of support these are the
main extensors of the talocrural (tarsal) joint.  The flexors of the digits
(m. fl. digitorum longus, m. fl. hallucis longus, m. tibialis posterior)
will be the extensors proper of the tarsal (talocrural) joint with ventral
support.  However, if as a result of knee joint extension and under the influence of their own contraction, the gastrocnemial muscles act to extend
the talocrural joint, raising the tuber calcanei a little, which favors transmitting propulsive thrust to the body, the flexors of the digits must have
an influence in drawing the proximal end of the tibia backward, which does
not favor transmitting propulsive thrust to the body.  This is why the angle
of origin of the digital flexors from the tibia is very small, and their pull
spreads almost along the long axis of the tibia, which makes their tension

* Exceptions are species adapted to the mediportal form of running: B os  ind ic us, 9.5%;  Phoëphagus
  mu tus, 8.9%;  B ison  am er ic a nus, 10.2%.

act almost exclusively to resist excessive extension of the digital joints in the preparatory period of the stage of hind support and to flex the digits in the stage of free transit of the limbs.

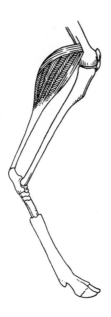

FIGURE 109. Structure of the gastrocnemial muscle in the giraffe

The development of the gastrocnemial muscles in ungulates depends to a large extent on the degree of interaction between the flexor-extensor movements in the knee and talocrural joints. In the case of a deep automatic connection of these movements, the gastrocnemial muscles become reorganized into what are known as ligament muscles, and this usually incurs a decrease in their relative weight. For example, in the Equidae, which have a markedly elongated foot in connection with cursorial specialization, whereby the force tending to bend the talocrural joint is also enhanced, the gastrocnemial muscles are not strengthened, as we should expect, but are even slightly weakened in comparison with those in the tapir (they account for 3.2% of the weight of the hind limb muscles as against 3.3% in the tapir). A similar slight weakening of the gastrocnemial muscles is observed in the cursorial artiodactyls as well (Table 8). In the gnu, for instance, their weight makes up 3.5% of that of the hind limb muscles, in the saiga 4.1%, while in saltatorial ungulates (sheep and goats) they constitute from 4.8 to 5.7%. In the giraffe the straightened position of the knee joint throughout the stage of hind support removes the influence of the knee joint extension upon talocrural joint extension. This is why in the giraffe we observe a strengthening of the gastrocnemial muscles, which account for 8.4%

of the weight of the hind limb muscles as against 3.2—6.0% in other ungulates (Table 8).   At the same time, in the giraffe these muscles undergo a
profound structural reorganization:  they become multipinnate, which
accelerates the speed of their contraction and enhances their force
(Figure 109).


## TYPES OF SEGMENT RATIOS IN THE FORELIMB SKELETON

We will now examine the role of the forelimbs in ungulates.  With the
light lateral and the diagonal gallop, after the stage of free flight the forelimbs take the weight of the body while the muscles of the shoulder girdle
draw the trunk between the legs, lending it the speed necessary to begin the
stage of crossed flight; the latter occurs also under conditions of slower
motion (heavy and slow gallop).

In the phase of support, there are, as with the hind limbs, preparatory
and starting periods between the moment the forelimbs touch the ground
and the moment they leave it.  In the preparatory period the digital and
metacarpal-digital joints, plus sometimes the wrist joint, are extended, the
elbow and shoulder joints are flexed, and the trunk moves forward and down
between the forelimbs.  In the starting period the shoulder and elbow joints
are extended and the digital, metacarpal-digital and wrist joints are flexed,
and the trunk is drawn forward and upward between the forelimbs.  The
movement of the trunk in the preparatory period tends to bend the shoulder
and elbow joints, so that the extensors of these joints are in a state of tension from the very beginning of the stage of front support, at first striving
to reduce the speed of flexion and then actively extending the joints.  Therefore, the forelimb muscles which bear the greatest load are the shoulder
and elbow joint extensors and the muscles drawing the trunk between the
legs (m. anconeus longus, m. endopectoralis, m. latissimus dorsi,
m. brachiocephalicus, and others).

Like the analysis of the length ratios of the hind limb segments, an analysis
of the relative dimensions of the segments in the forelimbs shows (Table 9)
a number of patterns of change related to adaptation to different forms of
running.  Comparing the forelimb segments in perissodactyls adapted to
the battering-ram and cursorial forms of running, we get an idea of the
phylogenetic pathways of their change.  As already mentioned (p. 130), the
structure of the tapir's skeleton is very similar to that of the precursors
of the Equidae.*  In Equus hemionus and E. quagga the hand and
forearm are markedly elongated in comparison with Tapirus americanus, and this leads to an overall increase in the relative length of the
limbs which even a certain shortening of the shoulder does not prevent.  It
is interesting to note that the heavier zebra shows the reverse trend: shortening of the hand and forearm, this being characteristic for some of the extinct heavy Equidae (Gregory, 1912).

---

* For more details on the bone-muscle system in the Perissodactyla see the author's special paper (Gambaryan,
  1964).

TABLE 9. Changes of the relative length of the forelimb segments in connection with adaptation to different forms of running

| Species | Relative size of segments | | | | | Absolute length (a+b+c), mm |
|---|---|---|---|---|---|---|
| | shoulder (a) | forearm (b) | hand (c) | scapula | total (a+b+c) | |
| Tapirus americanus ···· | 29.1 | 35.8 | 28.0 | 33.1 | 92.9 | |
| Equus quagga chapmani | 27.4 | 39.6 | 41.0 | 33.0 | 108.0 | 970 |
| E. hemionus onager ····· | 28.0 | 43.8 | 45.0 | 31.7 | 116.8 | |
| Connochaetes gnou ····· | 26.0 | 42.2 | 40.0 | — | 108.2 | 860 |
| Taurotragus oryx ········ | 27.6 | 35.5 | 38.2 | 36.7 | 101.3 | 1,280 |
| Saiga tatarica··········· | 26.0 | 39.3 | 44.0 | 28.8 | 109.3 | 572 |
| Gazella gutturosa ······ | 24.6 | 35.1 | 43.6 | — | 103.3 | 603 |
| G. subtutturosa ·········· | 27.4 | 41.5 | 56.3 | 31.0 | 125.2 | 750 |
| Ovis ammon polii ······ | 29.0 | 45.0 | 48.0 | 33.2 | 122.0 | 680 |
| Capra aegagrus ········· | 30.6 | 42.2 | 39.0 | 33.0 | 111.8 | 645 |
| Bison bonasus ··········· | 28.6 | 36.8 | 31.8 | 39.8 | 97.2 | 1,152 |
| B. americanus ··········· | 28.0 | 35.5 | 30 0 | 38.1 | 93.5 | 1,140 |
| Cervus elaphus sibiricus | 28.5 | 42.6 | 43.3 | 34.0 | 114.4 | 1,028 |
| C. elaphus elaphus······ | 31.2 | 45.3 | 45.0 | 34.6 | 121.5 | 922 |
| C. dama ················ | 29.0 | 42.0 | 47.0 | 32.0 | 118.0 | 751 |
| Capreolus capreolus ··· | 33.2 | 45.4 | 53.0 | 31.3 | 131.6 | 674 |
| Moschus moschiferus ··· | 34.8 | 43.3 | 52.2 | 27.3 | 130.3 | 550 |
| Rangifer tarandus······· | 29.4 | 42.4 | 43.3 | 31.8 | 115.1 | 836 |
| Giraffa camelopardalis | 48.5 | 83.0 | 94.0 | 56.0 | 225.5 | 2,320 |

Note. The length ratio of the segments is given as a percentage of the sum of lengths of the lumbar and thoracic regions of the spine.

Looking at the ratios of the segments in the forelimb skeleton in the Artiodactyla, we see a close resemblance in the types of changes to the above-analyzed ratio types in the hind limbs. Thus, adaptation to cursorial running leads to a shortening of the forelimb segments. This is observed in the group of deer, in which the reindeer (Rangifer tarandus) displays the shortest forearm and hand: 115.1% of the length of the lumbar and thoracic vertebrae as against 117.1—131.6% in other deer. Like the foot, the hand undergoes a secondary reduction in size when the body mass increases. The maral, which is 1.5 times as large as the reindeer, has a relatively smaller hand than Rangifer tarandus (Table 9). Similar changes in the ratios of the segments with adaptation to cursorial running are observed in the Cavicornia. Comparing the Mongolian gazelle, saiga and goitered gazelle (the first two adapted to cursorial running, while the goitered gazelle shows elements of saltatorial-cursorial running), we see that in the goitered gazelle the forelimbs are relatively much longer than in the other two forms. The indexes of the hand are especially large (in Gazella subgutturosa the length of the hand amounts to 56.3% of the length of the lumbar and thoracic regions of the spine, while in G. gutturosa it amounts to 43.6% and in Saiga tatarica 44.0%). Cavicornia adapted to cursorial running also exhibit a shortening of the hand with an increase in body mass.  In the gnu the index of the hand equals 40.0% of the length of the lumbar and thoracic vertebrae, and in the eland even 38.2%, whereas in the Mongolian gazelle and saiga these indexes are substantially higher (Table 9).  A further secondary shortening of the forelimb segments is observed also in animals adapted to the mediportal form of running, in which the sum of lengths of the forelimb segments is less than 100% of the lumbar and thoracic regions, returning to the values close to those of the primitive Tapiridae.  Gregory (1912) showed that in natural phylogenetic series of ungulates we perceive a lengthening of the hand and its subsequent shortening as the body mass increases;  this is analogous to what we described in comparative anatomical series.

The very large increase of all the forelimb segments in the giraffe came about through adaptation to feeding on leaves high up in trees;  the mechanics of motion of this animal undergoes reconstruction in view of this lengthening of the segments, and not vice versa, as in the other ungulates, in which the mechanics of motion governs the type of ratio of the segments.

## WORK OF THE FORELIMB MUSCLES

To recap, the muscles which bear the greatest load in the stage of front support are those drawing the trunk between the legs and the extensors of the shoulder and elbow joints.

The muscles drawing the body between the legs belong to a series of groups: pectoral — m. m. ectopectoralis, endopectoralis, sternoscapularis (Figure 67,1—3);  m. serratus ventralis pars thoracalis (Figure 67,8);  m. latissimus dorsi (Figure 67,4);  m. spinotrapezius (Figure 67,5).  The most advantageously placed of the pectoral muscles is the deep pectoral muscle, which originates on the caudal segments of the sternum and ensiform process and ends on the proximal part of the crest of the tuberculum majus, on the tuberculi majus and minus humeri, and often on the coracoid process.  Owing to this distribution, the bundles of this muscle are directed almost horizontally forward.  The superficial pectoral muscle is usually divided into two parts: cranial and caudal.  The cranial part begins from the manubrium and ends on the antebrachial fascia.  The caudal part begins on the first two segments of the sternum and ends on the crest of the tuberculum majus humeri.  In accordance with the direction of the bundles, both parts, and particularly the second, are arranged almost diagonal to the trunk, and therefore cannot exert an appreciable effect on drawing the body between the legs.  Their main function is to prevent the trunk from sagging. M. sternoscapularis always originates on the cartilage of the first rib and sometimes also on the manubrium.  The muscle generally terminates in the region of the fascia of m. supraspinatus, but when it is not strongly developed it ends in the region of the tendinous inscriptions between the clavicular parts of the deltoid and trapezoid muscles.  If strongly developed, it can act very effectively to extend the shoulder joint, and when the limbs move forward and out (preparatory period of the phase of support) it can also take part in drawing the trunk between the legs.

Individual bundles of m. serratus ventralis pars thoracalis have their
origin on the ribs: from III to IX (X). All these teeth converge toward the
vertebral margin of the scapula, where they terminate. The further cranially the bundles are situated, the more they take part in keeping the trunk
between the legs. And the more caudally the teeth are arranged, the greater
their role in drawing the trunk between the legs. M. latissimus dorsi begins on the spinous processes from thoracic vertebra V to the last thoracic
vertebra and also on the lumbar fascia; it has several costal teeth on ribs
VIII—X. The muscle ends on two peduncles: on the crest of the tuberculus
minor humeri together with m. teres major and on the crest of the tuberculus major humeri together with m. panniculus carnosus. The more caudally the bundles of this muscle are inserted, the more the horizontal component of its pull dominates over the vertical. The dorsal part of the trapezoid muscle originates on the spinous processes from thoracic vertebrae
V—XIII and ends on the tubercle of the spina scapulae. The most caudal of
its bundles are especially useful for drawing the trunk between the legs.
Interaction of the above muscles during simultaneous flexor-extensor movements in the shoulder and elbow joints in the phase of support ensures that
the body move forward smoothly between the legs. The extensors of the
shoulder joint comprise the joint extensors proper — m. m. supraspinatus
and infraspinatus* — and the muscles acting indirectly to extend the shoulder
joint: m. acromiotrapezius, m. rhomboideus cervicalis, m. serratus ventralis pars cervicalis and m. sternoscapularis. The last of these muscles
not only helps draw the trunk forward between the legs, but at the same time
facilitates extension of the shoulder joint. The elbow joint is usually
extended at the same time as the shoulder joint, and when this occurs
m. biceps brachii (the main flexor of the elbow joint) is also involved
in extending the shoulder joint.

M. supraspinatus (Figure 67,*14*) begins in the prespinous fossa, sometimes also penetrating partly into the anterior part of the infraspinous fossa,
for example in the tapir (Gambaryan, 1964); it ends on the proximal end of
the tuberculus major humeri. M. infraspinatus (Figure 67,*15*) begins in
the infraspinous fossa and ends just distal to the preceding muscle on the
tuberculus major humeri. M. acromiotrapezius (Figure 67,*6*) originates
on the ligamentum nuchae and the first 2—4 spinous processes of the thoracic vertebrae and ends on the spina scapularis on the area between the
tubercle of the spina and the acromion. M. serratus ventralis pars cervicalis (Figure 67,*9*) begins on the transverse processes of cervical vertebrae
II—VII and on the first two or three ribs and ends at the cranial angle of the
scapula. The rhomboid muscles (m. m. rhomboideus cervicalis and
thoracalis — Figure 67, *7*) begin under m. acromiotrapezius on the
ligamentum nuchae and on the first 3—5 spinous processes of the
thoracic vertebrae; they end on the vertebral margin of the scapula.
M. sternoscapularis was described above. M. biceps brachii (Figure 67,*16*)
originates on the tubercle of the scapula and ends on the proximal end of the
ulna.

* In many manuals it is considered that m. infraspinatus is an abductor, and not an extensor of the shoulder
joint (Klimov, 1955; Akaevskii, 1961; and many others). But while this is true in man, in the Ungulata
this muscle ends above the axis of the shoulder joint and therefore will be in the main an extensor, not
an abductor.

The extensors of the elbow joint include the extensors proper and muscles indirectly serving to extend the joint.  The extensors proper comprise all three capita of m. triceps brachii and also m. dorsoepitrochlearis and m. m. epitrochleoanconeus.  The indirect extensors of the elbow joint are the complex m. brachiocephalicus and m. latissimus dorsi.

M. triceps brachii (Figure 67, *11—13*) begins as a long caput (m. anconeus longus) on the caudal margin of the scapula, a lateral caput (m. anconeus lateralis) on the crest of the tuberculus majus humeri and a medial caput (m. anconeus medialis) on the medial crest of the shoulder and its volar surface.  All three capita end on the olecranon.  M. dorsoepitrochlearis begins on the caudal margin of the elbow and m. latissimus dorsi and ends almost entirely on the apex of the olecranon and only partly descends from it along the antebrachial fascia.  M. epitrochleoanconeus originates either on the medial epicondyle of the humerus (m. epitrochleoanconeus medialis) or on the lateral epicondyle (m. epitrochleoanconeus lateralis); it ends on the apex of the olecranon.

M. brachiocephalicus (Figure 67, *10*) begins on the cranial third of the ligamentum nuchae and the occipital crest in the form of m. clavotrapezius, on the mastoid process of the temporal bone in the form of m. clavomastoideus, and on the transverse processes of cervical vertebrae II—IV in the form of m. atlantoscapularis.*  It ends on a connective-tissue septum, a homologue of the clavicle, on which m. clavodeltoideus takes its origin, ending in turn on the crest of the tuberculus major humeri and the dorsal margin of the shoulder.  M. atlantoscapularis, intersecting the bundles of m. clavotrapezius, ends partly on the fascia of the shoulder joint in the region of the acromion.  M. latissimus dorsi was described above.

As we mentioned above (p. 83), we must examine the work of the forelimb and shoulder girdle muscles in two phases of motion: free transit and support.  And each of these phases is to be divided into two periods: the first into drawing-up and adjustment and the second into preparatory and starting (Figure 54).  During the course of these two phases all the joints of the free forelimb are extended and flexed twice.  Three moments of transition from flexing to extending movements occur in the phase of free transit and only one moment in the phase of support.  In the phase of free transit the whole limb with the humeral girdle moves from the extreme back position to an anterior position.  Thus, in this phase there are two extreme positions of the limb with respect to the trunk, posterior and anterior, since after the end of the phase of support the forelimb continues for a short while to be drawn back in the air before it is brought forward, and prior to coming down to the ground the leg manages to draw back somewhat from the line of the body. The advantage of this activity is clear, as up to the last moment of the phase of support the leg actively propels the body forward, while before landing it somewhat slows down its own forward progression and moves behind and under relative to the trunk, thereby making for a smoother touchdown.

The heaviest load is borne during the phase of support, throughout which the main working elements are, as stated above, the muscles drawing the body between the legs and the extensors of the shoulder and elbow joints.

---

* This is not generally singled out as a separate muscle, as it is strongly intergrown with m . clavotrapezius.
  However, according to both its origin and its termination it is certainly homologous to m. atlantoscapularis.

Distinctive working conditions for the forelimb muscles arise in the giraffe. Here the proximal end of the scapula hardly changes its position relative to the trunk throughout the cycle of motion, this being due to the activity of m. serratus ventralis. As a rule, in ungulates this muscle has 9—10 costal bundles and 5, less often 6, cervical teeth. The teeth depart from the ribs at the level of the costal cartilages, and their insertion on the ribs extends to a third, more rarely to a half, of the total length of the rib. A more extensive area for insertion of the costal teeth is generally situated on ribs II—IV. It is this middle part of m. serratus ventralis which in ungulates is most heavily strained in keeping the trunk between the legs, while the anterior cervical bundles work to extend the shoulder joint and the posterior thoracic bundles act to draw the body between the legs. The strength of the cervical part of this muscle in ungulates is usually little inferior to that of its thoracic part, 35—45% of the weight of the muscle as a whole. The muscle ends on the medial surface of the vertebral margin of the scapula, forming on its cranial and caudal angles markedly widened endings, which promote two functions: extension of the shoulder joint and dragging the body between the legs. In the giraffe, m. serratus ventralis has 11 costal bundles and only two cervical. The cervical part of the muscle in the giraffe accounts for at most 12% of the entire weight of the muscle. The costal bundles begin 11—37 cm proximal to the costal cartilages and extend along the rib as far as the costal tubercle. The denticles fall just short of the tubercle only on ribs IX, X, and XI. Hence, in the giraffe, unlike all the other ungulates, the costal denticles of m. serratus ventralis occupy not the distal but the proximal half to two thirds of the rib. Tendinous attachments extend from the costal tubercles to the vertebral margin of the scapula. They are especially well developed on the surfaces from the third to the seventh costal teeth. On the vertebral margin of the scapula the insertion of m. serratus ventralis does not form two thickened parts on neither the cranial nor the caudal side. Because the tendinous attachments are very short, they impede forward and backward movement of the vertebral margin of the scapula with respect to the vertebra, but do not prevent the distal end of the scapula from being deflected to the side, which is essential to the giraffe when it is watering or snatching up food from the ground. Due to the relative immobility of the vertebral margin of the scapula in the giraffe, the active work in drawing the trunk between the legs which is characteristic for this muscle in ungulates is replaced in this animal by a passive role, that of fixing the trunk between the legs. Its relative weight in the giraffe therefore constitutes 7.5% of the weight of the forelimb muscles, whereas in other ungulates it constitutes from 12.3 to 16.0%.

Fixation of the trunk between the scapulae in the giraffe changes the working conditions of the forelimb muscles, since during saltatorial gaits the body is maintained between the limbs by the tendon-muscle apparatus of m. serratus ventralis. The series of shoulder girdle muscles can therefore act exclusively to drag the body between the legs. The overall load on them thereby diminishes. For instance, m. endopectoralis, the main muscle drawing the trunk between the legs, is somewhat weakened in the giraffe. Its weight is 6.9% of that of the forelimb muscles as against

7.6—14.2% in the other ungulates.   At the same time a greater load is placed
on m. ectopectoralis, which works to raise the trunk after the animal has
eaten grass or taken water.   This muscle is therefore more strongly developed in the giraffe than in other ungulates.   Its weight is 9.5% of that
of the forelimb muscles, only 2.6—5.3% in the other species.

In ungulates with a nonfixed scapula in the phase of front support a
heavy load is placed on the muscles supporting and drawing the trunk
between the legs to an equal extent when there is a need for swift movement and when the animal runs with steep trajectories of jumps.   This
is why it is hard to distinguish the specific differences in their development in ungulates adapted to different forms of running.

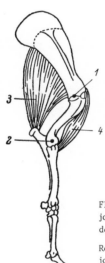

FIGURE 110.  Work of the shoulder and  elbow
joints of the horse as a factor of the activity of
double-jointed antagonists of these joints.

Rotation points in:  1) shoulder joint, 2) elbow
joint;  3) m. anconeus longus;  4) m. biceps
brachii.

Extension and flexion of the shoulder and elbow joints are to a large
degree interrelated in the Ungulata.   The relation is brought about both
by antagonistic action of the long caput of m. triceps brachii and m. biceps brachii and by a number of other muscles.   Tension of m. biceps
brachii simultaneously flexes the elbow and extends the shoulder joint,
while tension of the long caput of m. triceps brachii, on the contrary,
flexes the shoulder and extends the elbow joint (Figure 110).   However,
these two joints are generally flexed and extended together, which is

TABLE 10. Relative weight of the forelimb muscles in ungulates (% of total weight of forelimb

| Muscle | Tapirus americanus | Equus quagga | Equus hemionus | Connochaetes gnou | Strepsiceros cudu | Taurotragus oryx | Saiga tatarica | Gazella sub- gutturosa | Rupicapra rupicapra | Ovis ammon |
|---|---|---|---|---|---|---|---|---|---|---|
| brachiocephalicus[1] | 7.1 | 8.3 | 9.9 | 7.1 | 6.6 | 4.5 | 4.5 | 7.1 | 6.2 | 5.5 |
| trapezius . . . . | 3.9 | 2.3 | 1.8 | 4.9 | 3.7 | 2.1 | 2.3 | 3.3 | 2.0 | 2.0 |
| rhomboideus . . | 2.1 | 2.6 | 2.6 | 5.8 | 2.4 | 2.1 | 3.1 | 2.9 | 3.2 | 1.8 |
| latissimus dorsi | 5.0 | 5.6 | 5.7 | 7.7 | 6.7 | 5.3 | 9.4 | 9.3 | 4.3 | 5.9 |
| sternomastoideus | 3.8 | 2.3 | 2.4 | 1.5 | 1.0 | 1.8 | 1.5 | 2.8 | 0.9 | 1.6 |
| ectopectoralis . . | 3.3 | 4.3 | 5.9 | 4.8 | 4.4 | 5.3 | 2.9 | 3.5 | 3.3 | 4.3 |
| endopectoralis . . | 7.6 | 7.9 | 8.9 | 14.2 | 8.6 | 10.1 | 11.2 | 9.9 | 11.4 | 10.2 |
| sternoscapularis . | 6.3 | 6.0 | 4.2 | — | — | 0.5 | 0.2 | — | 0.2 | 0.1 |
| serratus ventralis | 11.8 | 11.4 | 11.3 | 12.3 | 14.9 | 13.2 | 13.6 | 14.5 | 14.2 | 15.3 |
| deltoideus . . . | 0.9 | 1.7 | 1.7 | 1.3 | 1.8 | 2.1 | 1.3 | 1.3 | 1.5 | 1.3 |
| supraspinatus . . | 6.6 | 4.5 | 4.7 | 4.5 | 7.4 | 7.2 | 7.2 | 6.6 | 8.2 | 8.9 |
| infraspinatus . . | 5.7 | 5.1 | 5.1 | 4.6 | 7.2 | 6.7 | 7.5 | 6.7 | 6.9 | 8.8 |
| subscapularis . . | 3.7 | 3.1 | 3.9 | 3.2 | 2.6 | 3.7 | 3.9 | 3.6 | 3.9 | 3.0 |
| teres major . . | 2.2 | 1.8 | 1.8 | 1.8 | 2.1 | 1.7 | 2.6 | 2.3 | 2.3 | 2.3 |
| teres minor . . | 0.3 | 0.4 | 0.5 | 0.2 | 0.6 | 0.7 | 0.3 | 0.3 | 0.5 | 0.5 |
| coracobrachialis . | 0.3 | 0.3 | 0.6 | 0.4 | 0.6 | 0.7 | 0.3 | 0.4 | 0.5 | 0.4 |
| anconeus longus . | 10.5 | 11.8 | 10.0 | 11.8 | 9.3 | 10.4 | 11.0 | 10.1 | 9.4 | 9.8 |
| anconeus lateralis | 3.2 | 2.2 | 2.6 | 2.2 | 3.2 | 3.4 | 2.4 | 2.5 | 4.3 | 3.2 |
| anconeus medialis[2] | 1.8 | 1.6 | 0.6 | 0.3 | 0.7 | 2.0 | 0.9 | 0.7 | 2.0 | 1.3 |
| dorsoepitrochlearis | 2.4 | 1.5 | 0.9 | — | 0.5 | — | 0.9 | 1.1 | 0.4 | 0.9 |
| biceps brachii . . | 1.6 | 2.5 | 2.3 | 2.1 | 2.7 | 3.0 | 2.0 | 1.6 | 2.3 | 1.8 |
| brachialis . . . . | 2.2 | 1.7 | 1.8 | 1.4 | 1.9 | 2.5 | 1.5 | 1.2 | 1.8 | 1.5 |
| ext. carpi radialis | 1.9 | 2.6 | 2.6 | 2.0 | 2.7 | 3.0 | 2.2 | 1.6 | 2.9 | 2.1 |
| ext. digitorum[3] . . | 1.3 | 1.0 | 1.4 | 0.8 | 1.0 | 1.4 | 1.0 | 0.9 | 1.4 | 0.9 |
| ext. carpi ulnaris | 0.7 | 0.8 | 1.2 | 0.8 | 1.4 | 1.5 | 1.6 | 1.2 | 1.6 | 1.5 |
| fl. carpi ulnaris | 0.3 | 0.8 | 1.0 | 0.5 | 0.6 | 0.8 | 0.7 | 0.8 | 0.9 | 1.1 |
| fl. carpi radialis | 0.4 | 0.5 | 0.1 | 0.2 | 0.5 | 0.5 | 0.3 | 0.2 | 0.3 | 0.4 |
| fl. digitorum[4] . . | 2.8 | 2.6 | 2.9 | 3.1 | 4.7 | 4.8 | 3.5 | 2.9 | 4.6 | 3.2 |

[1] m.m. clavotrapezius + omotransversarius + clavodeltoideus + clavomastoideus.
[2] m. anconeus    medialis + m.  epitrochleoanconeus.
[3] m.ext. digitorum communis + m. ext. digitorum lateralis.
[4] m. digitorum sublimis+ m. fl. digitorum profundus.

muscles)

| Ovis orientalis | Capra sibirica | Capra caucasica | Capra caucasica | Capra falconeri | Phoëphagus mutus | Bison bonasus | Bison americanus | Capreolus capreolus | Cervus elaphus | Cervus nippon | Cervus dama | Rangifer tarandus | Giraffa camelo-pardalis |
|---|---|---|---|---|---|---|---|---|---|---|---|---|---|
| 7.3 | 8.1 | 7.3 | 6.9 | 7.0 | 6.0 | 5.4 | 6.3 | 8.5 | 6.0 | 6.1 | 7.7 | 6.2 | 5.1 |
| 2.6 | 2.5 | 2.3 | 2.2 | 2.3 | 6.6 | 3.6 | 3.5 | 2.2 | 3.0 | 2.0 | 1.7 | 2.6 | 0.3 |
| 1.9 | 2.7 | 2.8 | 2.7 | 2.3 | 5.5 | 5.4 | 5.9 | 3.4 | 2.5 | 2.6 | 3.4 | 4.2 | 1.3 |
| 5.8 | 4.5 | 6.0 | 5.3 | 5.5 | 5.7 | 6.5 | 7.3 | 6.9 | 7.3 | 7.4 | 8.6 | 6.3 | 5.5 |
| 1.6 | 1.9 | 1.4 | 1.3 | 1.5 | 0.7 | 0.7 | 0.9 | 3.3 | — | 1.4 | 1.5 | 2.0 | 0.8 |
| 4.0 | 3.7 | 4.5 | 4.0 | 3.7 | 4.7 | 3.5 | 5.3 | 2.7 | 2.6 | 4.1 | 3.0 | 3.6 | 9.5 |
| 11.0 | 10.7 | 9.9 | 10.3 | 10.2 | 11.1 | 11.7 | 12.1 | 10.6 | 13.0 | 9.6 | 10.7 | 12.1 | 6.9 |
| 0.1 | 0.2 | 0.1 | 0.2 | 0.1 | — | 0.1 | 0.1 | — | — | — | — | 0.2 | — |
| 15.4 | 15.6 | 14.1 | 15.8 | 16.2 | 15.8 | 16.0 | 15.0 | 14.1 | 14.0 | 14.4 | 12.6 | 14.8 | 7.5 |
| 1.5 | 1.8 | 1.6 | 1.7 | 1.8 | 1.7 | 1.7 | 1.7 | 1.1 | 1.6 | 1.3 | 1.6 | 1.4 | 2.5 |
| 8.2 | 7.7 | 8.2 | 7.9 | 7.8 | 5.8 | 5.9 | 5.3 | 6.9 | 7.6 | 7.7 | 7.7 | 5.9 | 4.5 |
| 8.1 | 7.1 | 7.0 | 7.8 | 7.3 | 6.7 | 6.8 | 6.1 | 6.5 | 8.2 | 6.7 | 7.8 | 6.5 | 4.6 |
| 3.7 | 3.2 | 3.3 | 4.0 | 3.3 | 4.9 | 3.3 | 3.3 | 3.0 | 3.3 | 3.0 | 4.5 | 4.0 | 3.5 |
| 1.9 | 1.8 | 1.8 | 1.9 | 2.1 | 1.5 | 1.8 | 1.6 | 2.3 | 1.5 | 1.8 | 1.5 | 1.8 | 0.9 |
| 0.6 | 0.6 | 0.4 | 0.6 | 0.6 | 0.6 | 0.7 | 0.6 | 0.2 | 0.6 | 0.5 | 0.4 | 0.6 | 0.6 |
| 0.5 | 0.5 | 0.5 | 0.5 | 0.7 | 0.4 | 0.4 | 0.4 | 0.2 | 0.6 | 0.5 | 0.4 | 0.6 | 1.1 |
| 9.2 | 7.7 | 8.1 | 8.0 | 8.3 | 9.5 | 11.1 | 11.1 | 8.7 | 10.3 | 9.0 | 9.3 | 10.9 | 14.6 |
| 3.6 | 3.2 | 3.2 | 3.5 | 3.8 | 2.2 | 2.7 | 2.0 | 3.3 | 2.9 | 3.5 | 2.8 | 2.5 | 5.2 |
| 1.2 | 1.9 | 1.3 | 1.3 | 1.0 | 0.7 | 0.5 | 0.7 | 0.6 | 1.0 | 0.9 | 1.2 | 1.0 | 1.4 |
| 0.9 | 0.7 | 0.7 | 0.5 | 0.6 | 0.3 | 0.2 | 0.2 | 0.7 | 0.7 | 0.7 | 1.2 | 0.4 | 1.1 |
| 1.8 | 1.9 | 1.9 | 1.6 | 2.0 | 2.1 | 1.9 | 1.9 | 2.1 | 2.1 | 2.0 | 1.9 | 1.8 | 5.0 |
| 1.5 | 1.8 | 1.5 | 1.5 | 1.4 | 1.5 | 1.8 | 1.7 | 1.8 | 1.5 | 1.4 | 1.1 | 1.8 | 2.7 |
| 2.5 | 2.9 | 2.5 | 2.9 | 2.8 | 1.8 | 2.4 | 2.2 | 1.7 | 2.5 | 2.5 | 2.0 | 2.1 | 5.5 |
| 1.1 | 1.3 | 1.2 | 1.4 | 1.3 | 1.4 | 1.2 | 1.1 | 1.0 | 1.1 | 1.3 | 1.0 | 1.0 | 1.6 |
| 1.4 | 1.5 | 1.7 | 1.6 | 1.8 | 1.0 | 1.0 | 0.9 | 2.0 | 1.2 | 1.1 | 1.4 | 1.0 | 2.6 |
| 1.0 | 0.8 | 0.8 | 0.8 | 0.7 | 0.4 | 0.5 | 0.4 | 0.3 | 0.6 | 0.4 | 0.6 | 0.4 | 2.2 |
| 0.4 | 0.3 | 0.3 | 0.3 | 0.3 | 0.2 | 0.2 | 0.2 | 0.3 | 0.4 | 0.5 | 0.3 | 0.2 | 0.8 |
| 4.1 | 3.5 | 3.5 | 3.5 | 3.5 | 2.3 | 3.5 | 2.9 | 4.6 | 4.1 | 2.6 | 4.3 | 4.1 | 3.4 |

achieved as follows.  In the preparatory period of the phase of support, under the influence of the body's weight, which tends to push the trunk forward between the legs, the elbow and shoulder joints are flexed with a force which prevails over the action of their extensors.  The extensors of these joints gradually inhibit their flexion, and active extension of the joints begins in the starting period.  The force of the shoulder joint extensors (m. acromiotrapezius, m. rhomboideus cervicalis, m. serratus ventralis pars cervicalis, and m. m. supraspinatus and infraspinatus) is here greater than the force of tension of m. anconeus longus of m. triceps brachii, while resistance of m. biceps brachii to extension of the elbow joint diminishes in connection with extension of the shoulder joint.

The double-jointed work of m. anconeus longus depends on combined movement in both joints.  In the preparatory period of the phase of front support this muscle is contracted by 11—47% in all ungulates.  This causes the most proximal bundles to become maximally shortened, and they become much shorter than the distal bundles.  M. anconeus longus of m. triceps brachii becomes most contracted of all in the Perissodactyla, and also in the saltatorial-cursorial and cursorial Artiodactyla.  In ungulates adapted to the mediportal form of running this muscle becomes only 11—12% contracted in the preparatory period, and in the giraffe just 8%, as against 23—47% in other ungulates.  In the starting period the length of this caput of m. triceps brachii remains unchanged in the giraffe, while in other ungulates it increases by 6—33%.

The effect of a double-jointed muscle is profitable if in the preparatory period of the phase of support the muscle contracts negligibly and in the starting period is not elongated.  Of the ungulates, this working condition is fulfilled only in the giraffe.  It is natural to expect that this caput will be more strongly developed in the giraffe, and indeed its weight amounts to 14.6% of that of the forelimb muscles in this animal, while only 8.1—11.8% in other ungulates.

The shoulder joint is extended by its extensors proper (m. m. supraspinatus and infraspinatus) and by muscles indirectly acting to extend it (those of the shoulder girdle).  The force with which the shoulder joint is extended may depend on the degree of development of various muscle components.  For example, in the gnu and Equidae, m. m. supra- and infraspinatus are not powerful.  Their weight together is 9.1—9.8% of that of the forelimb muscles while in other ungulates it is 12.3—17.7%.  Instead of these muscles being developed, in the Perissodactyla m. sternoscapularis is strengthened, and in the gnu the rhomboid muscles (Table 10).

## PATHS AND CAUSES OF REORGANIZATION OF THE LOCOMOTORY ORGANS IN UNGULATES

As we have already discussed, the most profitable of the asymmetrical gaits from the mechanical point of view are the various forms of the gallop. At present it is hard to determine why the ancestors of the Artiodactyla and Perissodactyla took up the gallop.  It seems likely that they developed these gaits in view of their being the most economical means of leaping across obstacles in the forest habitats which were the original home of all the Ungulata.

The characteristics of the locomotory organs of recent ungulates were laid down by historical paths of specialization for swiftest progression in the existing habitat conditions.  Differences of body size are also of considerable significance for them, and the structure of some representatives of these orders is sometimes governed chiefly by the mode of obtaining food.  Even with all this, however, adaptation to fast running proves to be quite an important factor in working out a specific mechanics of running.

The variety of habitats, body dimensions and modes of life makes it possible, even when investigating a small number of perissodactyls and artiodactyls (about 30 species), to discern a series of trends in cursorial specialization, leading to the formation of different types: battering-ram, cursorial, saltatorial-cursorial, saltatorial, mediportal and stilt.

The relict battering-ram form was adopted by the tapirs, dwellers of shrub and reed thickets in humid tropical forests; here were preserved the primary low speeds of running with large vertical fluctuations of the center of gravity, brought about by the presence of numerous obstacles.

The cursorial form of running was developed in inhabitants of open spaces, where speed and endurance are the only safety devices.  The absence of obstacles enabled the animals to economize on vertical fluctuations of the center of gravity (Equidae, gnu, eland, saiga, Mongolian gazelle, reindeer).  The increase in body mass which is typical for several phylogenetic branches of the Artiodactyla and Perissodactyla places heavier demands on economy of running, leading in the extreme variants to the appearance of the mediportal form of running.  The latter is characterized by a further straightening of the trajectory of jumps, a decrease of the pace angles and other features of economical movement (large Bovidae).

When the ungulates secondarily returned to the forests and shrub thickets they developed the saltatorial-cursorial form of running, in which the trajectory of jumps became again steeper, stamina diminished and speeds were somewhat reduced, while at the same time maneuverability became enhanced. The preferred means of escaping from enemies now was leaping from one thickly overgrown area to another, speed and endurance being less important (forest deers, goitered gazelle, some Ovidae).  Saltation is a specific adaptation to life among the rocks, where large leaps are more valuable than speed and endurance.

Feeding on leaves on the tops of trees led to a huge increase in the length of the giraffe's legs, and its cursorial mechanics was appropriately adjusted (progression by means of relatively straight legs (stilts), small pace angles, etc.).

But with all the diverse trends of changes in the locomotory organs of individual branches of artiodactyls and perissodactyls, they all show some common mechanical characteristics.  In all these groups of mammals both the hind limbs and the forelimbs are locomotory and the spine is kept as rigid as possible.  In view of this we propose to name their typical mode of running *dorsostable dilocomotory*.

The forms of running we have distinguished provoke specific reorganizations of the locomotory organs in the Artiodactyla and Perissodactyla. Still, even in the presence of specialization for a similar form of running, arising independently in separate groups of animals, changes in the mechanics of motion of the skeleton and muscles are diverse.

Selectivity with regard to habitats often determines the nature of movement with uniform gaits. In forest dwellers the trot shows smaller fluctuations of the center of gravity and greater amplitudes of movement in the joints than the gallop, while for inhabitants of open spaces the opposite is true.

Rigidity of the spine, common to all the Artiodactyla and Perissodactyla, is achieved by marked expansion of the spinous processes and development of the supraspinous ligament, all of which restricts vertical flexion. Resistance to lateral flexures is brought about by the specific hinges of the articular processes and by the horizontally situated transverse processes with the intertransversal ligament. In both these cases the processes serve as beams of compression and the ligaments as beams of expansion on the principle of lattice girders.

It is useful in this type of study to consider some relationships between the indexes of force developed by a limb and the ratio of the size of the body and the locomotory mechanics of the limbs. The average forces developed in the phase of support (formula (9) on p. 69) decrease in proportion to the elongation of the limb and the sine of half the pace angle, and increase in proportion to the sine of twice the angle of departure and the squares of the speed and the weight of the body. The advantages gained from longer legs and larger pace angles are in turn limited. With a lengthening of the limbs and a simultaneous increase in body weight the load on the legs increases as the cube, whereas the elongation increases linearly. Therefore at a certain moment the relative length of the legs begins even to be reduced. The limits of pace angle increase are set by the increment in the geometrical progression of the vertical fluctuations of the proximal end of the limb as it increases. This is why the pace angle usually starts to decrease in animals with a large body weight.

The low running speeds typical for the relict perissodactyls (tapirs) lead to the maintenance of primitive short legs. The legs become longer in the Equidae in connection with adaptation to swift running in open spaces. The large angles of departure in the saltatorial-cursorial ungulates (forest deers, goitered gazelle) necessitate greater relative forces than in the cursorial animals, despite the fact that their speeds are relatively lower. It is thus not surprising that they have longer legs than the cursorial forms living in open areas (saiga, Mongolian gazelle, gnu, eland, reindeer). The greater body mass is reflected in a secondary reduction in the relative length of the limb segments, especially the distal (in the gnu and eland in comparison with the Mongolian gazelle and saiga and in the zebra in comparison with the wild ass); finally, the segments are even shorter in the large Bovidae adapted to the mediportal form of running. Adaptation to one-time large leaps, as is characteristic for animals dwelling among rocks, may take place with two main systems of flexor-extensor movements. In the first system there is synchronized extension of the hip, knee and talocrural joints, and in the second system the talocrural joint is extended later than the knee and hip joints. Activity according to the second system leads to greater loads on the foot, which allows for the final moment of thrust, and as a result of this the foot is relatively shortened. The elongated appendicular skeleton of the giraffe is explained by adaptation to feeding from the top branches of trees. Along with this, all the joints are straightened, and therefore the growth of the distal segments does not produce an increase of flexing moments acting on the joints.

Even though there are only a few systems of ratios in the limb segments and they show a direct relation to the mechanics of running and the weight of the body (the specific link with the mode of feeding is an exception in the giraffe), it is very difficult to define the reasons for muscle changes, because one type of movement can be attained in the presence of different ratios of development in various groups of muscles. In the phase of hind limb support the heaviest loads are borne by the hip, knee and talocrural joint extensors. Three groups of muscles take part in extending the hip joint: gluteal and the long and short postfemoral. The action of the long postfemoral group depends on flexor-extensor movements simultaneously in the knee and hip joints and on the topography of their origin and end. If we analyze the systems of movement in the pushing-away phase in animals adapted to all six forms of dorsostable dilocomotory running, we can perceive the causes of the strengthening and weakening of these muscles and the factors leading to a structural reorganization of their separate components. The favorable working conditions for these muscles are a small degree of contraction in the preparatory period and continued contraction in the starting period of the phase of support. These conditions are met by m. semitendinosus in the Equidae and in artiodactyls adapted to mediportal running, whereas in other ungulates the development of this muscle is unprofitable. It is therefore strengthened in the former group of animals and weakened in the latter. We examined the work of m. biceps femoris and m. m. gracilis from this point of view. The proximal part of m. biceps femoris anterior, together with the tensors of the femoral fascia and m. biceps femoris posterior, strain the anterior and posterior tendons to promote activity of the most advantageously situated distal part of m. biceps femoris anterior, which accounts for 80% of the total weight of the muscle. The role of m. gracilis anterior is to produce additional strain on the anterior margin of m. gracilis posterior, which extends the hip joint.

Apart from directly extending the hip joint, the short postfemoral muscles and the gluteal group act indirectly to extend the knee joint. The gluteal muscles are particularly conveniently arranged for this purpose. It is therefore easily understandable that a strengthening of these muscles in the gnu among the artiodactyls and in the perissodactyls leads to a relative weakening of the knee joint extensors proper.

Because the gastrocnemial muscles are inserted on the femur at an angle almost perpendicular to its axis, a clearly defined dependence is observed between extension of the knee and talocrural joints. Specialization for running therefore leads to structural reorganizations of these muscles rather than to changes of their strength. Owing to the straightened position of the knee joint during the phase of support in the giraffe, there is an independent reinforcement of the gastrocnemial muscles.

In the phase of forelimb support, of greatest importance are the functions of drawing and keeping the trunk between the legs. This calls for simultaneous development of the shoulder and elbow joint extensors. In the giraffe the scapulae are rigidly connected to the trunk and, in contrast to the other ungulates, a number of muscles take the load only for drawing the trunk between the legs. As a result, m. serratus ventralis and m. endopectoralis, which are the most powerful muscles of the shoulder girdle in the Artiodactyla and Perissodactyla, are weakened in the giraffe. At the same time, in this animal the elbow joint extensors proper are markedly strengthened. Extension of the shoulder joint may take place due to development both of the shoulder joint extensors proper (m. supraspinatus and m. infraspinatus) and of a number of muscles of the anterior girdle (m. serratus ventralis pars cervicalis and m. rhomboideus). Both these variants of strengthening the function of shoulder joint extension may be found in the Ungulata.

The structural features of the skeleton and muscles linked with various trends of specialization for running or with independent pathways of adaptation to similar forms of running enable us in a number of cases to clear up certain points concerning the phylogenetic relations in several groups of ungulates. Conclusions of this kind are usually of a "negative" character, since an analysis of the movements of a very small number of species in a group already shows us that certain forms cannot be grouped together satisfactorily into natural taxonomic units. In fact, it is still premature to judge their natural position in the system; more specialized and broader studies would be needed for this. The two systems of activity of the hind limb joints in the phase of hind support, one of which characterizes the genera Capra and Ovis and the other the genus Rupicapra, suggest that these groups adapted to life in the mountains in independent ways. We can thus hardly include them in the same tribe (Sokolov, 1953). The limb segments (especially the distal) are shorter in the cursorial animals of the open spaces than in the saltatorial-cursorial forms. This is confirmed by the presence of shorter segments in the tundra-dwelling reindeer in comparison with forest deers of the genera Cervus and Capreolus, and in the cursorial gnu, eland, Mongolian gazelle and saiga in comparison with the saltatorial-cursorial goitered gazelle. The theory propounded by Sokolov et al. (1964) on the mountain mode of life of the saiga's ancestors, based on a certain degree of shortening of the distal limb segments in comparison with the goitered gazelle, therefore seems invalid.

*Chapter 5*

## ADAPTATION TO RUNNING IN THE PROBOSCIDEA

### MECHANICS OF RUNNING

It is due to Osborn (1936, 1942) that the phylogeny of the Proboscidea has been very thoroughly studied. Still, as is the case with the Ungulata, the origin of these animals is rather obscure. Actually, the origin of the Ungulata is even clearer, since it is thought that their roots evolved into the polymorphous group Condylarthra, and the only thing which is not certain is the genus to which their direct precursors belonged. All we know about the ancestors of the Proboscidea, on the other hand, is that they probably derived from a particular branch of the Condylarthra. This branch almost certainly gave rise to, apart from the Proboscidea, the Hyracoidea, the aardvarks (O r y c t e r o p u s) and the Sirenia. It is very hard to say to which family and genus of the group Condylarthra the general ancestor of these four recent orders of mammals was related. Even more problematic are the mode of life and habits of this unknown ancestor of the elephant's recent parents. What we do know about the mode of life and habits of the recent hyraxes, and especially the fossorial aardvarks and natatorial sea cows, is of little help in clarifying the original type of locomotion of the Proboscidea. Of their recent parents, the hyraxes may be the closest in this respect to their precursors. However, perfection of land locomotion in the Hyracoidea was accompanied by the reduction of the digits which is common to the mammals, whereas in the Proboscidea the number of digits did not undergo reduction. Preservation of a five-toed structure of the hand and foot in the elephants is all the more paradoxical in that their limbs, like those of the ungulates, have only locomotory function.

In reconstructing the mode of life of a number of extinct groups of the order Proboscidea, Osborn noted that the representatives of many branches were characterized by a littoral mode of life and feeding on moisture-loving succulent vegetation.*

Littoral and swamp vegetation was often pulled up by the roots, which bore nutritive swellings. It is probably in view of this mode of pulling up plants that the tusks developed, the neck muscles became stronger and the neck itself became shorter. This shortening of the neck apparently led to the appearance of the proboscis or trunk, with which the animal introduces food into the mouth.

* The transition to feeding on semiaquatic and then aquatic vegetation led to the development of the true aquatic phytophagous mammals, the Sirenia.

Progression over swampy ground and littoral areas necessitated a widening of the hand and foot to increase navigability; this is why the number of rays was not reduced.   The need for this was all the greater that in the evolution of this group there was a progressive increase in the weight of the animals.   The direct ancestors of the elephants, having a proboscis, tusks and five-rayed limbs, gradually went over to an arboreal mode of feeding and became increasingly entrenched in forest and savanna habitats.   Under these conditions it was profitable for them to pluck leaves and young shoots growing high up, which were out of reach of the small Artiodactyla and Perissodactyla.   As a result, the increase in the size of the body, which was already one of their features, became one of the leading trends of adaptive phylogenesis peculiar to the elephants.   The tusks acquired greater importance as a weapon, while the long proboscis enabled the animals to take food from very high branches without extending the neck (Figure 111).   Exclusively locomotory limbs come more and more to lose an external division into rays, and in the recent Proboscidea they are a monolithic supporting column composed of a dermofascial sac sheathing the five inner rays right up to the distal phalanges.

FIGURE 111.  Elephant picking food off branches (from Grzimek, 1966)

The primary gait of the elephant's ancestors, which were no larger than the recent Hyracoidea, was probably some sort of asymmetrical gait. But the recent Proboscidea secondarily lost the ability to move by means of asymmetrical saltatorial gaits, and their high-speed gait became the fast walk, during which at least one leg was always on the ground.

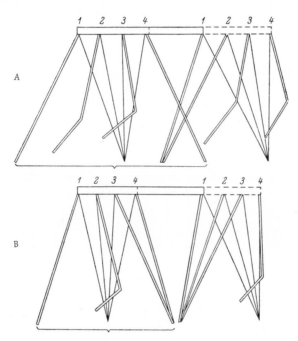

FIGURE 112.   Appearance of racklike gaits in long-legged animals as an adaptation to an increased pace length:

A) slow racklike walk; B) very slow walk. 1—4) movement of limbs during half a cycle. A regular line denotes the left leg, a double line the right leg.

Lengthening of the limbs in the Proboscidea led to a change in the type of symmetrical locomotion, especially during slow forms of movement. This is because a forelimb standing on the ground prevented the hind limbs from moving forward in long-legged animals (Figures 52 and 112). A change in the rhythm of locomotion therefore took place in the direction of racklike gaits, during which the interval between the moments at which the forefeet and the hind feet left the ground became shorter and shorter (Figure 112). As a result, the hind limb could freely move farther than the spot where the forelimb had first stepped.  However, as we showed above (p. 23), when movement is speeded up, the support/transit ratio, that is, the rhythm of leg activity, decreases.  This means that also with a 1:1 rhythm of loco-motion a lateral stage of support arises, so that even with this rhythm of

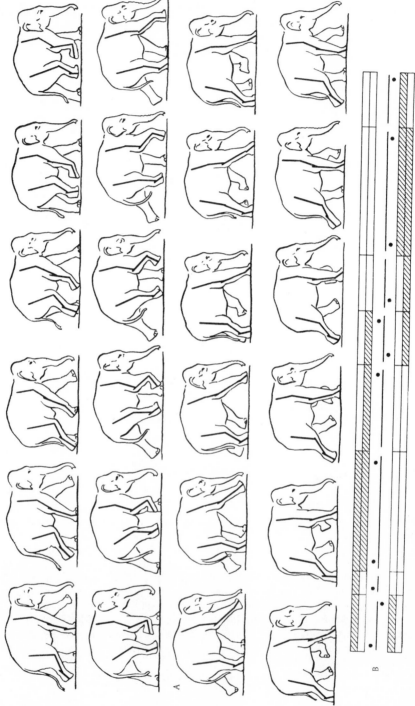

FIGURE 113. Movement of the appendicular skeleton of the elephant during a fast walk (A) and its support graph (B)

locomotion, with no shift toward racklike gaits, the hind limbs are carried farther from the place where the forelimbs of the same side stepped (Figure 52).  Thus, when movement was accelerated, there became no point in shifting the rhythm of locomotion in the direction of racks.  In addition, the stage of free flight appears all the sooner the more closely the rhythm of locomotion approaches a trot or a rack (Figure 22). At rack and trot rhythms the stage of free flight appears when the acceleration is such that the support/transit ratio of limb activity is 1:1.  Meanwhile it is known that the decrease in the rhythm of limb activity is almost directly proportional to the speed of motion, and with a fast walking rhythm of loco-motion the stage of flight appears only after the support/transit ratio has reached 1:3.  Thus, the stage of free flight appears at a speed three times lower with the rack than with the fast walk.

A change in the rhythm of locomotion toward a rack is characteristic for elephants during slow movement, while a reverse switch of the rhythm is typical during accelerated motion.

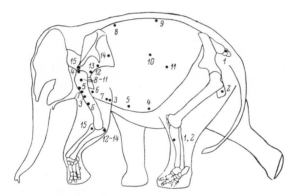

FIGURE 114. Appendicular skeleton of the elephant, sketched in its outline, with the points of insertion of muscles:

1—2) m. semitendinosus: 1) pars vertebralis, 2) pars ischiadicus; 3—5) m. endopectoralis: 3) margo cranialis; 4) margo caudalis; 5) linia mediana; 6—7) m. ectopectoralis: 6) margo cranialis, 7) margo caudalis; 8—11) m. latissimus dorsi: 8—9) pars verte-bralis; 10—11) pars costalis, 8, 10) margo cranialis, 9, 11) margo caudalis; 12—14) m. anconeus longus; 15) m. biceps brachii.

With the fast walk, the main load on any leg comes to bear in the very middle of that leg's phase of support.  This is because one of the limbs accepting part of the load lands almost immediately after another leg has passed through the middle of its phase of support.  When this latter limb leaves the ground, the phase of support of the leg which previously landed is approaching its middle (Figures 113 and 120).  Because of this distribu-tion of loads on the limbs the leg which carries the main weight during a fast walk is that which is in the vertical position.  The functional importance of such a distribution of loads is very great where running is concerned. We mentioned earlier (p. 75) that minimum flexing moments are exerted on the knee and talocrural joints when the arrangement of the limbs is pillarlike and their distal elements (foot, tibia) are shortened. This distribution of moments, which is typical for the statics of the animal, in fact changes little also during running in view of the fact that in elephants the main load on the limbs is brought to bear during the moments they are arranged vertically.

The specific running* of elephants, with the secondary disappearance of a stage of flight in the cycle and with a perfectly even distribution of loads on all the limbs, represents the means by which these very heavy animals adapted to running. We may therefore call their type of running heavy, or graviportal (Gregory, 1912). A characteristic feature of both the elephants and the ungulates is that the thoracic and lumbar regions of the spine remain rigid throughout the cycle of movement. Hence, as for the Ungulata, we may describe the elephant's mode of running as *dorsostable dilocomotory*.

This mode and form of running are characterized by many specific features of locomotory mechanics, which we can most conveniently examine by looking at the changes in the angles of the limb joints of a skeleton drawn in the outline of successive frames of a running elephant (Figures 115 and 116)

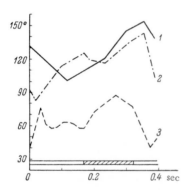

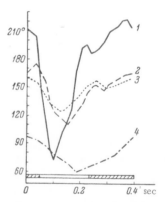

FIGURE 115. Changes in the angle of the hind limb joints during a cycle of fast walking in Elephas indicus:

1) hip joint; 2) knee joint; 3) talocrural joint (hatched part of bar denotes the phase of support). The abscissa gives the time in seconds, the ordinate the angles in degrees.

FIGURE 116. Changes in the angle of the forelimb joints during a cycle of fast walking in Elephas indicus:

1) wrist joint; 2) elbow joint; 3) shoulder joint; 4) angle between horizontal plane and axis of scapula (hatched part of bar denotes the phase of support). Abscissa and ordinate as in Figure 115.

---

* According to the definition given above, running is a form of high-speed motion with a stage of flight in the air. This does not apply to elephants, but still we think it is worthwhile calling the fastest form of this animal's locomotion running.

The hind limbs are characterized by a different type of activity in the hip joint and knee joint.  For the hip joint the change in the size of the angle shows only two moments of direction switching, while for the knee joint there are four such moments.  In the knee joint, as in other mammals, only a transition from flexion in the preparatory period to extension in the starting period takes place in the phase of support.  The other three transitional moments of the change in the size of the joint belong to the phase of free transit of the limbs (Figure 115).  The hip joint is extended throughout the phase of support.  Extension is continued at the very beginning of the phase of free transit, but then the joint is flexed, and the flexion lasts almost up to the middle of this phase, whereupon the joint is again extended (Figure 115).

In elephants, as in other mammals, the cycle of motion contains two transitions from flexion to extension and two from extension to flexion of the shoulder and elbow joints.  But whereas in other mammals only one of these four moments occurs in the phase of support, in the elephant two moments occur in the phase of support and two in the phase of free transit. In the phase of support the shoulder and elbow joints at first continue to be extended, then are slightly flexed, and toward the end of the phase are again extended.  This extension continues into the beginning of the phase of free transit, then quickly changes to flexion, and then back to extension toward the end of this phase (Figure 116).  The angle between the axis of the scapula and the horizontal plane decreases markedly in the phase of free transit and continues to decrease at the very beginning of the phase of support; it then increases almost up to the end of this phase, but before the end it once more starts to decrease, in other words, the phase of support shows two direction switches in the change of the angle (Figure 116).  It is interesting to note that in the hip joint the analogous changes of angle size occur in the phase of free transit (cf. Figures 115 and 116).

The talocrural and wrist joints of the elephant work under special conditions.  Support spreads to the joint itself through a cushion of fat, and therefore the load tending to break these joints is quite negligible.  The enormous amplitude of flexor-extensor movements in the wrist joint during the phase of free transit (152°) is far above that in the talocrural joint (37°).  The reason for this is that support on the forelimbs involves overextension of the wrist joint, while during the transfer the forefoot lifts off the ground on account of its flexion.  On the other hand, during transfer of the hind foot, the talocrural joint is extended just as it is during the time of support on the foot.

The huge weight of elephants, as we have said, accounts for the pillarlike arrangement of the limbs.  This means that there will be a much greater amplitude of movements in the joints in the phase of free transit than in the phase of support (Table 11).

Vertical movements of the hip joint and apex of the scapula are very limited in elephants (Figure 117), but slightly less so in the case of the hip joint.  If, however, we take into account that the center of gravity has to undergo still less pronounced vertical fluctuations than the proximal ends of the limbs, we can more or less consider that the center of gravity moves horizontally in the cycle of motion.

TABLE 11. Amplitudes of flexor-extensor movements in the joints in Elephas indicus (degrees)

| Joint | Phase of support | | | Phase of free transit | | |
|---|---|---|---|---|---|---|
| | maximum | minimum | difference | maximum | minimum | difference |
| Hip ........ | 150 | 111 | 39 | 154 | 100 | 54 |
| Knee ....... | 140 | 116 | 24 | 146 | 83 | 63 |
| Talocrural .. | 90 | 58 | 32 | 76 | 39 | 37 |
| Shoulder .... | 168 | 146 | 22 | 155 | 109 | 46 |
| Elbow ...... | 172 | 150 | 22 | 165 | 110 | 55 |
| Forearm-wrist | 218 | 186 | 32 | 224 | 72 | 152 |

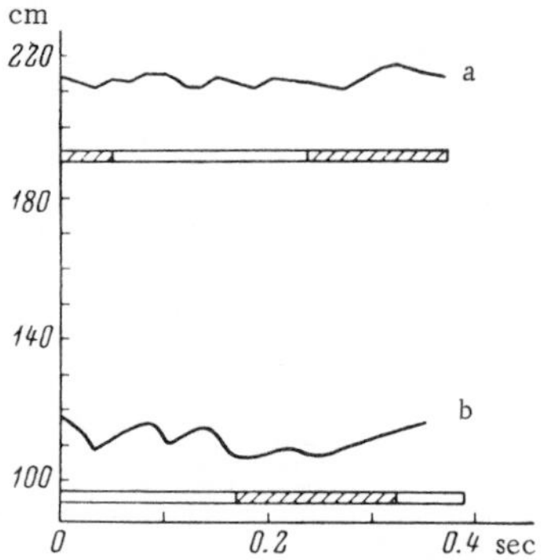

FIGURE 117. Vertical movements of the apex of the scapula (a) and the hip joint (b) during a cycle of fast walking in Elephas indicus:

Hatched part of bar denotes the phase of support. The abscissa gives the time in seconds, the ordinate the length in centimeters.

Hence, the mechanical features of locomotion characteristic for the elephant are: 1) rigidity of the spine during the entire cycle of running; 2) a relative straightening of the joints in the phase of support; 3) two moments of direction switching in the changes of hip joint size, instead of the four in other mammals; 4) two moments of angle change in the shoulder and elbow joints during the phase of support instead of one moment as in other mammals; 5) much greater flexor-extensor movements in the joints during the phase of free transit than during the phase of support; 6) virtual absence of vertical fluctuations of the center of gravity in the cycle of motion; 7) immobility of the proximal end of the scapula with respect to the trunk.

## STRUCTURAL FEATURES OF THE SPINE

We mentioned in the preceding section that mobility of the spine is very limited in the elephant's cycle of motion.  The analogous decrease in spinal mobility in the Ungulata is due to a system of rigid adaptations of the vertebral column: the withers in the thoracic section, the transverse processes in the intertransversal ligament, and the spinous processes with the supraspinous ligament and the hinges of the articular processes in the lumbar region.  Rigidity of the spine is even more complete in elephants.  In the thoracic region it is achieved more or less in the same way as in ungulates, but in the lumbar region quite differently.

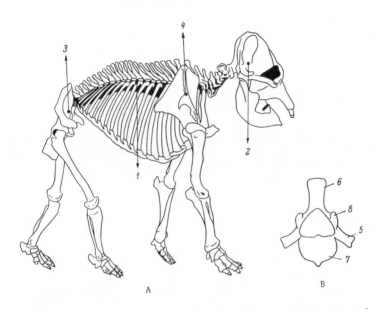

FIGURE 118.  Skeleton of E l e p h a s  i n d i c u s showing the points of application of the forces of gravity and the reactive forces of the limbs to the spine (A) and the structure of lumbar vertebra II (B):

1) weight of the body;  2) weight of the head;  3) reactive force of the pelvis;
4) scapulae;  5) transverse process;  6) spinous process;  7) body of the vertebra;
8) articular process.

Two oppositely directed loads act on the spine of an elephant.  In the middle between the two supports (forelimbs and hind limbs) the spine tends to sag under the weight of the body.  In the region of attachment of the forelimbs the spine is extended owing to the force of support on the limbs, which is equal to the sum of the forces formed by the weight of the body behind and the weight of the neck and head in front of the place where the scapulae are attached to the spine (Figure 118, A). Sagging of the thoracic and lumbar regions is counteracted by the vaulted position of the spine. The bodies of all the thoracic and lumbar vertebrae have a wedge-shaped widening from bottom to top. This widening is not easy to see on an individual vertebra, but taken as a whole it forms a regular arch, the bases of which lie on the scapulae and on the ilia. The powerful extending influence of the head and body in the region of the scapulae is damped by the rigid formation of the thoracic region of the spine. This rigidity is due to expansion of the spinous processes and supraspinous ligament and the very rigid thorax. Interestingly, the spinous processes of the elephant slope backward beginning from the last cervical vertebra. The angle of the slope progressively increases up to thoracic vertebrae XV—XVI, after which it decreases. The elephant has few lumbar vertebrae, usually less than four. Because of this the last rib is situated very close to the wing of the ilium, which expands anteriorly and broadwise (Figure 118, A).

The transverse processes of the lumbar vertebrae are considerably weakened (Figure 118, B), while their articular processes do not form any hinges in the intervertebral joints. Resistance of the lumbar region to lateral flexures is achieved differently than in ungulates. A bilateral fibrous connection of the last rib with the wing of the ilium creates sufficient resistance to lateral bends in the very reduced lumbar region of the elephant's spine. At the same time, a certain degree of mobility in the ribs allows for the very limited lateral flexures which are needed during the impacts imparted to the body by the hind limbs one after the other. Owing to the position of the limbs, the force of these impacts is directed at a small angle to the axis of the spine. The similar mobility of the loin in the ungulates does not permit the transverse processes to be knitted into a single bony unit.

The relatively small height of the spinous processes in the region of the withers in the elephant, which even in absolute measurements are smaller than the spinous processes of the thoracic vertebrae in most of the large ungulates, probably has two explanations. First, being unable to move by asymmetrical saltatorial gaits, elephants are spared the heavy impacts typically felt by ungulates when they land on the forefeet. And secondly, the elephant has a very short neck, which leads to a decrease of the moment acting to extend the spine in the region of scapular attachment.

The apexes of the spinous processes of the lumbar and last thoracic vertebrae widen markedly anteroposteriorly, greatly decreasing the interspinal spaces of the supraspinous ligament. The spinous processes of the anterior thoracic vertebrae are strongly thickened in both the anteroposterior and transverse directions. The apexes of these vertebrae are also markedly widened in both directions. The most powerful spinous process is that of the third thoracic vertebra, on which is inserted the thickest fibrous band from the scapula, and it is apparently this process which takes the main extending load on the vertebra.

## RATIO OF THE LIMB SEGMENTS

Adaptation to taking food from the tops of trees has given the elephant very long legs. The ungulates with the longest legs are the giraffes, and the reason is the same. But the ratios of the limb segments are diametrically opposite in these two animals. In the recent and some of the extinct Proboscidea the total length of all the segments is usually not less than in the giraffe, but it is the hand and foot which are elongated in the giraffe, while in the elephant it is the shoulder and femur, the hand and foot being on the contrary greatly reduced (Tables 12 and 13).

We see from Tables 12 and 13 that the hand and foot are markedly shorter in elephants than in ungulates. However, if we take into account that in the elephant support is transmitted from the distal margin of the forearm and tibia via a callosity on the sole of the foot directly to the ground, we may consider the moment acting on the talocrural and wrist joint to be equal to 0, while a shortening of the tibia and forearm diminish also the moments acting on the elbow and knee joints.

Thus, despite the overall lengthening of the limbs in elephants, reduction of the distal segments is a characteristic feature. We described a similar process of reduction of the distal segments for the ungulates adapted to the mediportal form of running (p. 151).

## CHARACTERISTICS OF THE LEG FASCIAE AND MUSCLES

The muscles of the Proboscidea have been studied very superficially. Nevertheless, a detailed description of all the leg muscles of elephants will not be given in this section,* but only those muscles and fasciae which are of interest for understanding how the proboscidians became adapted to their distinctive type of running will be examined. With this in mind let us discuss the functions of the muscles separately in the phases of support and free transit.

### Fasciae and muscles of the hind limbs

As mentioned above, the fast walk is the elephant's typical high-speed gait. During this gait, at the moment the hind feet land the phase of front support is coming to an end, and the body is affected by the following: inertia, of motion, the force of the body's weight, the work of the forelimb and the just-beginning activity of the hind limb. Despite the fact that the hip and knee joint extensors enter into action, the force of gravity and the inertia

---

* The muscles of two species will be treated in detail in a special paper (Gambaryan and Rukhkyan, 1972). All the data in this chapter are based on original studies. Only one work has appeared on a morphofunctional analysis of the Indian elephant's forelimb muscles (Druzhinin, 1941), but the very badly damaged specimen used by the author detracts greatly from the value of his conclusions.

TABLE 12.  Absolute length of the limbs and the relative sizes of their segments (% of sum of segment lengths in the relevant limb without the girdles)

| Species | Length of hind limb, mm | Index | | | Length of forelimb, mm | Index | | | |
|---|---|---|---|---|---|---|---|---|---|
| | | femur | tibia | foot | | shoulder | forearm | hand | scapula |
| Loxodonta africana ......... | 2,240 | 53.5 | 32.4 | 14.1 | 2,100 | 46.0 | 37.6 | 16.4 | 42.6 |
| Elephas indicus ............. | 2,230 | 55.2 | 30.9 | 13.9 | 2,160 | 43.0 | 37.6 | 21.9 | 37.1 |
| Archidiscodon meridionalis | 2,970 | 51.6 | 33.3 | 15.2 | 2,800 | 47.1 | 37.5 | 15.4 | 34.2 |
| Mammonteus primigenius .. | 2,340 | 55.6 | 29.4 | 15.0 | 2,200 | 44.0 | 35.9 | 20.1 | 32.2 |
| Giraffa camelopardalis .... | 2,080 | 23.1 | 26.5 | 50.4 | 2,320 | 21.3 | 36.9 | 41.8 | 26.3 |

TABLE 13.  Relative sizes of the limb segments in proboscidians and the giraffe (% of sum of lengths of the lumbar and thoracic regions of the spine)

| Species | Relative size of segments and of legs as a whole | | | | | | | | |
|---|---|---|---|---|---|---|---|---|---|
| | femur (a) | tibia (b) | foot (c) | sum (a+b+c) | shoulder (d) | forearm (e) | hand (f) | sum (d+e+f) | scapula |
| Loxodonta africana ......... | 59.2 | 35.8 | 15.4 | 110.4 | 48.2 | 39.4 | 17.3 | 104.9 | 44.7 |
| Elephas indicus ............. | 57.7 | 32.2 | 14.4 | 104.4 | 45.1 | 36.9 | 23.0 | 105.0 | 39.0 |
| Archidiscodon meridionalis | 56.1 | 36.2 | 16.5 | 108.8 | 50.2 | 40.0 | 16.5 | 106.7 | 38.0 |
| Mammonteus primigenius .. | 56.0 | 29.6 | 15.0 | 100.6 | 46.2 | 37.6 | 21.0 | 104.8 | 40.0 |
| Giraffa camelopardalis .... | 48.5 | 55.5 | 106.2 | 210.2 | 48.5 | 83.0 | 94.0 | 225.5 | 56.0 |

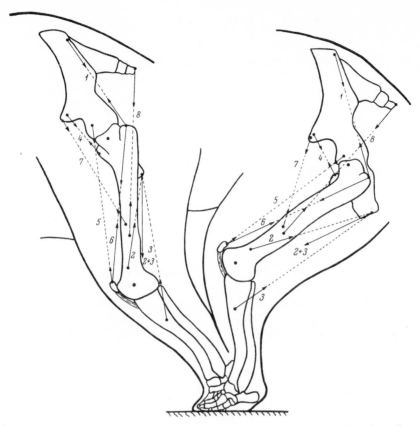

FIGURE 119. Diagram showing the action of the hind limb muscles at the end of the phase of support during the fast walk of an elephant:

1) gluteal; 2) short postfemoral; 3) long postfemoral; 4) m. iliopsoas; 5) m. rectus femoris; 6) knee joint extensors; 7) m. tensor fasciae latae + m. gluteus superficialis; 8) m. biceps anticus.

cause the knee joint to be flexed, in connection with which the hip joint moves forward almost horizontally. In elephants as opposed to ungulates, movement in the talocrural joint is apparently passive; at any rate it does not play a decisive role in transmitting thrust to the trunk. The work of the hip joint extensors and flexors together with the knee joint extensors and the activity of the forelimb helps move the body forward. In order to understand the action of these muscles let us first take a simplified scheme of their work (Figure 119). The position of the skeleton of the pelvic girdle and hind limbs is shown at the beginning and end of the phase of support, and the resultants of the main flexors and extensors of the hip and knee joints are plotted. The hip joint extensors are represented in the form of resultants of three groups: gluteal (1) and short and long postfemoral (2, 3).

From the short postfemoral group we have singled out m. biceps femoris anterior *(8)*, as its activity is not equivalent at the beginning and end of the phase of support. The activity of the hip joint flexors is shown by m. iliopsoas *(4)*, m. tensor fasciae latae *(7)*, and m. rectus femoris *(5)*. The knee joint extensors proper comprise the three femoral capita of m. quadriceps femoris *(6)*.

We can assess the general trend of movement associated with the work of the above muscles by calculating in any scale the lever arms of the application of their force on the corresponding joints and by multiplying them by the relative weight of the relevant muscle or group of muscles. Although such a calculation does not give us an idea of the actual forces developed by a limb in the phase of support, it does allow us to objectively evaluate their sum effect on the joint in question. In the phase of support apparently all the extensors of these joints are in action and possibly some of the flexors morphologically related to the extensors (m. tensor fasciae latae). However, even if we arbitrarily accept that all the above muscles act together, the relative force of extension of these joints will all the same prevail over the force of their flexion. Taking a moment of flexion to be negative and one of extension to be positive, we see that at the beginning of the phase of support the lever arm for m. iliopsoas is −1.5, for m. tensor fasciae latae −3.5, for m. rectus femoris −0.8 and for m. biceps femoris anterior −0.3. The product of the relative weight of these muscles (Table 14) times the lever arms in all yields the value −38.6. At the same moment the effect of the long and short postfemoral muscles has a lever arm of +4.0 while the gluteal group has one of +1.2. Their product times the relative weight in the sum gives 143.6, that is, the force of extension of the hip joint is 3.7 times greater than the force of its flexion. Analogous measurements of lever arms at the end of the phase of support yield the following values: m. iliopsoas −1.3, m. tensor fasciae latae −2.3, m. rectus femoris −1.0, in all the moment of flexion equaling −27.2. M. biceps femoris anterior has a value of +2.0, the long and short postfemoral muscles +2.8 and the gluteal muscles +1.8. In all the moment of extension is +136.4, i. e., the relative force of extension in the hip joint is five times greater than the force of flexion. Hence, the force of extension of the hip joint during the entire phase of support is 3—5 times greater than the force of its flexion, even taking into account the simultaneous work of all the muscles acting on this joint.

Apart from implementing movement in the hip joint, the gluteal, short postfemoral and iliopsoas muscles and m. tensor fasciae latae indirectly affect movement in the knee joint. To determine this effect, their resultants are resolved into components perpendicular and parallel to the axis of the femur. Considering their extending effect on the knee joint to be positive and their flexing effect to be negative and taking into account the work of the three femoral capita of m. quadriceps femoris, we see that the sum relative force of extension is 2.5—2.7 times greater than the relative force of flexion in the knee joint. This causes a thrust to be imparted to the body due to the work of the hip and knee joint extensors during the whole phase of support (Figure 120). The thrust is directed forward and upward along a line situated a little in front of the line connecting the overall center of gravity with the point of support of the limb.

TABLE 14. Relative weight of muscles and groups of muscles in elephants (% of the sum of weights of the hind limb muscles)

| Muscles | Elephas indicus | Loxodonta africana | Muscles | Elephas indicus | Loxodonta africana |
|---|---|---|---|---|---|
| m. psoas minor ............. | 0.4 | 0.5 | m. gracilis anterior ....... | 2.1 | 1.4 |
| m. iliopsoas ............... | 6.6 | 6.1 | m. gracilis posterior ...... | 1.0 | 0.7 |
| m. gluteus superficialis + | | | m. popliteus ............. | 0.1 | 0.1 |
| + m. tensor fasciae latae ... | 5.9 | 6.1 | m . rectus femoris ........ | 4.8 | 3.9 |
| m. gluteus medius ......... | 11.0 | 14.3 | m. vastus lateralis ........ | 11.8 | 10.3 |
| m. gluteus minimus ........ | 2.0 | 1.6 | m. vastus medialis ........ | 4.3 | 3.8 |
| m. obturator externus ...... | 2.2 | 2.1 | m. vastus intermedius ..... | 0.4 | 2.2 |
| m. obturator internus ........ | 0.7 | 0.6 | m. gastrocnemius lateralis | 0.3 | 0.1 |
| m. gemelli ...............ı | 0.4 | 0.5 | m. gastrocnemius medialis | 0.4 | 0.5 |
| m. pectineus ............... | 1.2 | 2.3 | m. plantaris ............. | 0.1 | 0.5 |
| m. adductor brevis .......... | 5.3 | 4.8 | m. soleus ............... | 0.4 | 0.6 |
| m. adductor maximus ....... | 6.5 | 5.7 | m. tibialis anterior ........ | 0.4 | 0.2 |
| m. biceps anticus .......... | 14.0 | 12.0 | m. ext. digitorum longus .. | 1.6 | 1.1 |
| m. biceps ⎧ pars femoralis ⎫ | 5.8 | 1.6 | m. peroneus longus ······· | 0.5 | 0.4 |
| posticus ⎩ pars tibialis ⎭ | | 5.0 | m. peroneus brevis ....... | 0.6 | 0.1 |
| m. semitendinosus .......... | 5.1 | 5.7 | m. peroneus digiti quinti .. | 0.6 | 0.6 |
| m. semimembranosus anterior | 1.0 | 1.9 | m. tibialis posterior ...... | 0.6 | 0.3 |
| m. semimembranosus posterior | 0.9 | 0.9 | m.m. flexores digitorum .. | 1.2 | 1.2 |
| Gluteal ................. | 13.0 | 15.9 | Knee joint extensors ...... | 16.5 | 16.3 |
| Long postfemoral .......... | 14.1 | 13.7 | Hip joint flexors ......... | 17.3 | 16.1 |
| Short postfemoral .......... | 32.1 | 31.5 | Gastrocnemial .......... | 1.1 | 1.2 |
| Hip joint extensors ........ | 59.2 | 61.1 | | | |

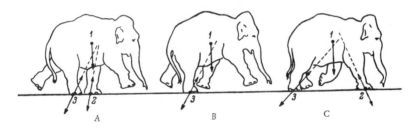

FIGURE 120. Three moments of the phase of hind support during a fast walk of the elephant, and the forces acting on the body:

A) beginning, B) middle, C) end of the phase of support. 1) force of gravity; 2) force of thrust by the forelimbs; 3) force of thrust by the hind limbs.

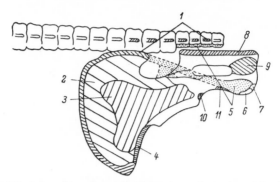

FIGURE 121. Points of insertion of the hind limb muscles on the pelvic girdle in Loxodonta africana, dorsal:

1—3) m. gluteus: 1) superficialis, 2) medius, 3) minimus; 4) m. tensor fasciae latae; 5, 6) m. biceps femoris: 5) anterior, 6) posterior; 7) m. semitendinosus; 8, 9) m. obturator: 8) externus, 9) internus; 10, 11) m. gemelli: 10) cranialis, 11) caudalis.

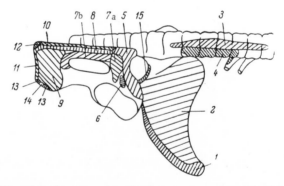

FIGURE 122. Points of insertion of the hind limb muscles on the pelvic girdle in Loxodonta africana, ventral:

1) m. tensor fasciae latae; 2) m. iliacus; 3) m. psoas minor; 4) m. psoas major; 5) m. pectineus; 6) pubic stem of superficial medial fascia; 7a) anterior part of m. obturator externus; 7b) posterior part of m. obturator externus; 8, 9) m. adductor: 8) maximus, 9) brevis; 10) m. gracilis; 11, 12) m. semimembranosus: 11) anterior, 12) posterior; 13) m. semitendinosus; 14) m. biceps femoris posterior; 15) m. biceps anterior.

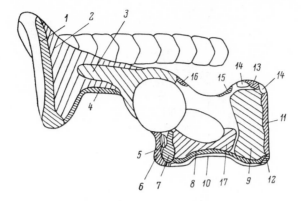

FIGURE 123. Points of insertion of the hind limb muscles on the pelvic girdle in Loxodonta africana, lateral:

1–3) m. gluteus: 1) superficialis, 2) medius, 3) minimus; 4) m. tensor fasciae latae; 5) m. pectineus; 6) pubic stem of superficial medial fascia; 7) anterior part of m. obturator externus; 8, 9) m. adductor: 8) maximus, 9) brevis; 10) m. gracilis; 11, 12) m. semimembranosus: 11) anterior, 12) posterior; 13) m. biceps femoris posterior; 14) m. semitendinosus; 15, 16) m. gemelli: 15) caudalis, 16) cranialis; 17) posterior part of m. obturator externus.

Before going on to a more detailed scrutiny of the work of the different hind limb muscles, we must point out that due to the special working conditions of the foot, which is supported as far as the talocrural joint by the fat cushion, the talocrural joint extensors have little influence on the animal's movement. It is therefore not surprising that the relative weight of these muscles is markedly lower in the elephant than in the ungulates. For instance, in the elephant the weight of the gastrocnemial muscles is 1.1–1.2% of the weight of the hind limb muscles, whereas in ungulates it is at least 4.0% (Tables 14 and 8).

The deep functional bond between the muscles situated in front of and behind the hip joint results in a morphological union of m. tensor fasciae latae, m. gluteus superficialis and m. biceps femoris anterior into a single whole. These three muscles originate linearly for a large distance, being inserted on the pubis along the entire ventral margin of the ilium (this part is probably homologous to m. tensor fasciae latae), along the crest of the ilium and the spinous processes of the sacral vertebrae (this part is probably homologous to m. gluteus superficialis) and along the spinous processes of the first two caudal vertebrae and the sacroischial ligament (this part is homologous to m. biceps femoris anterior) (Figures 121–123). The first part ends on a fibrous plate of the broad fascia of the femur, which is inserted on the patella. The second and third parts end on the lateral lip of the femur. The bundles of m. gluteus superficialis run anteroposteriorly, those of m. biceps femoris anterior posteroanteriorly (Figure 124).

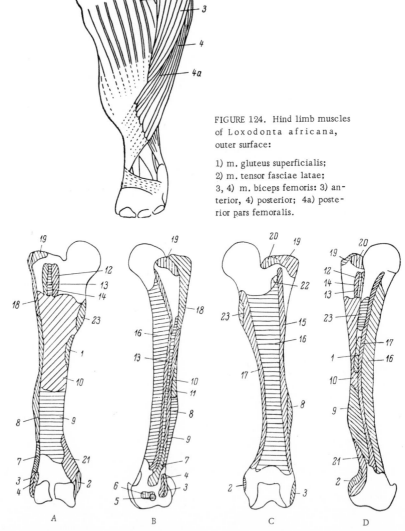

FIGURE 124. Hind limb muscles of Loxodonta africana, outer surface:

1) m. gluteus superficialis; 2) m. tensor fasciae latae; 3, 4) m. biceps femoris: 3) anterior, 4) posterior; 4a) posterior pars femoralis.

FIGURE 125. Points of insertion of the hind limb muscles on the femur in Loxodonta africana:

A) posterior view; B) lateral view; C) anterior view; D) view from the medial side. 1) m. pectineus; 2, 3) m. gastrocnemius: 2) medialis, 3) lateralis; 4) m. plantaris; 5) m. popliteus; 6) m.ext. digitorum longus + m. peroneus longus + m. peroneus tertius; 7, 8) m. biceps femoris: 7) posterior, 8) anterior; 9, 10) m. adductor: 9) maximus, 10) brevis; 11) m. tensor fasciae latae; 12, 13) m. obturator: 12) internus, 13) externus; 14) m. gemelli cranialis + caudalis; 15–17) m. vastus: 15) lateralis, 16) intermedius, 17) medialis; 18–20) m. gluteus: 18) superficialis, 19) medius, 20) minimus; 21) m. semimembranosus anterior; 22) m. capsularis; 23) m. iliopsoas.

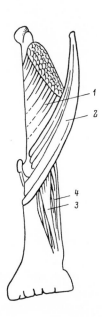

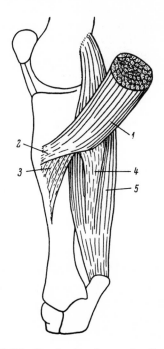

FIGURE 126. M. biceps femoris
in Loxodonta africana:

1, 2) m. biceps femoris: 1) anterior,
2) posterior; 3) tendinous band to
the calcaneal tubercle; 4) m. gastrocnemius.

FIGURE 127. Termination of m. semitendinosus
in Loxodonta africana:

1) m. semitendinosus; 2) its tendon; 3) elastic
band to the tibia; 4) fibrous band to m. gastrocnemius; 5) m. gastrocnemius.

M. tensor fasciae latae and m. gluteus superficialis prevent the hip
joint from becoming overextended in the phase of support. A similar
function is performed by m. rectus femoris, which begins on a special
tuberosity on the body of the ilium and ends on the patella; likewise with
m. iliopsoas. The weight of these muscles in all constitutes 16.1—17.3%
of that of the hind limb muscles (Table 14).

The postfemoral muscles show the opposite action to these. The short
group consists of the already described m. biceps femoris anterior and the
adductors of the femur (m. adductor maximus and m. adductor brevis), plus
m. pectineus and m. semimembranosus anterior. All these muscles originate on the vertical branch and body of the ischium (Figures 121—123) and
end on the plantar side of the femur (Figure 125). In addition, the tuber
ischiadicum is the site of origin of m. biceps femoris posterior, one part
of which ends on the lateral lip of the femur directly under the termination
of m. biceps femoris anterior (Figure 125), while the second part ends on
the crest of the tibia and can therefore be considered to belong to the group

of long postfemoral muscles (Figure 126). A little-elastic tendon stretches between the second part of m. biceps femoris posterior to the calcaneum and the fascia of the foot.

The short postfemoral muscles together make up 31.5—32.1% of the weight of the elephant's hind limb muscles (Table 14). A similar effect for propelling the body is exerted by the long postfemoral muscles, whose activity depends not only on flexor-extensor movements simultaneously in the knee and hip joints but also on the topography of their attachment on the pelvis and tibia. These muscles include the capita of m. gracilis, m. semimembranosus posterior, m. semitendinosus and the above-described m. biceps femoris posterior. Apart from ending on the tibia, all these muscles have a connection with the foot, which makes an analysis of their functions all the more complicated.

M. semitendinosus originates as two capita: vertebral and ischial, the first originating on the spinous process of the second caudal vertebra and the second on the proximal tuber ischiadicum. The two capita become fused at the level of the distal tuber ischiadicum, and a well developed tendon stretches from the point of their fusion to m. biceps femoris anterior. M. semitendinosus ends on the medial lip of the tibia, almost at its middle. From its tendon to the medial malleolus extends a triangular tendon (Figures 124 and 127). In addition, a little above the terminal tendon a nontensile tendon extends to the calcaneal tubercle and to the fascia of the foot (Figure 127). A graph showing the changes in the length of the ischial caput during a cycle of fast walking in the Indian elephant (Figure 128) helps us understand the function of this muscle. The figure shows two different graphs for the ischial caput of m. semitendinosus. Curve 1 shows the change in the length of the posterior margin of the ischial caput and curve 2 the influence that the nontensile tendon stretching to the foot exerts on its length, in other words, how it affects the tension of m. semitendinosus. It is seen from the graphs that the amplitude of change in the length of the muscle increases appreciably in relation to the effect of the bracing band which stretches to the calcaneal tubercle. Apart from this, the working conditions for the muscle are improved in the phase of support, this being connected in the first place with the increased tension of the muscle in the phase of free transit due to its artificial elongation, and secondly, with the fact that its tension is utilized to a large degree in the phase of support, since the bracing band prevents it from contracting excessively. Thus, in the phase of support, due to the action of the brace, the muscle contracts by only 9%, while if we artificially exclude the effect of the brace, it would contract by 16—17% (Figure 128). These working conditions of m. semitendinosus are further complicated by the influence exerted by the vertebral caput and the tendinous band to m. biceps femoris anterior. By actively pulling up the ischial caput, the vertebral caput serves as a buffer enhancing the work of the ischial caput, while the band to m. biceps femoris anterior prevents the place where the vertebral and ischial capita meet from being diverted backward. This last effect is especially important in the last moments of the phase of support, when a straight line from the place of origin of the vertebral caput to the end of the whole muscle is situated caudal to the place of union of the two capita (Figure 114).

A similar function is performed by m. biceps femoris posterior, the femoral caput of which has a similar influence to that of the vertebral caput of m. semitendinosus, while the tendon to the caudal part of this muscle fully repeats the role of the brace of m. semitendinosus.

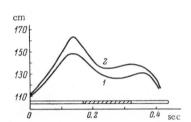

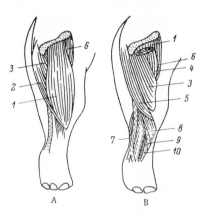

FIGURE 128. Changes in the length of m. semitendinosus during a cycle of fast walking in Loxodonta africana:

1) excluding the influence of the tendinous band; 2) with the influence of the tendinous band. Hatched part of bar denotes the phase of support. The abscissa gives the time in seconds, the ordinate the length in centimeters.

FIGURE 129. Hind limb muscles of Loxodonta africana from the medial side:

A) surface layer; B) deep layer. 1) m. gracilis; 2) m. semimembranosus posterior; 3, 4) m. adductor maximus: 3) posterior bundles, 4) anterior bundles; 5) m. semimembranosus anterior; 6) m. pectineus; 7) fibrous band to m. gastrocnemius from m. semitendinosus and m. gracilis; 8) tendon of m. semitendinosus; 9, 10) its elastic bands to the tibia.

M. m. gracilis anterior and posterior begin as a common tendon from the pelvic symphysis, falling short of the anterior margin by a quarter of its length. The tendon is considerably thinner in its middle part, so that we can arbitrarily divide the muscle into anterior and posterior parts. M. gracilis anterior ends on the medial epicondyle of the tibia and along its median lip. On the inner side of the tibia the fibers of the posterior muscle intertwine, so that in its topography the ending of these parts is very reminiscent of the union of m. gracilis anterior and m. gracilis posterior in the Ungulata (Figures 107 and 129). All this taken together suggests that there are two not yet fully separate gracilis muscles in the Proboscidea. In addition, m. semimembranosus posterior is entwined in this complex of m. gracilis. This muscle originates along the whole caudal margin of the ischium. The bundles then gradually converge and descend toward the tibia, ending in a thick tendon which is intergrown with the common tendon of m. m. gracilis (Figure 129). From the caudal margin of the common tendon and m. semimembranosus posterior begins an inextensible tendon which stretches to the fascia of the foot and the calcaneal tubercle.

Tracing the change in the length of m. m. gracilis and m. semimembranosus posterior during a cycle of fast walking, we see that the overall contraction of m. m. gracilis is not great and that the anterior margin of the muscles acts as an extension onto the posterior margin in the phase of support (analogously to the action of m. gracilis anterior in ungulates; see p. 146). The change in length of m. semimembranosus posterior during a cycle of fast walking is of a very similar nature to that observed in m. semitendinosus. We have to infer that the tendon stretching from the caudal margin of m. m. gracilis and m. semimembranosus posterior to the calcaneal tubercle also acts analogously to that of m. semitendinosus.

As a whole, the group of long postfemoral muscles together with the short postfemoral muscles and the well developed gluteal muscles acts to extend the hip joint with a force which markedly exceeds that of all the flexors of this joint. Forceful extension of the hip joint is extremely important to the elephant because during the fast walk there is a stage of support on one hind limb. At this time the weight of the body tends to bend the hip joint. A rectilinear movement of the trunk is under these conditions possible only in a case where the force of the hip joint extensors can counteract the flexing moment of the body's weight. Remembering that the elephant's center of gravity is situated nearer the forelimbs, not the hind limbs, the need for strongly developed hip joint extensors becomes evident. The flexing effect of the center of gravity on the hip joint is particularly marked toward the end of the one-leg stage of support, when the center of gravity is especially far removed from the point of contact between the leg and the ground (Figure 120). We can thus understand why predominance of the total relative force of extension of the hip joint over the relative force of its flexion is enhanced in the course of the phase of support (for more details back to pp. 173, 175).

In elephants the weight of the hip joint extensors accounts for 60.2—61.6% of the total weight of the hind limb muscles as against 51.0—63.9% in the Ungulata (Tables 8 and 14). Hence, the main working groups of hind limb muscles (the extensors of the hip joint) are in proboscidians even more strongly developed than in many ungulates highly specialized for running.

The extensors proper of the knee joint are also strengthened in the elephant, that is, the three femoral capita of m. quadriceps femoris. The straightened position of the limbs (with strongly extended knee joint) during the whole load-bearing phase of support leads to a very small amplitude of contraction of the femoral capita of this muscle. Therefore, apart from the strengthening deriving from the increased weight of the muscle (cf. Table 14 with Table 8), there is a structural reorganization of the lateral caput in proboscidians. It becomes 5-plumose. The tendinous interlayer forming the first plume of this muscle, which is closest to the femur, is somewhat concave. Therefore, when the animal is standing, it can work on the principle of extensor muscles (Gambaryan, 1957). In general, its plumose structure leads additionally to a considerable strengthening of the knee joint extensors proper.

Owing to the marked extension of the knee joint toward the end of the phase of support (Figure 115), considerable tension is exerted in the long postfemoral muscles and the muscles situated in front of the hip joint: m. tensor fasciae latae, m. gluteus superficialis and the superficial fascia of the hind limb. All this together causes sharp flexion of the hip and knee

joints from the very beginning of the phase of free transit.  We must especially note the work of the superficial fascia of the hind limb, which promotes
automatic flexion of the knee and hip joints together with the muscles which
were extended at the end of the phase of support.  The lateral layer of the
outer fascia of the hind limb is weakly expressed.  It covers m. gluteus
superficialis with a thin elastic sheet 2—3 mm thick.  It ends anteriorly on
the little-extensible fibrous broad fascia of the femur, passing posteriorly
into the fascia of the tibia (Figure 130).  The middle layer of the superficial
fascia, which begins as two teeth, is more differentiated and well developed.
The first tooth is inserted along the whole ventral crest of the ilium right
up to the anterior margin of the pubis, covering m. tensor fasciae latae from
the medial side.  The second tooth is in the form of a strong fibrous tendon
about 40 mm in diameter; it begins on the pubis and is situated on a special
tuberosity between the origin of m. pectineus and m. adductor maximus
femoris.  This tendon becomes gradually wider and passes into an elastic
layer 12—14 mm thick, covering all the femoral muscles medially.  On the
upper third of the femur the bundles of both teeth crisscross, unite with each
other and pass in a common layer into the fascia of the tibia, which is densely
attached on the foot and the median lip of the tibia.  It is this layer of the
superficial fascia which produces the automatic pull of the tibia in the phase
of free transit of the limbs.  In the adjustment period of the phase of free
transit the hip and knee joint extensors again straighten the leg, which is
then placed on the ground to start a new cycle of motion.

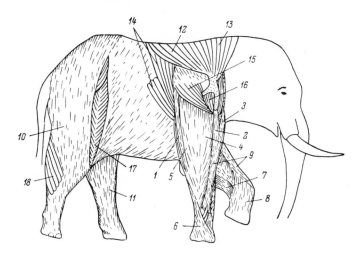

FIGURE 130. Superficial fasciae and muscles in Loxodonta africana:

1) band from caudal angle of scapula to olecranon; 2, 3) origin of the acromioradial
ligament; 2) from the acromion, 3) from the scapular tubercle; 4, 5) intermediate
bundles between the acromioradial ligament and the band from the caudal angle of
the scapula; 6—8) fasciae of the hand: 6) lateral, 7) medial deep, 8) medial superficial; 9) acromioradial ligament; 10, 11) superficial fasciae of the hind limb:
10) lateral, 11) medial; 12) m. spinotrapezius; 13) m. acromiotrapezius; 14) m.
latissimus dorsi; 15) m. spinodeltoideus; 16) m. acromiodeltoideus; 17) m. vastus
lateralis; 18) m. biceps femoris posterior.

## Fasciae and muscles of the forelimbs

In the phase of support the locomotory organs of both the forelimbs and the hind limbs cause the body to move forward. Even a simplified scheme of the work of the forelimbs is much more complex than the one we have given for the hind limbs. During the fast walk of an elephant, at the moment of landing and before it leaves the ground the forelimb is acted upon by the following forces: the continuing or beginning thrust of the hind limbs, inertia of motion, the weight of the body and the activity of the skeleton and muscles of the forelimbs. But in the middle of the phase of support, when only one leg is left on the ground, the hind limb no longer exerts any influence on the animal's movement.

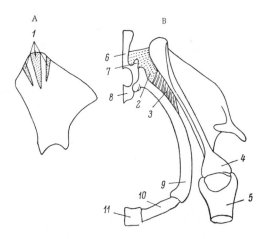

FIGURE 131. Diagram showing the attachment of tendinous bands from the thoracic vertebrae to the scapula (A) and the areas of their insertion on the scapula (B) in Loxodonta africana:

1) area of their attachment to the scapula; 2) tendinous band; 3) m. serratus ventralis; 4) scapula; 5) humerus; 6) spinous process of thoracic vertebra III; 7) its articular process; 8) its body; 9) rib; 10) its cartilage; 11) segment of sternum.

Before discussing the work of the shoulder girdle muscles it is useful to compare the relative mobility of the scapula in the Proboscidea and Ungulata. In ungulates the thorax is movably attached between the scapulae by means of the teeth of m. serratus ventralis, which begin on the distal ends of the ribs and end on the vertebral margin of the scapula. Only in the giraffe does m. serratus ventralis originate on the proximal ends of the ribs and even of the costal tubercles. Well expressed tendinous attachments extend from these tubercles to the vertebral margin of the scapula.

This structure of m. serratus ventralis in the giraffe greatly restricts
the mobility of the proximal margin of the scapula with respect to the
thorax. In proboscidians from the medial surface of the scapula in the
region of the tubercle of the vertebral margin to the tubercle of the third
rib, the base of the spinous process and the mastoid process of thoracic
vertebra III extends a strong fibrous band 5—6 cm thick. Less well expressed tendinous bands no more than 1 cm thick reach ribs II and IV and
the thoracic vertebrae from the above-mentioned tubercle on the vertebral
margin of the scapula (Figure 131). These three tendinous bands practically immobilize the proximal end of the scapula with respect to the
thorax. As a result, the axis of motion of the scapula relative to the thorax
corresponds in elephants to its proximal end in the region of the tubercle
and the base of the spinous process of the thoracic vertebra.

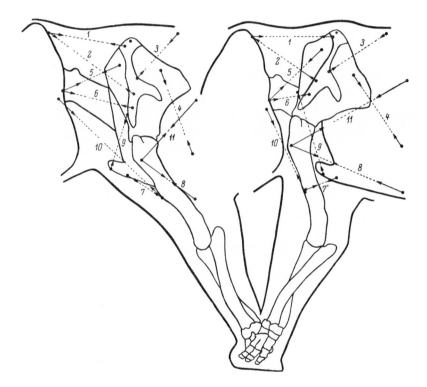

FIGURE 132. Diagram showing the activity of the main shoulder girdle muscles at the beginning
and end of the phase of support during the fast walk of an elephant:

1) m. rhomboideus capitis; 2) m. acromiotrapezius; 3) m. spinotrapezius; 4) m. serratus
ventralis pars thoracalis; 5) m. serratus ventralis pars cervicalis; 6) m. omotransversarius inferior; 7) m. ectopectoralis; 8) m. endopectoralis; 9) m. sternoscapularis; 10) m. brachiocephalicus; 11) m. latissimus dorsi.

In the preparatory period of the phase of support the shoulder and elbow joints are flexed to cushion the impact, while in the starting period they are extended (Figure 116), imparting an active thrust to the body. Owing to these movements of the elbow and shoulder joints and the change in the form of the hand, the apex of the scapula is able to move almost in the horizontal plane (Figure 117).

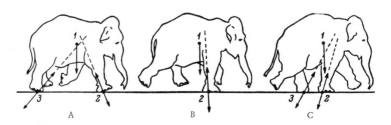

FIGURE 133. Three moments of the phase of support on the forelimbs during the fast walk of an elephant, and the forces acting on the body:

A) beginning; B) middle; C) end of the phase of support. 1) force of gravity; 2) force of thrust by the forelimbs; 3) force of thrust by the hind limbs.

With respect to the apex of the scapula the elephant's body has a tilting moment (clockwise rotation in Figure 132). If we take this to be positive and the moment of reverse rotation to be negative, we can estimate the influence of the shoulder girdle muscles on these movements. In the phase of support the action of these muscles directs the resultant of the thrust of the forelimbs in front of the common center of gravity, enhancing the tilting moment. At the beginning of the phase of support (Figure 133, A) the thrust imparted to the trunk is directed against the motion (shock absorption and inhibition). But at this time the thrust imparted by the hind limb neutralizing this negative effect for propelling the body is still in force. Toward the middle of the phase of support (Figure 133, B) when the center of gravity has still not intersected the vertical passing through the point of support, the shoulder girdle muscles, creating a tilting moment, already lend pull, thereby promoting some forward acceleration. At the end of the phase of support (Figure 134, C), when the thrust of the forelimb comes to have a favorable direction for propulsion and the following hind limb is already entering into action, the tilting moment becomes unimportant.

The force lever at the beginning of the phase of support is +11.8 in m. ectopectoralis, +11.5 in m. endopectoralis, +5.3 in m. sternoscapularis, +1.0 in m. serratus ventralis pars thoracalis, +2.6 in m. spinotrapezius and +7.1 in m. latissimus dorsi. The total moment of positive rotation of the body (the product of the lever arm times the relative weight of these muscles, Table 15) equals +216.5. The force lever of m. brachiocephalicus (m. cleidomastoideus + m. clavodeltoideus) is −9.0, of m. serratus ventralis pars cervicalis −0.3, of m. omotransversarius inferior −5.5, of m. acromiotrapezius −5.0 and of m. rhomboideus −0.5, the total moment being −68.5. Hence, at the beginning of the phase of support the tilting moment is 3.2 times greater than the relative force of its inhibition due solely to the influence of the shoulder girdle muscles.   The inhibiting effect of these muscles gradually increases, and toward the end of the phase of support it becomes practically equal to their tilting effect. Thus, the lever of m. ectopectoralis becomes −10.6, of m. sternoscapularis −1.8, of m. brachiocephalicus −8.8, of m. serratus ventralis −1.8 and −1.7, of m. omotransversarius inferior −6.0, of m. acromiotrapezius −3.8 and of m. rhomboideus −0.6;  the total relative moment of motion is −114.6.   A positive force lever is preserved for m. latissimus dorsi, +8.3, m. endo-pectoralis, +8.3 and for m. spinotrapezius, +2.9, while their relative total moment of motion (+156.3) is only about 1.3 times greater than the inhibit-ing effect.

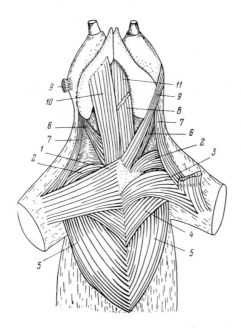

FIGURE 134.  Shoulder girdle muscles of L o x o d o n t a   a f r i c a n a, ventral:

1) m. ectopectoralis;  2) m. sternoscapularis;  3—5) m. endopectoralis;
6) m. sternomastoideus;  7) m. cleidomastoideus;  8) m. sternohyoideus;
9) m. sternomaxillaris;  10) m. sternomandibularis;  11) m. digastricus.

Therefore, whereas with the hind limbs, which play the leading propulsive role in elephants, we observe increasing predominance of the effect of ex-tension of the hip joint over its flexing effect up to the end of the phase of support, the tilting moment caused by the forelimbs gradually dies out.

The structural features of the elephant's shoulder girdle muscles enable us to expand on the general scheme of their work.  In elephants the trape-zoid muscle does not have a typical clavicular part, but nevertheless, its

origin spreads onto the head (Figures 130 and 135).  The cephalic part of
m. acromiotrapezius ends on the base of the metacromion and acromion
and is the strongest part of the whole muscle.  The other part of the muscle
begins on the midline of the neck and ends on the spina scapulae between
the metacromion and the tubercle of the spina scapulae (Figure 135).  Contraction of the cephalic part of m. acromiotrapezius affects the drawing
forward of the shoulder girdle in the phase of free transit and inhibits movement of the body in the phase of support.

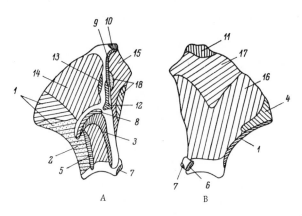

FIGURE 135. Points of insertion of the forelimb muscles on the scapula
in Loxodonta africana:

A) from the lateral side; B) from the medial side. 1) m. anconeus longus;
2) m. spinodeltoideus; 3) m. acromiodeltoideus; 4) m. teres major;
5) m. teres minor; 6) m. coracobrachialis; 7) m. biceps brachii; 8, 9) m.
omotransversarius: 8) inferior, 9) superior; 10, 11) m. rhomboideus:
10) capitis, 11) thoracalis + cervicalis; 12) m. acromiotrapezius; 13) m.
spinotrapezius; 14) m. infraspinatus; 15) m. supraspinatus; 16) m. subscapularis; 17) m. serratus ventralis; 18) m. sternoscapularis.

In elephants there are two parts to m. omotransversarius, inferior and
superior.  Both start from the wing of the atlas, but m. omotransversarius
superior ends on the vertebral margin and the proximal third of the spina
scapulae while m. omotransversarius inferior ends on the metacromion
(Figures 132 and 135).  Because the center of movement of the scapula is
situated in the region of the tubercle of its vertebral margin, contraction of
both muscles promotes forward movement of the shoulder joint in the phase
of free transit and inhibits movement of the body in the phase of support.
These muscles inhibit movement of the trunk with respect to the limbs.

M. cleidomastoideus begins on the mastoid process of the temporal bone
and ends on the clavicular part of the deltoid muscle, which ends on the distal
part of the continuation of the crest of the tuberculum majus humeri
(Figure 136).  Contraction of these muscles pushes forward the distal part
of the scapula and the humerus, and together with them also the shoulder

joint in the phase of free transit while it impedes forward movement of the body in the phase of support.

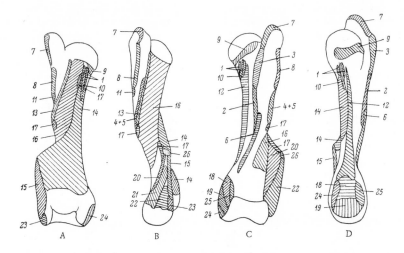

FIGURE 136. Points of insertion of the forelimb muscles on the humerus in Loxodonta africana:

A) from the volar side; B) from the lateral side; C) from the dorsal side; D) from the medial side. 1) m. latissimus dorsi; 2) m. ectopectoralis; 3) m. endopectoralis; 4, 5) m. spino-, acromiodeltoideus; 6) m. clavodeltoideus; 7) m. supraspinatus; 8) m. infraspinatus; 9) m. subscapularis; 10, 11) m. teres: 10) major, 11) minor; 12) m. coracobrachialis; 13, 14) m. anconeus: 13) lateralis, 14) medialis; 15) m. epitrochleoanconeus; 16) m. brachialis; 17) m. brachioradialis; 18) m. palmaris longus; 19) m. fl. digitorum profundus; 20) m. ext. carpi radialis longus et brevis; 21, 22) m. ext. digitorum: 21) communis, 22) lateralis; 23) m. ext. carpi ulnaris; 24, 25) m. fl. carpi: 24) ulnaris; 25) radialis; 26) tendinous band to the acromioradial ligament.

The pectoral muscles of proboscidians are clearly differentiated into three parts: 1) m. sternoscapularis originates on the anterior margin of the first rib and ends on the cranial margin of the scapula and the fascia of m. supraspinatus (Figures 132 and 135); 2) m. ectopectoralis begins on the manubrium and the first segment of the sternum and ends as two stems. The first stem, covering the bundles of m. endopectoralis, passes to the crest of the tuberculum majus humeri and ends just distal to the end of m. endopectoralis on its medial tuberosity (Figure 136). The second stem ends on the superficial fascia of the shoulder and together with this extends to the medial epicondyle of the humerus, ending on the tendon which is the degenerated m. pronator teres; 3) m. endopectoralis begins from the second segment of the sternum to the base of the ensiform process, but in Loxodonta africana its origin extends along this process and even to the cranial part of the white line of the abdomen; it ends on the crest of the tuberculum majus humeri, descending along it down to the upper boundary of the termination of m. ectopectoralis. The most caudal bundles of this muscle

end the most proximally and the cranial the most distally, forming a crisscrossing of bundles of m. endopectoralis. The effects exerted by these
three parts of the pectoral muscles are well illustrated by a graph of the
length changes of their fibers during a cycle of fast walking (Figure 137).
Before going into details about their activity, we should note that for the
anterior margin of m. endopectoralis, m. sternoscapularis and m. ectopectoralis the changes in the length of the fibers were determined not according to their position in the skeletons which were sketched in the successive frames, but with calculation of the necessary corrections. The
problem is that with movement of the humerus with respect to the thorax,
we can measure only the changes in the length of the projection of the
muscle on the horizontal plane, not its true length. For each case, therefore, the true length of the muscle was calculated according to its projection.
Here the length of one leg of the triangle was constant (the distance from
the thorax to the humerus) while that of the second was equal to the length
of the muscle's projection. Tracing the length changes of individual parts
of the pectoral muscle, calculated in this manner, during a cycle of fast walking, we see that from the very beginning of the phase of support all these
parts are contracted in unison, promoting movement of the trunk between
the legs. Then gradually m. sternoscapularis and m. ectopectoralis, and
later also the anterior parts of m. endopectoralis, attain their minimum
length and again begin to be elongated (Figure 137). Only the caudal part
of m. endopectoralis continues to shorten right up to the end of the phase
of support. M. sternoscapularis, m. ectopectoralis and the anterior bundles
of m. endopectoralis begin to contract from the very beginning of the phase
of free transit, promoting accelerated movement of the shoulder joint forward, while toward the end of this phase they are again elongated. Thus, in
the most crucial period of the phase of support all the pectoral muscles together help speed up the movement of the trunk between the legs. All these
muscles taken in toto represent one of the main forces exerting an effect to
produce the tilting moment of the trunk. Subsequently m. m. ectopectoralis
and sternoscapularis promote extension of the shoulder and elbow joints, facilitating the work of a number of muscles in the free forelimb imparting the necessary thrust to the body. On the other hand, at the beginning of the phase of free
transit the anterior parts of the pectoral muscles act as synergists of m. acromiotrapezius, m. omotransversarius superior and inferior and m. cleidomastoideus, speeding up the forward movement of the shoulder joint. Toward the end
of the same phase the whole pectoral muscle inhibits movement of the proximal
parts of the forelimb so that the limb is adjusted for the next phase of support.

Among the muscles acting to draw the trunk between the legs, apart from
the pectorals, of great importance is m. latissimus dorsi and partly
m. spinotrapezius. In the African and Indian elephants m. latissimus
dorsi has seven costal teeth from ribs V to XI. In Loxodonta africana the tooth from rib V begins almost a third of the length of the rib
away from the rib cartilage, while the tooth from rib XI is situated a little
above half the length of the rib. In Elephas indicus the tooth from
rib V begins almost at the level of the rib cartilage, while that from rib XI
begins above $\frac{4}{5}$ of its length, in other words, this part is much more compact
in the African species. Apart from the ribs, this muscle begins also from

the spinous processes of the thoracic vertebrae: from IV to VII in the
Indian elephant and from V to XI in the African form.   The two parts
become fused and end as two stems on two sides of the terminal tendon
of m. teres major.   M. spinotrapezius originates on the spinous processes
of the thoracic vertebrae and ends on the tubercle of the spina scapulae.

We obtain quite a similar picutre when we examine a graph of the length
changes of individual parts of m. latissimus dorsi during the cycle of fast
walking.   To study the changes in the length of this muscle's bundles we
took four extreme points: from ribs V and XI and from the spinous processes of thoracic vertebrae V and XI (Figure 138).   The graph shows that
the bundles display their greatest length at the very beginning of the phase
of support, becoming steadily shorter during the course of this phase right
up to its end.   In the phase of free transit the bundles lengthen at first a
good deal and then less intensively.   The maximum amplitude of contraction of the bundles from rib V is 41% of its maximum length;  this value is
28% for the bundles of rib XI, 16% for those from the fifth spinous process
and 17% for those from the eleventh.

The bundles of m. spinotrapezius change little in length, since they end
on the tubercle of the spina scapulae near the center of movement of the
limb.   As a result, their change in length does not exceed 6—7% of the
muscle's maximum length and is difficult to discern by the method of sketching in the muscles.

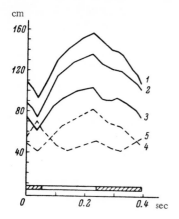

FIGURE 137.  Changes in the length of
the pectoral muscles during a cycle of
fast walking in Loxodonta africana:

1—3) m. endopectoralis: 1) caudal,
2) middle, 3) cranial bundles; 4—5) m.
ectopectoralis: 4) caudal, 5) cranial
bundles. The abscissa gives the time
in seconds, the ordinate the length in
centimeters (hatched part of bar denotes
the phase of support).

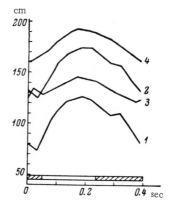

FIGURE 138.  Changes of different
parts of m. latissimus dorsi during a
cycle of fast walking in Loxodonta africana:

1, 2) costal bundles; 3, 4) vertebral
bundles. Abscissa and ordinate as
in Figure 137.

The above analysis of the work of the shoulder girdle muscles does not give a clear enough picture of the activity of the forelimbs without taking account of the activity of the most important muscles of the free limb. The most interesting of these are the extensors of the shoulder and elbow joints, which together with the muscles of the shoulder girdle work to impart a thrust to the body during the phase of support. The shoulder joint extensors include two single-jointed muscles: m. supraspinatus and m. infraspinatus, which begin respectively on the prespinous and infraspinous fossae of the scapula and end on the tuberculum majus humeri. The elbow joint extensors must be considered to include both single-jointed muscles: the lateral and medial capita of m. triceps brachii (m. anconeus lateralis and medialis) and m. epitrochleoanconeus, which begin on the humerus (Figure 136) and end on the olecranon, and a double-jointed one, the long caput of m. triceps brachii (m. anconeus longus), which originates on the wide area of the caudal surface of the scapula (Figure 135) and ends on the apex of the olecranon. All the single-jointed muscles are contracted and extended together with the extension and flexion of the shoulder and elbow joints (Figure 116). The work of m. anconeus longus, however, depends simultaneously on the ratios of flexor-extensor movements in both these joints and on the topography of the origin of its fibers. In order to study its activity, therefore, let us look at the relationship between the change in the length of its fibers during a cycle of a full pace and the position occupied by the origin of individual bundles. For this we present graphs showing the length changes of the bundles distributed most distally, most proximally and in the middle (Figure 139). As seen from the figure, in the phase of support they are at first lengthened together with the extension of the shoulder and elbow joints and then shortened, this contraction lasting slightly longer than flexion of the elbow and shoulder joints (Figures 116 and 139). They then begin to lengthen, to contract again in the phase of free transit together with flexion of the shoulder and elbow joints. But whereas flexion of these joints continues into half the phase of free transit (of the elbow joint slightly longer), the fibers at first become shorter, then longer and then shorter again up to the middle of this phase, finally lengthening again to the very end of it (Figure 139). Under these conditions the apex of the tooth of fiber elongation appears increasingly later the further distally the bundles are attached to the scapula. The overall amplitude of contraction of this muscle fluctuates in the limits of 10% of the maximum length for all parts of m. anconeus longus. It is interesting to note that the more or less uniform amplitude of contraction of all the fibers of this muscle is explained by the distinctive form of the scapula, in which the caudal surface is markedly widened and hangs down. The small percentage of contraction of the fibers promotes optimum conditions for the work of the muscle throughout the cycle of motion.

Before studying the activity of the forelimbs as a whole, we must look at the structure of the superficial fascia. The common fascia of the elephant's forelimb may be divided into lateral and medial layers. The superficial layer, which is common for the lateral and medial surfaces, lies directly under the subcutaneous fat. This very thin layer (at most 2 mm thick) consists of loose fibrous tissue with scarcely any admixture of elastic fibers and it covers the whole musculature of the trunk and limbs and also the deeper-lying layer. The lateral superficial fascia is divided from the medial only from the second.layer. It begins on a broad surface and is thickened in three places in the form of powerful teeth. The first of these teeth (Figure 130, *1*) begins on the caudal angle of the scapula, the second on the acromion (Figure 130, *2*) and the third on the tubercle of the scapula (Figure 130, *3*). Between the teeth of the acromion and the caudal angle of the scapula the fascia is inserted on the distal line of the deltoid muscle. Elastic fibers, derivative fasciae of the biceps and humeral muscles, reach from the crest of the tuberculum majus humeri to the inner side of the superficial fascia in the region of the shoulder.

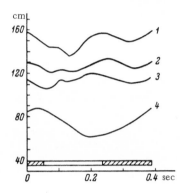

FIGURE 139. Changes in the length of muscles during a cycle of fast walking in L o x o d o n t a  a f r i c a n a:

1–3) m. anconeus longus: 1) proximal, 2) middle, 3) distal bundles; 4) m. biceps brachii. Abscissa and ordinate as in Figure 137.

The tooth from the caudal angle of the scapula stretches to the olecranon, and the teeth from the tubercle of the scapula and the acromion become intergrown in a layer 100–120 cm wide and 25 cm thick. Between this layer and the band from the caudal angle of the scapula, crisscrossing elastic fibers of the intermediate layer of the superficial fascia of the shoulder region extend obliquely downward and to the side (Figure 130, *5*). Around the proximal end of the lateral epicondyle of the humerus strong elastic bundles merge into a band from the acromion; from here on the forearm the band reaches a thickness of up to 65 mm, and, remaining at this thickness, it stretches to the distal end of the radius, where it is inserted as two stems, admitting between them the tendon of the lateral extensor of the digits (the acromioradial ligament) (Druzhinin, 1941). From the caudal angle of the scapula, apart from the tooth of the superficial fascia, a deeper band begins which penetrates between m. anconeus longus and m. anconeus lateralis and ends on the lateral crest of the olecranon. The superficial lateral fascia of the hand (Figure 130, *6*) begins on the olecranon independently of the fascia of the humerus.

The first layer of the superficial medial fascia begins on the inner surface of the acromioradial ligament. The second layer of the medial fascia originates as a strong tooth from the medial margin of the crest of the tuberculum majus humeri (Figure 140, *2*). To it from the medial layer of the fascia of m. biceps (Figure 140, *1*) and from the medial epicondyle of the humerus (Figure 140, *3*) reach powerful elastic bands which receive additional bundles from the olecranon (Figure 140, *4*).

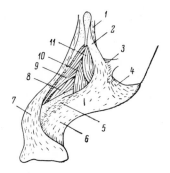

FIGURE 140. Fasciae of the forearm and hand of L o x o d o n t a  a f r i c a n a:

1) band from m. biceps brachii; 2) band from the crest of the tuberculum majus humeri; 3) band from the medial epicondyle of the humerus; 4) band from the olecranon; 5—7) medial fascia of the forearm: 5) common, 6) deep, 7) super-ficial; 8) m. ext. carpi radialis longus and brevis; 9) m. brachioradialis; 10) m. brachialis; 11) m. biceps brachii.

All these formations pass to the distal half of the medial lip of the radius and the fascia of the hand as the inner layer of the forearm fascia (Figure 140, *5*, *6*). They consist almost entirely of elastic fibers which are stretched by the weight of the body or by active tension of muscles. In the absence of tension, therefore, all these fasciae automatically contract. In the phase of free transit the band from the caudal angle of the scapula raises the olecranon and together with it flexes the shoulder joint. The acromioradial ligament sharply flexes the elbow joint, drawing the radius up, while the medial and lateral fasciae of the hand flex the wrist joint. This means that without any muscular efforts, the action of all the bands of the forelimb outer fascia carry the limb upward and facilitate its free progression.

Taking a round view of the activity performed by the elephant's forelimb, we note that fixation of the scapula in the region of the tubercle on its vertebral margin results in completely distinctive functional relations and features in the structure of its muscles and fascial cover.

To make their effect clearer, let us first study the less loaded phase of free transit. In the preparatory period of this phase the limb is brought upward, the shoulder joint is pushed forward and the limb joints are flexed. This period lasts from the time the foot leaves the ground to the maximum point of flexion of all the joints. At the moment the foreleg pushes away it is almost completely straight with maximally extended elbow and shoulder joints and with the axis of the scapula at an angle of 95° to the horizontal plane. At this stage the acromioradial ligament and the medial and lateral fasciae of the hand are maximally extended. So are m. acromiotrapezius, especially its cephalic part, m. omotransversarius, especially its distal part, m. sternoscapularis, m. ectopectoralis, m. brachioradialis, m. brachialis, m. biceps brachii and also the flexors of the digits. As soon as the limb leaves the ground the load on it is greatly lightened, and the elongation of all the layers of the fasciae and muscles helps accelerate the lifting of the limb. All the limb joints are now flexed

simultaneously.   The shoulder joint is flexed owing to the action of
m. acriomio- and spinodeltoideus, m. teres major and m. latissimus
dorsi.   Its flexion leads to contraction of m. anconeus longus, contraction
which cannot even be compensated by simultaneous flexion of the elbow
joint, which extends this muscle.

In the adjustment period the shoulder and elbow joints are extended.
Two main systems act to extend the shoulder joint.   The first consists of
the extensors proper:  m. supraspinatus and m. infraspinatus.   The second
is made up of m. cleidomastoideus which is intergrown with m. clavodeltoideus.   It pulls the distal part of the humerus forward, while m. latissimus dorsi and the group of pectoral muscles inhibit the proximal part of
the shoulder from moving forward.   Their simultaneous action leads to
the formation of a force couple which rotates the shoulder and extends the
shoulder joint.   In addition, m. omotransversarius inferior and the cephalic
part of m. acromiotrapezius continue to move the distal part of the scapula
forward.   The tension of m. anconeus longus arising as a result of the extension of the shoulder joint promotes extension of the elbow joint, while
m. brachioradialis, which ends on the radial ossicle of the carpus, extends
the wrist joint together with the extensors of the hand and digits.   This
activity of the muscles causes the leg to move forward, straighten, and come
to rest on the ground.

The phase of support lasts from the time the forefoot lands to the time
it pushes away again, and during this phase the limb takes part or all of the
body's weight.   At the beginning and end of the phase of support the load on
the limb constitutes a small percentage of the weight of the body, but in the
middle, when there is one-legged support, the whole weight of the body is on
this limb (for more details see p. 188).

We must also investigate the activity of the peculiarly structured hand
and foot of the elephant.   Under a load the hand and foot are vertically compressed and expanded along the diameter of the sole.   The callosity on the
sole now becomes markedly compressed.   In elephants this callosity is
represented by a cushion of fat embedded under the skeleton of the foot and
hand.   The stroma of this cushion is kept in place by strongly developed
fibrous strands on the subcutaneous fat of the sole and on the extreme rays
of the hand and foot (Figure 141).   From these main fibrous strands elastic
fibers, in whose complex network accumulations of fat cells are included,
penetrate into the sole in different directions.   These calluses make for a
smooth descent of the distal end of the tibia and forearm under the body's
weight.   At the same time as the load increases and the tibia and forearm
descend, the limb becomes more and more vertical, and therefore, to com-
pensate for the raising of its proximal end, the shoulder and elbow joints
are slightly flexed (Figure 116).

From the very beginning of front support, muscle activity imparts acceleration to the body.   This function is performed by the group of pectoral
muscles, m. latissimus dorsi and m. spinotrapezius.   As mentioned above,
only m. endopectoralis from the pectoral group promotes forward movement of the body throughout the phase of support (tilting moment).   The
most advantageously placed for this purpose is m. latissimus dorsi, especially its costal bundles.   It is therefore understandable why in proboscidians

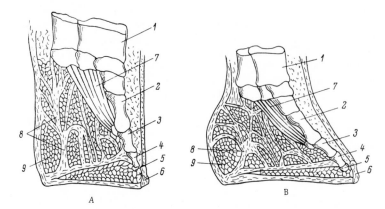

FIGURE 141. Diagram showing the activity of the hand callus in elephants in the phase
of transit (A) and in the phase of support (B):

1) bones of the forearm; 2) metacarpals; 3–5) phalanges of third digit: 3) first, 4) second,
5) third; 6) hoof of third digit; 7) m. interossei; 8) tendinous interlayers and callous bands;
9) fatty tissue of callus.

TABLE 15. Relative weight of individual muscles in the Proboscidea (% of the weight of all the
forelimb muscles)

| Muscle | Elephas indicus | Loxodonta africana | Muscle | Elephas indicus | Loxodonta africana |
|---|---|---|---|---|---|
| m. acromiotrapezius . . . | 1.9 | 1.7 | m. coracobrachialis . . . . | 0.5 | 0.5 |
| m. spinotrapezius . . . . . | 2.0 | 2.1 | m. anconeus longus . . . . | 13.1 | 12.0 |
| m. rhomboideus . . . . . | 1.0 | 1.0 | m. anconeus lateralis . . . | 2.0 | 4.3 |
| m. rhomboideus capitis . . | 3.2 | 3.0 | m. anconeus medialis . . . | 2.2 | 3.4 |
| m. latissimus dorsi . . . | 10.1 | 10.1 | m. dorsoepitrochlearis . . . . | 0.9 | 0.9 |
| m. sterno-mastoideus . . . | 2.0 | 2.0 | m. epitrochleoanconeus . . . | 0.6 | 0.8 |
| m. cleido-mastoideus . . . | 2.7 | 2.7 | m. biceps brachii . . . . . . | 2.0 | 1.8 |
| m. ectopectoralis . . . . | 2.9 | 2.9 | m. brachialis . . . . . . . . | 2.4 | 2.6 |
| m. endopectoralis . . . . . | 8.0 | 8.0 | m. brachioradialis . . . . . | 1.2 | 0.8 |
| m. sternoscapularis . . . | 1.7 | 1.8 | m. ext. carpi radialis . . . | 1.9 | 1.5 |
| m. serratus ventralis . . . | 8.0 | 8.4 | m. ext. digitorum communis | 1.7 | 1.3 |
| m. omotransversarius superior . . . . . . . . . . | 0.9 | 0.9 | m. ext. digitorum lateralis | 0.6 | 0.5 |
| m. omotransversarius inferior . . . . . . . . . | 2.2 | 2.2 | m. ext. pollicis longus . . | 0.2 | 0.2 |
| m. deltoideus . . . . . . | 4.3 | 3.8 | m. abductor pollicis longus | 1.0 | 0.8 |
| m. clavodeltoideus . . . . | 2.1 | 2.1 | m. ext. carpi ulnaris . . . . | 1.2 | 1.3 |
| m. supraspinatus . . . . | 3.4 | 3.5 | m. fl. carpi ulnaris . . . . | 1.0 | 1.2 |
| m. infraspinatus . . . . . | 4.5 | 4.0 | m. fl. carpi radialis . . . . | 0.5 | 0.7 |
| m. subscapularis . . . . | 2.7 | 2.5 | m. palmaris longus . . . . | 0.5 | 0.4 |
| m. teres major . . . . . | 1.3 | 1.1 | m. fl. digitorum sublimis . . | 0.3 | 0.4 |
| m. teres minor . . . . . | 0.8 | 0.6 | m. fl. digitorum profundus . | 1.5 | 1.4 |

this muscle accounts for 10.1% of the weight of the forelimb muscles as against 5.0—9.4% in the Ungulata (Tables 15 and 10).  In ungulates a greater load for drawing the body between the legs is placed on m. endopectoralis, which is often relatively better developed in these animals than in the Proboscidea.  This is connected above all else with the fact that the saltatorial gaits typical for ungulates lead to increased loads on the pectoral muscles and m. serratus ventralis, which keep the trunk between the legs.  In the giraffe and the proboscidians the scapulae are fixed on the body, and so these muscles are more weakly developed.  Thus, in elephants m. serratus ventralis accounts for 8.0—8.4% of the weight of the limb muscles, in the giraffe 7.5% but in the other ungulates 11.3—16.0% (Tables 10 and 15).

The shoulder girdle muscles can drag the body between the legs only when there is simultaneous movement of the whole limb forward with respect to its distal part.  Under these conditions there is a constant load on the extensors of the shoulder and elbow joints.  Because the scapula is fixed, m. anconeus longus can be kept tensed during the whole phase of support without reinforcing the actual function of extending the shoulder joint.  Reinforcement of this function is of secondary importance for a certain promotion of the scapula.  It is thus natural to expect that the shoulder joint extensors proper will be relatively less developed in the elephant and giraffe than in the other ungulates.  Indeed, the relative weight of m. supraspinatus is 3.4—3.5% of that of the forelimb muscles in the elephant as against 4.5% in the giraffe and 4.5—8.9% in other ungulates.  The weight of m. infraspinatus constitutes 4.0—4.5% in elephants, 4.6% in the giraffe and higher indexes in the other ungulates (Tables 10 and 15).  The important part played by elbow joint extension in the phase of front support in ungulates and proboscidians leads to greater development of the extensors of this joint.  However, the need for as rectilinear movement of the body as possible where the heavy elephant is concerned calls for special development of these muscles.  Thus, the sum of the relative weights of the elbow joint extensors is 20.8—21.4% in the elephants but only 12.7—17.9% in ungulates.  Only in the giraffe is the index even higher than in the elephant, 22.3%.  The only explanation for this is that in the giraffe the distal segments of the limbs are very elongated, and therefore there is a large lever arm for the application of force.

## PATHS AND CAUSES OF REORGANIZATION OF THE LOCOMOTORY ORGANS IN PROBOSCIDIANS

The ancestors of the Proboscidea lived in littoral thickets in swampy areas, as a result of which they retained a five-toed structure of the hand and foot which was inherited by the elephants.  The evolution of the group is characterized by adaptation to feeding on leaves and branches, in connection with which the animals developed long legs.  Another characteristic of the elephant is the secondary loss of the ability to move by means of asymmetrical gaits, which can probably be attributed to two factors: the increase in body mass and the elongation of the limbs.  The very heavy elephant is ill-suited to jumping, since in saltatorial gaits there is a stage of free flight.  Without a stage of flight sufficient speed can be mustered only in long-legged forms in which speed is achieved by the length of the stride.  For example, typical gaits of the rhinoceros are the gallop and fast trot, which have a stage of free flight, and yet these animals are hardly any lighter than the elephant.

Owing to their long legs, elephants came to adopt some typical symmetrical gaits which were little used by other mammals.  With slow movement the gait rhythm changed in the direction of racks; elephants employ the slow racklike walk in which a hind limb and then immediately afterward the forelimb on the same side push away, this increasing the length of the stride.  With accelerated motion the gait rhythm changes in the reverse direction, and the animal goes over to the fast walk, during which maximum speeds are attained in the absence of a stage of flight.

The increased height of elephants was promoted not only by the elongation of the limbs but also by their pillarlike arrangement and straightened joints.  Lengthening of the legs is observed in other mammals too, but usually the hand and foot are lengthened while the femur and shoulder remain of the same length.  In the elephant, on the other hand, the femur and shoulder become longer and the hand and foot shorter.  This pattern of elongation gave the animal maximum resistance to bending of the limbs, because the moments of the forces acting on the talocrural, knee, wrist and elbow joints are reduced in proportion to the shortening of the distal segments of the limbs.

The peculiar ratio of the segments, the long legs, their pillarlike arrangement, maintenance of the inner five-toed structure of the hand and foot, plus the enormous weight of the body govern the specific features of the locomotory mechanics of these animals.  The most characteristic of these are the following: 1) more or less steady rectilinear movement of the center of gravity in the horizontal plane, achieved by minimum vertical movements of the hip joint and of the apex of the scapula, which is immovably fixed on the thorax;  ·2) the almost straight position of the limbs in the shoulder, elbow and knee joints in the phase of support, making it possible for the limbs to refrain from bending and to preserve their pillarlike arrangement with the minimum waste of energy; 3) the distinctive strongly compressing activity of the hand and foot when they are expanded during support and straightened during free transit;  4) the smoothness of all movements;  5) the even distribution of the load on all four limbs during fast forms of motion;  6) the extreme rigidity of the spine.

Because of the spinal rigidity in the cycle of motion and the even distribution of the load on all four limbs, we may consider that elephants show adaptation to dorsostable dilocomotory running.  We may call this form of running graviportal, defined by all the above-described mechanical features of motion.

Although the form of running of elephants and ungulates is determined by a similar mechanism of active thrust by the hind limbs and forelimbs with the spine rigid, this form arose independently in these groups of mammals. Good proof of this are the distinctive pathways by which the spine achieved its rigidity in these two groups. The special hinges of the articular processes and the strongly developed transverse processes which are so typical for the Ungulata are absent in elephants in the part of the spine which is most subject to bending influences. Rigidity in this case is achieved by reduction of the number of lumbar vertebrae, by the closeness of the last ribs, by the greatly widened crests of the wing of the ilium, and also by the connection between the ribs and the pelvis provided by the little-elastic fibrous layer.

With specialization for fast walking, when only one leg supports the body in the middle of the phase of support, the load on the limbs is very great, considering the enormous weight of the animal. The main working groups of leg muscles are therefore powerfully developed. For the hind limbs this applies to the hip and knee joint extensors. Thus, according to their relative weight, the hip joint extensors are almost as strongly developed as in the ungulates most highly specialized for running. In the phase of support the angle of the knee joint changes negligibly, and therefore, apart from the strengthening due to the increased muscle mass, a profound structural reorganization of the knee joint extensors proper takes place. The lateral caput of m. quadriceps femoris becomes 5-plumose, and its strength is thereby enhanced.

The absence of a stage of flight in the fast walking cycle diminishes the load for keeping the body between the legs. Also promoting this is the firm fixation of the scapula on account of the strong fibrous bands attached between the tubercle of the vertebral margin of the scapula and ribs II, III and IV and the thoracic vertebrae. As a result, m. serratus ventralis and the group of pectoral muscles are markedly weakened in elephants as opposed to ungulates. Movement of the body between the legs is promoted by the muscles of the shoulder girdle situated caudal to the limbs. The most advantageously placed of these are m. latissimus dorsi and the pectoral group. Since the pectoral muscles do not keep the trunk between the legs, they are no better developed in elephants than in ungulates. As for m. latissimus dorsi, it is much more powerfully developed in elephants than in ungulates. Apart from the strong development of the shoulder girdle muscles, bringing about a tilting moment of the body with respect to the limbs, a number of muscles of the free limbs are well developed, particularly the elbow joint extensors. It is these muscles which bear the main load in keeping the legs straight in the phase of support.

The peculiar work of the hand and foot is due to the very broad callosity comprising a complex network of elastic and fibrous bands, which contain accumulations of fat cells. The result is that under pressure the hand and foot spread markedly, the interradial space is increased, and all this leads to a larger area of support. As the pressure diminishes the hand and foot are automatically compressed under the influence of the same callosity and their diameter diminishes. This decreases the chances of the animal's sinking into swampy ground as it can easily extract its wedgelike narrowed foot from viscous mud.

The smooth movement of elephants is apparently due to the distinctive structure of the superficial fascia, which is represented almost all along by a layer of elastic fibers.  This fascia is especially well differentiated on the lateral side of the forelimb and the medial side of the hind limb. In the forelimb a strong elastic band (the acromioradial ligament) stretches from the acromion to the distal part of the radius.  This band receives bundles from the caudal angle of the scapula and epicondyles of the humerus. The deep and superficial layers of the hand fascia branch off from this ligament.  In the hind limb the most powerful layer is attached to the symphysis and wing of the ilium and extends to the tibia and foot.  In the phase of support these layers of the fascia are markedly extended, while in the phase of free transit they automatically raise the legs, thereby allowing them to move smoothly.

A study of the mechanical features of movement and the structure of the skeleton and muscles in two species of proboscidians (the only ones that have survived from the one-time rich fauna) showed that adaptation to high-speed movement took a very distinctive path in these animals.  The rigidity of the spine and the unique shock-absorbing role of the limbs developed in a way quite different from that in the Ungulata.  It is enough to say that the seemingly primitive five-toed structure of the hand and foot in fact serves as a very effective shock-absorbing apparatus during running.  In addition, the spreading and narrowing of the hand facilitate navigation over swampy terrain.

*Chapter 6*

## ADAPTATION TO RUNNING IN THE CARNIVORA

### MECHANICS OF RUNNING

The methods of obtaining food may be considered to be the main factors governing the motion mechanics of predatory mammals. The long chase, hot pursuit of the victim at close quarters, prey-seeking in burrows, the catching of live fish, specialization for hunting small and large animals, group or solitary hunting, feeding on carrion, and the primarily phytophagous type of feeding — all these create special conditions dictating the motion mechanics of the Carnivora, and not only this but their morphology as well.

Carnivorousness was the characteristic feature of the creodonts, the group from which the recent carnivores descended. It is certain that in the ancestors of the recent carnivores a divergence must have taken place in the modes of hunting which brought about a splitting off of different groups. Although subsequent divergence within these groups led to the appearance of convergently similar ways of hunting, the original trend of specialization still exerted a specific influence on the habits and structure of the animals. Therefore, in analyzing the trends of specialization in various families, we will try to reconstruct the original mode of life and hunting typical for the group and show its influence on the diversity of habits and hunting techniques observed in the recent carnivores.

It is very likely that the ancestors of the recent Felidae adapted to stalking their prey which they then overcame as a result of one or several final large leaps. The victim was usually seized by the claws, which were put out only at the moment of capture. This assumption finds confirmation in the fact that this strategy is excellently developed in the recent cats, and all the known extinct representatives of the family had retractile claws (Gromova, 1962). If they do show any deviations in hunting habits, these clearly stem from the techniques described above. The usual high-speed gait of the Felidae is the half-bound, in which a stage of extended flight is promoted by the simultaneous pushing away of both hind limbs. The large springs which climax the act of stalking are more easily performed when both hind limbs leave the ground together, and therefore the half-bound characteristic for cats may serve as one more argument in favor of their primary mode of hunting.*

---

* The half-bound has become so firmly fixed in cats that automatic nerve paths controlling movement make it an unconditioned act. Thus, in experiments with a mesencephalic preparation (a cat with the prosencephalon and diencephalon removed) it was found that stimulation of a certain region of the mesencephalon elicits locomotion. When the intensity of the stimulus is raised, the form of locomotion goes over without fail to the half-bound, the typical high-speed gait of this family (Shik, Severin and Orlovskii, 1966).

Specializ. tion for this form of hunting gave rise to a monolithic group in which the main hunting technique was increasingly perfected.   However, this does not mean that the specialized recent members of the family cannot vary their methods of hunting.   Stalking is often replaced by patient lying in ambush.   And solitary hunting, which is typical for stalking animals, is often replaced by group hunting (the well known prides of lions).

Somewhat unusual techniques are used by the cheetah, in which the final jumps are transformed into hot pursuit at close quarters.   The animal seizes the prey in its front or hind paws or else flips it over with a blow of the paw. The final spurt instead of the leaps of other cats has come about due to the fact that cheetahs inhabit savanna or desert land, where prey easily detect a stalking predator.   During the spurt the usual half-bound is abandoned. But when cheetahs slow down, they take up this gait (Hildebrand, 1959), that is, they return to the high-speed gait typical for all the Felidae. This proves that it was the primary gait of their ancestors too.   On the whole it is clear that the cheetah's method of hunting is homologous to that of the other cat species and derives from the methods used by them (Randall, 1970).

In contrast to cats, dogs became adapted to seeking out hidden prey, puzzling out their tracks and nosing out their hiding places.   The probability that this type of hunting was the original one for the Canidae is demonstrated by observations of the habits of recent representatives of this group, which seek out prey mainly according to smell.   The typical gait of the dog is the light lateral gallop, which changes to the heavy lateral gallop when the animal slows down.   These forms of locomotion are far more economical than the cat's half-bound; all dogs use them, and even when hurdling obstacles they generally alternate the hind legs in pushing away.   We may conclude from this that prolonged pursuit was their primary mode of hunting. The pursuit of large, swift-footed prey called for considerable endurance and speed, a deficiency of which was often compensated by organized group hunting. Good examples of this may be seen by observing Cape hunting dogs (L y c a o n p i c t u s) or wolves in action.   Having scented a large antelope, the Cape hunting dogs give chase in a pack.   Several individuals chase it at top speed, while the others find the shortest path by orienting themselves according to the barking or turns made by the animal.   When the dogs in the lead get tired, the ones which were moving less swiftly before take their place.   After several such shifts even an ungulate with great stamina may become a victim. Wolves are experts at using the driving together and rounding-up technique. In the first case some of the predators conceal themselves, while the others, surrounding the prey, force it toward the hidden animals.   In the second case the chase is performed in a line of saclike shape, where the animals at the ends are in front of those in the middle, which are making straight for the prey.   Here any deviation from running along a straight line brings the victim closer to the animals at the ends.   The method of seizing the prey is another characteristic feature of dogs.   All the Canidae seize the prey in their teeth, not their claws.   The methods of hunting call for more endurance than speed.   The well developed sense of smell enables dogs to enjoy more often than cats not only the fruits of their own hunting but also other food (carrion).

These differences in the big-game hunting techniques between cats and dogs are somewhat smoothed out when it comes to catching small rodents. The Felidae hardly change their habits: just as when hunting large animals, they lie in wait for or steal up to a mouse, and with one or several jumps come upon it and crush it in their claws. On the other hand, the fox (V u l p e s v u l p e s) orients itself by hearing rather than by sight, and not by smell, felling the victim with its paw instead of grabbing it by the teeth, etc. However, when seeking a rodent's burrow, it uses its sense of smell in the way typical for the family. If the rodent is not hiding deep down in the earth, the fox digs it out, which a cat, with its retractile claws, cannot do.

When we compare the depth of specialization for running in different species of Felidae and Canidae we see that in fact no species of cat is highly specialized for fast running. Only the cheetah is equipped for short-distance spurts. Among the dogs, along with the little-specialized runners adapted to seeking small objects among thickets of littoral biotopes (C a n i s   a u - r e u s ,   N y c t e r e u t e s   p r o c y o n o i d e s), there are species which show a high degree of specialization and great stamina, capable of mustering speeds of 60 km/hr and more (L y c a o n   p i c t u s). Man has developed breeds of greyhound which can run nearly as fast as a cheetah. Interestingly, these breeds pursue their prey similarly to cheetahs, orienting themselves mainly by sight rather than by smell.

Hence, although the hunting habits of cats and dogs differ in their origin, and these differences become more pronounced as specialization is enhanced, it is not uncommon for either of them to employ strategies which are typical mainly for the other group. For instance, dogs often stalk prey or lie in wait for it. On the other hand, lions organize group hunting, and a lynx and jungle cat may hunt by following tracks, like hunting dogs. But these frequent deviations still do not destroy the nature of the group differences in their habits.

The most ancient of the known Mustelidae, found in the Lower Oligocene, are very similar in skeletal structure (Gromova, 1962) to the recent Siberian weasels (M u s t e l a   s i b i r i c u s). Like the recent representatives of the genus M u s t e l a, these small animals chased rodents right into their burrows. It seems probable that the diverse modes of hunting of the recent martens can be derived from their adaptation to pursuing rodents in their burrows. Proof of this is seen in both the morphology of the remains of ancient Mustelidae and the type of locomotion of the recent forms. We will discuss this in more detail later. The narrow galleries in rodents' burrows and the sharp turns along them must have led either to a marked decrease in the size of the predators or to a reduction in the length of their limbs and an increased mobility of the body. The latter trend is the more likely, since the predators had to kill their victims and therefore could hardly have become smaller; on the contrary, we should assume that it would be better for them to be bigger, and hence stronger, than their prey. The size increase of carnivores created additional difficulties during their movements along the galleries. Therefore, when they came to narrow parts, they widened them, this demanding the development of fossorial adaptation. While digging they often came upon root swellings and insect larvae, and these became more and more incidental "hors d'oeuvres" in their diet. This is why we should not be surprised that the descendants of the Carnivora which adapted to pursuing prey in burrows were able to go over to feeding on the subterranean parts of plants and insect larvae and became further specialized for digging — M e l e s   m e l e s,   M e l l i v o r a   i n d i c a (Gambaryan, 1960).   The catching of rodents, inhabitants of burrows in littoral areas, led to a continuation of the pursuit actually in the water (M u s t e l a   l u t r e o l a). The concomitant natatorial adaptations led to the possibility of catching live fish as well, and specialization for feeding on the fish became reflected in subsequent adaptation to swimming (L u t r a   l u t r a,   E n h y d r a   l u t r i s).*

The transition to forest habitats, where rodents do not have complicated burrows, led to the development of new hunting strategies: pursuit over land and in trees (representatives of the genus M a r t e s).

The battle inside the galleries of burrows required that the predator be stronger and more skillful than its victim, and so the former was generally bigger than the latter.   From another point of view its large size hampered its entry into the burrow, but this turned out to be an advantage in the case of capturing fairly large prey on the surface of the ground.   Adaptation to hunting large animals living above ground thus led to a further increase in the size of the body and loss of the ability to capture rodents in burrows (M e l l i v o r a   i n d i c a,   G u l o   g u l o).   Here the hunting strategies become similar to those employed by cats and dogs.   As we have already mentioned, hunting in burrows demanded an increased mobility of the backbone and a reduced length of the limbs.   Greater spinal mobility promotes an increase in the size of a full stride not only due to the flexor-extensor movements in the joints but also owing to the vertical flexures of the spine.   Preservation of this movement in the large terrestrial forms (G u l o   g u l o,   M e l l i v o r a   i n d i c a), natatorial species (L u t r a   l u t r a,   E n h y d r a   l u t r i s) and climbers (species of the genus M a r t e s) is further evidence of the primary nature of the fossorial mode of capturing food in the Mustelidae.

The dentition of bears indicates that these animals were long ago plant-feeders.**   Adaptation to feeding mainly on plants with the simultaneous increase in the size of the body on the one hand subjected the ancestors of bears to much less stiff competition from other predators and on the other served as good protection from enemies.   Because of the latter circumstance the importance of swift or sustained running diminished.   Still this group can display various degrees of cursorial specialization.   For example, the Malayan sun bear, which feeds practically exclusively on leaves, is certainly less adapted for swift running than the European bear.

As with the Ungulata, the gallop is the typical high-speed gait of the Carnivora.   The original reason that these animals adopted this gait is hard to determine.   We can assume, with the same degree of probability as for the ungulates, that the original carnivores inhabited forests.   The continual need to overcome obstacles in these surroundings must have led to their developing the ability to inhibit forward movement of the hind legs in the stage of extended flight, and it is this which made the gallop the main gait.

---

* The sea otter probably took to the sea when it switched from eating fish to feeding on other objects (sea urchins, mollusks, etc.).

** The exclusive carnivorousness of the white polar bears was certainly a secondary development, as the structure of their teeth is little different from that of the almost solely plant-feeding brown bears.

All carnivores are characterized during the gallop by vertical mobility of the vertebral column, which is extended before the stage of extended flight and flexed before the stage of crossed flight.  These movements of the spine promote an increase in the length of a full stride, thereby accelerating the speed of running.  In most Mustelidae, in fact, the running speed depends mainly on flexion and extension of the spine.

Since the hind limbs are most advantageously arranged for lending propulsion, it is naturally their thrust which is chiefly responsible for increasing the stride.  However, adaptation to climbing or to other considerable loads on the forelimbs, induced by the mode of obtaining food (bears), resulted in a strengthening of the forelimbs, which also during running could have a greater effect than the hind limbs on increasing the length of a pace.  Similarly, in the Mustelidae the length of a full stride depends more on the movements of the spine than on the thrusts of the limbs.  Nevertheless, specialization for running always leads to an enhanced role of hind limb thrust.  Therefore, a comparison of species of Mustelidae and Ursidae variously adapted to running shows that the force of hind limb thrust becomes increasingly important the more highly cursorial specialization is developed.

From what we have said above it is clear that we may call the modes of running of all carnivores *dorsomobile dilocomotory*, taking into account that hind limb thrust becomes increasingly vital for the cursorially specialized Mustelidae (wolverine) and for the Ursidae (European brown bear).

FIGURE 142.  Light lateral gallop of the dog (after Muybridge, 1887)

FIGURE 143.  Gallop of the cat (after Muybridge, 1887)

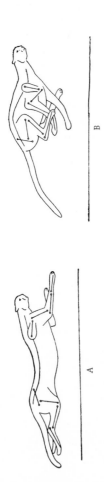

FIGURE 144.  Movement of the appendicular skeleton and spine in the cheetah (after Hildebrand, 1960):

A) extended flight;  B) crossed flight.

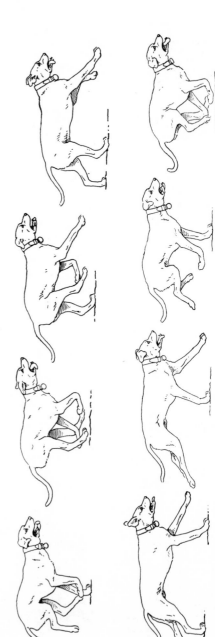

FIGURE 145.  Graviportal gallop of the dog (after Muybridge, 1887)

We pointed out on pp.203—205 that cursorial specialization proceeded along two paths in cats and dogs.  On the one hand short-distance spurts developed and on the other the long-distance, sustained form of running.

An analysis of the gaits used by the cheetah, which is the member of the cat family most highly specialized for short, energetic spurts, showed that at a speed of 80—99 km/hr (Hildebrand, 1961, fig. 1) the hind step constitutes 9%, the front step 16%, crossed flight 24% and extended flight 51% of the distance covered in an entire cycle of running.  An approximately similar distribution of stages is characteristic for the most energetic cycle — the light gallop of the dog (Figure 142), in which the hind step constitutes 10%, the front step 17%, crossed flight 18% and extended flight 55%.  The gallop of the domestic cat is very different from the above.  Here extended flight accounts for 80% and the front step 20% of the cycle.  Hind step and crossed flight are virtually absent (Figure 143).

Unfortunately, Hildebrand's analysis of movements does not contain data allowing us to assess the size of leaps.

FIGURE 146.  Position of the appendicular skeleton during the walk:

A) of a cheetah (from a photo taken from Das Tier, 1966, 4:15);  B) of a lion (from a photo taken from Bentley, 1961).

FIGURE 147.  Movement of the hind limb skeleton of the European brown bear (Ursus arctos) during the gallop

In contrast to the ungulates, a cat moving at a gallop shows a large amplitude of flexor-extensor movement in the limb joints and spine, as was noted by Hildebrand (1959, 1961, 1962) for the cheetah. Actually, a comparison of the movements in these joints in different representatives of the Felidae and Canidae did not reveal any substantial differences between them. We see that in the cat (Figure 143) and dog (Figure 142) the amplitude of movements may be larger than in the cheetah in certain joints (Figure 144) and smaller in others. It may be, however, that these features depend on the animal's degree of exertion during running, not on the specific movements. We can see this very clearly in running cycles of two breeds of dog that we reproduce here after Muybridge (Figures 142 and 145). Hence, until special studies are made of the differences in the amplitudes of movement in the joints and spine in the cheetah running at various speeds and in other species of cat and dog running under various degrees of exertion, we can only discuss the depth of specialization, not a concrete difference in the mechanics of their motion.

The apparently only real difference between the movement of the cheetah and that of the other cat species consists in the movement of the talocrural joint. In all cats, running fast or slowly, the foot comes down at a very acute angle (see the movement of the lion, tiger, leopard and domestic cat in Muybridge, 1887, pl. 124–138), whereas in the cheetah, even during the walk it is placed at a larger angle to the horizontal plane (Figure 146). Stepping in the cheetah, therefore, is analogous to that in most dogs, but not in cats.

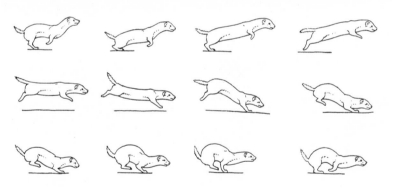

FIGURE 148. Gallop of the weasel (Mustela nivalis)

Quite a different ratio of the stages in the gallop cycle is observed in the bear (Figure 147), in which a stage of extended flight is practically absent even at the fastest speed of running, and the stage of crossed flight is strongly expressed, possibly due to the force of forelimb thrust. During the gallop of the Mustelidae, as we noted above, flexor-extensor movement of the spine assumes special importance (Figures 148 and 149).

Thus, the radical difference in cursorial motion mechanics between carnivores on the one hand and ungulates on the other lies in the work of the backbone. In ungulates and proboscidians the spine remains rigid

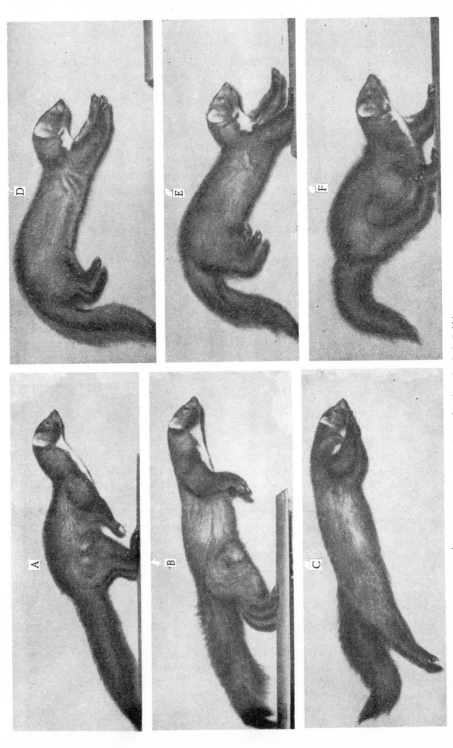

FIGURE 149.  Stages in the gallop of the stone marten (Martes foina).  Photo by R.G. Rukhkyan:

throughout the cycle of movement. In carnivores the vertical mobility of the spine promotes an increased stride and therefore greater speed.

## STRUCTURAL FEATURES OF THE SPINE

Active vertical mobility of the spine, which increases the full stride, is a property of all the Carnivora. The vertical and horizontal mobility of their spine lends it some specific structural features, which, as in the Ungulata, are manifested most fully in the lumbar region. But whereas in ungulates the vertebral column is maintained rigid owing to the spreading of the spinous processes with apexes widened anteroposteriorly and with a powerfully developed supraspinous ligament attached to them, in carnivores the apexes of the processes are narrowed and the supraspinous ligament is weakly developed or absent. Since the girder system (already discussed) of supraspinous ligament and strong spinous processes promotes rigidity of the spine, interspinous ligaments and interspinal muscles lose their importance for ungulates. In carnivores, on the other hand, a complicated system of interspinous ligaments develops. For instance, in cats, dogs and bears a fanwise spreading interspinous ligament runs from the caudal part of the apex of each spinous process in the lumbar vertebrae to the cranial margin of the adjacent vertebra. A similar interspinous ligament stretches from the cranial apex of the spinous process to the caudal margin of the next vertebra (Figure 150). These two ligaments together promote a smooth widening and narrowing of the distance between the spinous processes of adjacent vertebrae during flexion and extension of the spine. The particularly mobile spine in the Mustelidae leads to a reduction of all the interspinous ligaments, a considerable shortening of the spinous processes and to the development of the interspinal muscles, which are weakly expressed in cats, dogs and bears.

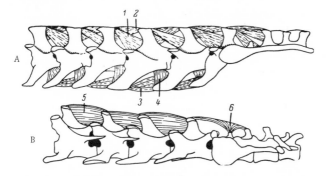

FIGURE 150. Structure of the spine in the Carnivora:

A) Felis pardus; B) Martes foina. 1, 2) interspinous ligaments; 3, 4) intertransversal ligaments; 5) interspinal muscle; 6) ligament from spinous process of last lumbar vertebra to spinous process of second sacral vertebra.

Reduced lateral mobility of the spine is achieved in ungulates by the spread of the transverse processes, which are directed perpendicular to the axis of the body of the vertebra in the horizontal plane. The apexes of these processes widen in the anteroposterior direction and are united with each other by the intertransversal ligament. In carnivores the transverse processes are directed obliquely forward and downward. Their apexes are negligibly widened or even become narrower. These processes are especially sloped in the Mustelidae, in which they are, furthermore, developed weakly and are united by a strongly developed intertransversal muscle. Because of the gradually increasing inclination of the transverse processes, the distance between their ends becomes greater, this lending extra freedom to lateral flexures of the spine (Figure 150).

The hinges on the articular processes in the Ungulata also inhibit spinal mobility. Such hinges are absent in the Carnivora.

We have discussed the structural features of the spine in general terms for the Carnivora, since the complex diversity of their hunting methods has had little effect on the development of the spine. The more complicated and differentiated role played by the limbs, on the other hand, compels us to study the corresponding changes in them in the framework of natural groups.

## STRUCTURAL FEATURES OF THE LIMBS

### Cursorial adaptation in the Felidae

As we showed above, cursorial adaptation among the Felidae is most clearly expressed in the cheetah, and therefore in this section we will examine the characteristics of this animal as a specialized form and will compare it with other cats less specialized for running. The cheetah is externally clearly distinguished from other species by its high-leggedness. Interestingly, this feature is not brought about by elongation of the limbs (Table 16) but by the peculiar type of their arrangement. We see from the table that according to the relative length of the hind limb segments the cheetah takes fifth place after F. lynx, F. caudata, F. manul and F. silvestris, and according to the length of the forelimb segments fourth place after F. manul, F. lynx and F. caudata.

In the preceding chapters we based our interpretation of the structural features of the skeleton and muscles on a study of the biomechanics of the main high-speed gaits characteristic for the animals. We assumed that less laborious forms of locomotion can be performed by the same organs with less exertion or with a different sequence of limb activity. Of course, this premise called for experimental proof, which presents a serious technical problem where large ungulates are concerned. We will therefore analyze here the biomechanics of two extreme variants of locomotion in the domestic cat: the pace or normal walk at a speed of 0.9 m per sec, which does not require special expenditures of energy, and a jump 170 cm long, which is pretty much the maximum leap a cat can make. We made a parallel study of the electrical activity of a number of hind limb muscles during the pace, trot and gallop in a mesencephalic preparation of a cat (Gambaryan, et al., 1970). Such a complex investigation of a model animal* enabled us to sound out the possibility of discovering the causes of the reorganization of the locomotory organs on the basis of biomechanical studies of high-speed gaits.

TABLE 16. Relative size of the limb segments in the Felidae (% of the total length of the lumbar and thoracic regions of the spine)

| Species | Relative length | | | | | | | |
|---------|-----------------|---|---|---|---|---|---|---|
| | femur (a) | tibia (b) | foot (c) | total (a+b+c) | shoulder (d) | forearm (e) | hand (f) | total (d+e+f) |
| Acinonyx jubatus | 43 | 43 | 42 | 128 | 38 | 45 | 26 | 109 |
| Felis pardus ..... | 44 | 42 | 41 | 127 | 39 | 40 | 28 | 107 |
| F. onza .......... | 42 | 36 | 36 | 114 | 38 | 39 | 24 | 101 |
| F. uncia ......... | 41 | 42 | 42 | 125 | 38 | 41 | 29 | 108 |
| F. concolor ...... | 42 | 40 | 42 | 124 | 37 | 37 | 30 | 104 |
| F. leo ........... | 40 | 36 | 40 | 116 | 38 | 41 | 27 | 106 |
| F. lynx .......... | 46 | 47 | 45 | 138 | 38 | 44 | 31 | 113 |
| F. serval ........ | 43 | 43 | 41 | 127 | 39 | 43 | 26 | 108 |
| F. manul ........ | 44 | 46 | 41 | 131 | 42 | 46 | 27 | 115 |
| F. chaus ........ | 41 | 42 | 41 | 124 | 36 | 41 | 24 | 101 |
| F. margarita ..... | 41 | 41 | 44 | 126 | 38 | 42 | 25 | 105 |
| F. euptilura ...... | 42 | 42 | 39 | 123 | 37 | 38 | 24 | 99 |
| F. silvestris ...... | 43 | 46 | 42 | 131 | 38 | 42 | 25 | 105 |
| F. caudata ....... | 43 | 45 | 46 | 134 | 40 | 44 | 25 | 109 |

As mentioned above (p. 77), during high-speed gaits the greatest load is carried during the phase of support, which is the task chiefly of the hip, knee and talocrural joint extensors. It is the hip joint extensors which take the main load, and therefore these are strengthened in an animal specialized for swift running. Some of the muscles of this group (the long postfemoral) may act as flexors of the knee joint during the phase of transit. Since this function demands an immeasurably smaller amount of energy than extension of the hip joint in the phase of support, we considered that the load on these muscles must correspond to the phase of support. With slow gaits, however, the strength developed by the muscles in the phase of support and phase of transit may be more similar, and in this case we can expect the activity of the long postfemoral muscles to shift from the phase of support to the phase of transit. We naturally wanted to check all these assumptions on a model.

* We deliberately chose the domestic cat as the model for the following reasons. First, of all domesticated animals cats have changed the least in comparison with their wild counterparts, and therefore changes due to domestication can hardly distort the results obtained. Secondly, cats are the favorite object of physiologists conducting experiments. A procedure for studying their locomotory features has been developed on a mesencephalic preparation (Shik, Severin and Orlovskii, 1966). Of course, locomotion induced by electrical stimulus of the brain stem in a cat whose cerebral hemispheres have been removed is not the natural movement of the intact animal. However, in a number of significant characters coordination of movements under conditions of such regulated locomotion does not differ from that under natural conditions (Shik, Orlovskii and Severin, 1966). But the technique of an experiment with a mesencephalic preparation is easier and makes it possible to obtain the necessary forms of locomotion with greater repetition. Since the electrical activity of the muscles studied in a preparation and investigated in intact animals proved almost the same (Engberg, 1964), we may hope that the work of other muscles, studied in a short-term experiment, is similar to the natural activity.

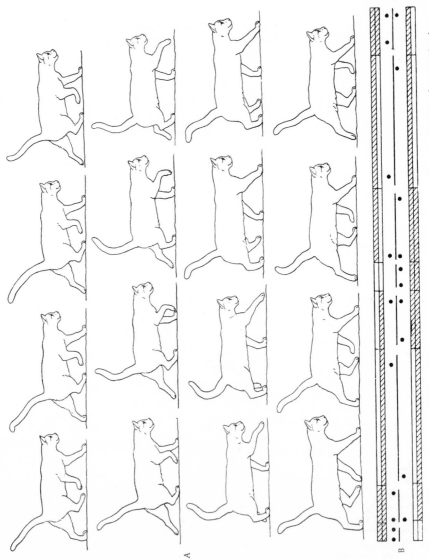

FIGURE 151. Successive frames of the normal walk of a cat (part of a cycle) showing a shift in the rhythm of locomotion in the direction of trots (A) and its support graph (B)

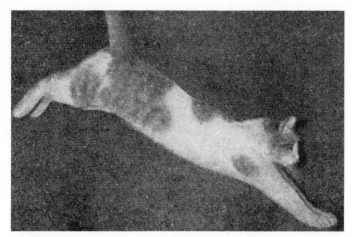

FIGURE 152. Cat landing on the ground (jump of 170 cm). Photo by G.P. Gambaryan.

It was pointed out on p. 91 that the ratio of the average speed of movement of the distal end of the limb in the phase of transit is determined by
the working rhythm of the limbs. In the case we took for analysis, a normal
walk with a rhythm of locomotion approaching that of a slow trot (Figure 151),
the rhythm of leg work was 1.2, that is, at a walking pace of 0.9 m/sec the
average speed of movement of the cat's hind feet and forefeet in the phase
of free transit was 2.1 m/sec (see the calculation method on p. 91). Taking
into account that at the beginning and end of the phase of transit the speed
of movement of the distal end of the limb is zero, we can assess the positive
and negative accelerations which must be developed by the leg muscles.
During the gallop (Figure 29) and jump, forward deflection of the hind limbs
begins only after the cat has landed on the forefeet (Figure 152) and occurs
mainly on account of the forces of inertia. The result is that the phase of
transit requires that less strength be developed during a leap or gallop than
during walking. We can estimate the comparative load on the limbs in the
phase of support from the following data. With slow pacing the load on one
limb reaches 40% of the weight of the body (Manter, 1938). The load on the
hind limbs during a jump may be calculated by studying the kinematics of
the center of gravity during the phase of support. For this, out of 100 frames
showing the phase of support of a jump, filmed at a speed of 263 frames/sec,
22 frames were enlarged. Every ninth frame was taken up to the 55th
frame and then every third frame. Natural-size cardboard models were
made from these frames and these were used to study the dynamics of movement of the center of gravity in the phase of support before a 170-cm jump
(Figure 153). The whole trajectory of the center of gravity was marked off
into six parts, in each of which we calculated the velocity and acceleration
of the center of gravity (Table 17).

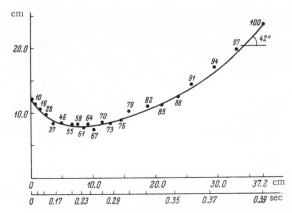

FIGURE 153. Kinematics of a cat's center of gravity in the phase of support before a 170-cm leap.

Figures on the curve denote the serial numbers of frames in which the center of gravity was determined. The angle of departure of the center of gravity (42°) is shown. The abscissa gives the time in seconds and the length in centimeters, the ordinate the height in centimeters.

TABLE 17. Analysis of the trajectory of a cat's center of gravity in the phase of support before a 170-cm leap

| Serial Nos. of frames (and their number) | Time, sec | Distance, cm | Speed, m/sec | Speed increment, m/sec | Average time in which acceleration developed, sec | Acceleration, m/sec² |
|---|---|---|---|---|---|---|
| 0−37 (37) | 0.144 | 4.9 | 0.34 | — | — | — |
| 37−58 (21) | 0.080 | 3.5 | 0.44 | 0.10 | 0.112 | 0.9 |
| 58−73 (15) | 0.057 | 5.8 | 1.02 | 0.58 | 0.068 | 8.5 |
| 73−88 (15) | 0.057 | 10.8 | 1.90 | 0.88 | 0.057 | 15.5 |
| 88−97 (9) | 0.034 | 13.1 | 3.85 | 1.95 | 0.045 | 43.3 |
| 97−100 (3) | 0.011 | 4.6 | 4.20 | 0.35 | 0.022 | 15.9 |

The table shows that the acceleration of the center of gravity increases almost up to the moment of pushing away, and toward the end of the phase of support reaches 43.3 m/sec². This means that the force developed by each limb in the phase of support is more than four times the weight of the body during the jump, and during the jump each leg develops 4−5 times more force than during walking.

The overall amplitude of flexor-extensor movements in the supporting phase of the walk is markedly smaller than during the jump (Figure 154), when the minimum size of the angle in the hip joint is 53° as against 58° in the supporting phase of the walk; the respective values for the knee joint are 43 and 85° and for the talocrural joint 40 and 83°.  But the maximum sizes of the angles for the hip joint are approximately the same for both types of motion, 135° for the jump and 130° for the walk;  the knee joint is extended to 158° during the jump and to 148° during the walk, and the talocrural joint to 153 and 138°, respectively.  If we take into consideration, moreover, that extension of these joints is completed in the jumping animal in 0.19 sec (hip joint) and 0.10 sec (knee and talocrural joints), and in the walking cat in 0.35 sec (hip joint) and 0.20 sec (knee and talocrural joints), we readily see that the rate of extension of the hind limb joints in the phase of support is much greater during the jump than during the walk.

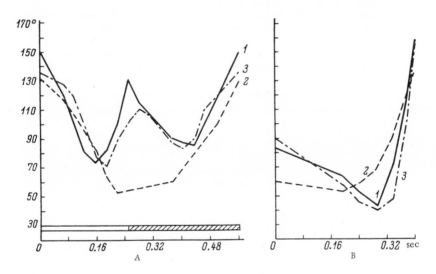

FIGURE 154.  Changes of the angles in the hind limb joints of the cat:

A) during a cycle of walking (cf. Figure 151) (hatched part of bar denotes the phase of support);
B) during the phase of support before a jump of 170 cm.  The abscissa gives the time in seconds,
the ordinate the angles of the joints in degrees.  1) knee, 2) hip, 3) talocrural joints.

In a phase of transit lasting 0.23 sec the knee joint is flexed from 148 to 73°, and the talocrural joint from 137 to 70°, after which they are extended to 100 and 130° (Figure 154, A).  This flexion and extension of the hind limb joints during the phase of transit promote the necessary accelerations and decelerations of the leg in the air.

In the adjustment period of locomotion each muscle develops activity in a certain phase of walking (Figure 155).  The single-jointed muscles do not alter the phase of their work in the cycle when the power developed by the animal increases, even when there is a shift to another gait.  But where double-jointed muscles are concerned the situation may be much more complicated.

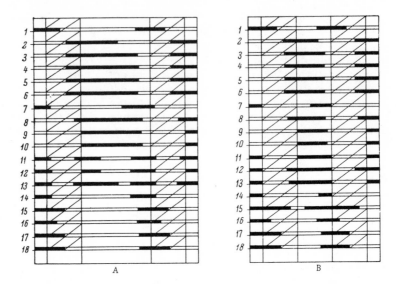

FIGURE 155. Diagram showing the electrical activity of the cat's hind limb muscles in a cycle of walking (A) and a cycle of galloping (B).

Hatched columns denote the phase of transit, nonhatched ones the phase of support; black parts of bars show muscles which are electrically active, light parts muscles which are electrically passive. 1) m. iliopsoas; 2) m. soleus; 3, 4) m. gastro-cnemius: 3) lateralis, 4) medialis; 5) m. plantaris; 6) m. vastus lateralis; 7) m. rectus femoris; 8) m. semimembranosus; 9) m. adductor; 10, 11) m. biceps femoris: 10) pars anterior, 11) pars posterior; 12) m. semitendinosus; 13) m. gracilis; 14) m. gluteus medius; 15) m. sartorius; 16) m. tensor fasciae latae; 17) m. ext. digitorum longus; 18) m. tibialis anterior.

With a constant intensity of brain stimulus the phase of transit is the standard element of the locomotory cycle. A change in the speed of the treadmill belt at a constant intensity of stimulus of the brain stem does not tell on either the kinematics of transit or muscle work in the phase of support (Figure 156, A and D). The figure shows that the amplitude of the EMG* during the phase of support does not depend on the speed of loco-motion either, but the duration of an EMG of extensor muscles changes in accordance with the duration of support. The latter, as we have already mentioned (p. 91), decreases substantially as the speed of locomotion increases.

Intensification of the stimulus enhances the power developed by the loco-motory apparatus. Even if the speed of motion is stabilized (actively in-hibiting the speed of movement of the belt), the amplitude of the EMG of all muscles increases, but not to an equal extent, especially for the muscles of

* Electromyogram, according to which muscular activity is assessed.

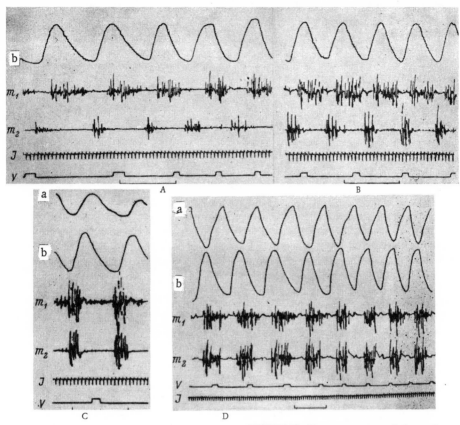

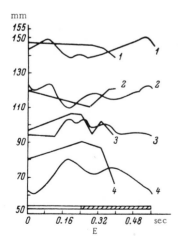

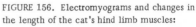

FIGURE 156. Electromyograms and changes in the length of the cat's hind limb muscles:

A) walk; B) trot; C) gallop. A, B) m. gastrocnemius lateralis ($m_1$) and m. tibialis anterior ($m_2$); C) m. ext. digitorum longus ($m_1$) and m. tibialis anterior ($m_2$); D) m. gastrocnemius lateralis ($m_1$) and m. soleus ($m_2$); a, b) longitudinal movements of the limbs: a) left, b) right limb (deviations upward correspond to the phase of transit, downward to the phase of support); $m_1$, $m_2$) electromyograms of leg muscles; $J$) marker of brain stimulus; $V$) velocity of the belt (interval between markers 0.5 m); the time marker is 0.5 sec; E) changes of muscle length (short lines: in the phase of support prior to jumping, long lines: during walking, full cycle). 1) m. ext. digitorum longus; 2) m. tibialis anterior; 3) m. gastrocnemius lateralis; 4) m. soleus. Hatched part of bars denotes the phase of support during the walk. The abscissa gives the time in seconds, the ordinate the length in millimeters.

different joints and the muscles belonging to different groups. The duration of the phase of transit under these conditions decreases by 10—20% and in accordance with this the time of activity of the muscles working in this phase of the walk is somewhat shortened (Figure 156, A, B).

*Hip joint extensors*

As noted above, the muscles extending the hip joint comprise three groups: gluteal and the long and short postfemoral. The gluteal and short postfemoral muscles are single-jointed, and their length (Figures 157, C, D and 158, E, F) changes with the change in the angle in the hip joint (Figure 154), increasing as this decreases and decreasing as it increases. During the phase of support of a jump, m. gluteus medius becomes 18% shorter, while during the same phase of a walk it becomes 14% shorter. The muscle contracts actively only toward the end of support (Figure 157, B, $m_1$), at the moment when the femur has already almost reached a posterior position. The participation of this muscle in creating propulsive thrust at the end of the phase of support is probably its important function during locomotion. In cats this muscle changes little in connection with cursorial specialization (Table 18).

The second group of single-jointed hip joint extensors includes two large muscles: m. semimembranosus (Figure 158) and m. adductor (Figure 159), and two smaller ones, the EMG of which was not studied. We can estimate the length of the first two muscles from the size of the proximal margin of m. semimembranosus and the distal margin of m. adductor which corresponds to it (Figure 158, E, F). As we see from the graphs, in the phase of support these muscles undergo a 25% shortening during the walk and a 20% shortening during a jump. M. semimembranosus and the femoral adductors are active during a walk or gallop from the end of transit almost to the end of support, i. e., the beginning of their activity (Figures 158, A, B, C, $m_1$ and 159, B, C, $m_2$) anticipates their shortening (work in a yielding regime). Specialization for running leads in cats in the first place to a strengthening of this group of muscles in particular. For example, in the cheetah the group of short postfemorals constitutes 16.2% of the total weight of the hind limb muscles, whereas in other Felidae it constitutes from 9.6 to 12.0%.

The double-jointed condition of the long postfemoral muscles and the numerous structural links that each of them has with other muscles make it especially difficult to analyze the work of this group. One of the most complex of them is m. biceps femoris pars posterior, which is divided into two portions of unequal strength: cranial and caudal. In some species of Felidae the mass of the cranial portion is 4—7 times that of the caudal. The pull of the cranial portion is transmitted through the posterior tendon of the muscle to extend the hip joint. The cranial portion is activated during both the walk and the gallop at the very beginning of the phase of support, and it remains active almost up to its end.

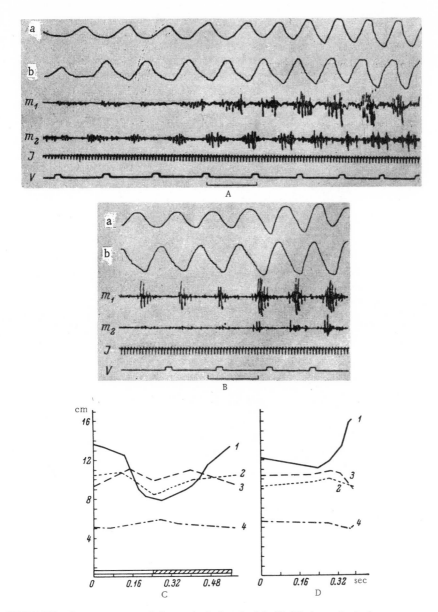

FIGURE 157. Electromyograms and changes in the length of the hind limb muscles in the cat:

A) m. vastus lateralis ($m_1$) and m. sartorius ($m_2$), intensity of brain stem stimulation gradually increasing (transition from a walk to a gallop); B) m. gluteus medius ($m_1$) and m. rectus femoris ($m_2$), transition from trot to gallop; C, D) change in muscle length during: C) phase of support before a 170-cm jump, D) the whole cycle during a walk. 1) m. sartorius; 2) m. rectus femoris; 3) m. vastus lateralis; 4) m. gluteus medius. Other notation as in Figure 156.

TABLE 18. Relative length of the hind limb muscles in the Felidae and Canidae (% of the total weight of the hind limb muscles)

| Muscles | Acionyx jubatus | Felis pardus | Felis panthera | Felis uncia | Felis onza | Felis concolor | Felis leo | Felis lynx | Felis chaus | Nyctereutes procyonoides | Vulpes vulpes | Canis lupus | Lycaon pictus |
|---|---|---|---|---|---|---|---|---|---|---|---|---|---|
| m. gluteus medius | 4.7 | 4.5 | 5.1 | 4.3 | 5.4 | 4.4 | 5.8 | 4.1 | 3.8 | 4.9 | 4.7 | 6.9 | 5.6 |
| m. gluteus minimus [1] | 1.4 | 1.2 | 1.3 | 1.6 | 2.1 | 1.4 | 2.8 | 1.1 | 1.8 | 1.5 | 1.1 | 1.2 | 1.5 |
| Gluteal | 6.1 | 5.7 | 6.4 | 5.9 | 7.5 | 5.8 | 8.6 | 5.2 | 5.6 | 6.4 | 5.8 | 8.1 | 7.1 |
| m. pectineus | 0.3 | 0.5 | 0.8 | 1.0 | — | 0.5 | 0.2 | 0.4 | 0.7 | 1.0 | 0.5 | 0.7 | 0.9 |
| m. quadratus femoris | 0.3 | 0.4 | 0.6 | 0.6 | 1.0 | 0.6 | 0.5 | 0.5 | 1.1 | 0.8 | 0.4 | 0.4 | 0.5 |
| m. adductores | 15.6 | 10.3 | 8.0 | 9.6 | 11.2 | 11.3 | 12.0 | 10.4 | 8.3 | 8.9 | 11.3 | 12.1 | 11.7 |
| Short postfemoral | 16.2 | 11.2 | 9.4 | 11.2 | 12.2 | 12.4 | 12.7 | 11.3 | 10.1 | 10.7 | 12.2 | 13.2 | 13.1 |
| m. biceps anticus | 2.5 | 2.5 | 2.7 | 10.1 | 2.0 | 1.9 | 1.6 | 0.7 | 1.0 | — | — | — | — |
| m. biceps posticus | 9.6 | 11.6 | 11.7 |  | 9.5 | 10.7 | 10.1 | 13.8 | 11.1 | 14.3 | 13.3 | 13.3 | 13.5 |
| m. semitendinosus | 4.2 | 4.2 | 5.1 | 4.8 | 4.5 | 4.4 | 4.2 | 3.6 | 3.8 | 4.0 | 4.5 | 5.4 | 4.6 |
| m. gracilis | 3.7 | 3.3 | 3.1 | 3.8 | 2.9 | 3.3 |  | 2.6 | 3.7 | 6.6 | 5.4 | 5.2 | 4.4 |
| m. semimembranosus | 11.7 | 11.2 | 9.0 | 10.2 | 10.8 | 12.9 | 9.9 | 11.6 | 9.0 | 7.8 | 10.8 | 8.4 | 7.9 |
| Long postfemoral | 31.7 | 32.8 | 31.6 | 28.9 | 29.7 | 33.2 | 29.7 | 32.3 | 28.6 | 32.7 | 34.0 | 32.3 | 30.4 |
| Hip joint extensors | 54.0 | 49.7 | 47.4 | 46.0 | 49.4 | 51.4 | 51.0 | 48.8 | 44.3 | 49.8 | 52.0 | 53.6 | 50.6 |
| m. vastus lateralis | 7.4 | 7.9 | 8.3 | 9.1 | 6.8 | 7.6 | 9.7 | 7.9 | 9.3 | 9.7 | 8.8 | 8.5 | 9.5 |
| m. vastus medialis [2] | 4.6 | 4.5 | 4.2 | 4.8 | 5.6 | 4.5 | 6.2 | 4.9 | 7.0 | 4.5 | 5.1 | 4.8 | 5.4 |
| Knee joint extensors | 12.0 | 12.4 | 12.5 | 13.9 | 12.4 | 12.1 | 15.9 | 12.8 | 16.3 | 14.2 | 13.9 | 13.3 | 14.9 |
| m. gastrocnemius [3] | 2.7 | 7.2 | 8.7 | 6.2 | 7.2 | 6.3 | 5.9 | 6.6 | 9.7 | 7.1 | 4.4 | 3.3 | 3.1 |
| m. psoas minor | 1.5 | 1.3 | 1.3 | 6.0 | 1.1 | 1.2 | 1.1 | 1.3 | 1.2 | 1.6 | 2.2 | 1.3 | 1.8 |
| m. iliopsoas | 6.1 | 3.9 | 3.6 |  | 3.9 | 5.1 | 4.2 | 3.6 | 3.5 | 3.7 | 3.1 | 4.0 | 4.6 |
| m. gluteus superficialis | 1.1 | 1.4 | 1.3 | 4.2 | 2.7 | 4.0 | 1.3 | 2.7 | 0.7 | 0.9 | 1.5 | 1.4 | 1.2 |
| m. tensor fasciae latae | 3.1 | 2.9 | 2.4 |  |  |  | 2.8 | 2.9 | 2.0 | 2.1 | 3.2 | 3.0 | 3.1 |
| m. obturator [4] | 2.0 | 1.5 | 1.7 | 2.0 | 2.4 | 1.8 | 2.0 | 1.7 | 2.4 | 1.9 | 2.0 | 2.0 | 2.2 |
| m. gemelli | 0.1 | 0.2 | 0.3 | 0.2 | 0.4 | 0.3 | 0.3 | 0.1 | 0.4 | 0.3 | 0.1 | 0.1 | 0.2 |
| m. popliteus | 0.5 | 0.4 | 0.5 | 0.7 | 0.5 | 0.6 | 0.7 | 0.6 | 0.8 | 0.7 | 0.5 | 0.7 | 0.6 |
| m. rectus femoris | 3.3 | 4.6 | 4.2 | 5.1 | 4.7 | 3.8 | 4.4 | 4.6 | 4.8 | 5.4 | 4.8 | 3.9 | 4.9 |
| m. sartorius | 5.6 | 4.1 | 5.9 | 4.6 | 5.0 | 4.7 | 5.4 | 3.9 | 3.4 | 3.2 | 2.3 | 4.2 | 4.2 |
| m. plantaris | 2.1 | 2.4 | 2.5 | 2.0 | 1.9 | 2.1 | 2.1 | 3.4 | 2.7 | 2.4 | 2.7 | 2.7 | 1.8 |
| m. tibialis anterior | 1.4 | 2.4 | 2.8 | 2.9 | 2.3 | 2.2 | 2.3 | 2.1 | 2.1 | 1.9 | 1.5 | 1.5 | 1.6 |
| m. ext. digitorum longus | 1.3 | 1.3 | 1.4 | 1.7 | 1.7 | 1.3 | 1.5 | 1.2 | 1.4 | 1.2 | 1.3 | 1.4 | 1.2 |
| m. peroneus [5] | 0.8 | 0.8 | 1.1 | 1.2 | 0.9 | 1.1 | 1.1 | 1.2 | 1.3 | 1.0 | 1.0 | 0.7 | 0.5 |
| m. flexor digitorum [6] | 2.5 | 2.5 | 1.7 | 3.3 | 2.2 | 3.1 | 2.8 | 2.1 | 2.9 | 2.6 | 2.4 | 2.4 | 1.8 |
| m. sacrospinalis [7] | 23.8 | 13.5 | 12.0 | — | 18.4 |  | 10.2 | 20.8 | 22.0 | 15.0 |  | 19.0 | 17.0 |

1) m. gluteus minimus + piriformis.   2) m. vastus medialis + vastus intermedius.   3) m. gastrocnemius lateralis + medialis + soleus.   4) m. obturator externus + internus.   5) m. peroneus brevis + longus + digiti quinti.   6) m. tibialis posterior + fl. digitorum longus + fl. hallucis longus.   7) For this muscle, an extensor of the spine, the percentage ratio to the total weight of the fore and hind limb muscles is given.

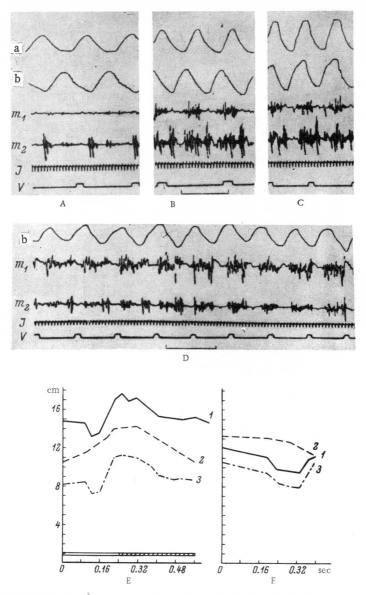

FIGURE 158. Electromyograms and changes in the length of the hind limb muscles in the cat:

A, B, C) m. semimembranosus ($m_1$) and m. semitendinosus ($m_2$): A) walk, B) trot, C) gallop; D) m. gracilis ($m_1$) and m. semitendinosus ($m_2$), transition from a walk to a trot; E, F) changes in muscle length during: E) phase of support before a 170-cm jump, F) whole cycle during a walk. 1) m. semitendinosus; 2) m. semimembranosus; 3) m. gracilis. Other notation as in Figure 156.

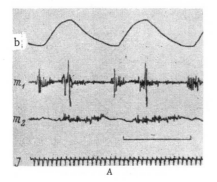

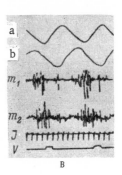

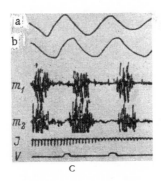

FIGURE 159. Electromyograms of
the hind limb muscles in the cat:

A) m. biceps femoris pars posterior ($m_1$)
and m. quadriceps femoris ($m_2$), walk;
B, C) m. biceps femoris pars anterior
($m_1$) and m. adductor ($m_2$): B) walk,
C) gallop. Other notation as in Figure 156.

Normal activity of the cranial portion of m. biceps femoris requires
that its anterior and posterior tendons be permanently strained.  Otherwise, instead of extending the hip joint the contraction of its fibers would
only cause these tendons to be deflected.  The strain of these tendons is
particularly important when the phase of support is coming to an end, as
then, due to the increased angle of plumosity, the likelihood of their being
deflected is considerably greater.

From the posterior tendon of the cranial portion of m. biceps femoris
originate the bundles of the caudal portion of the muscle, which strain this
tendon.  During the walk these bundles have their minimum length in the
middle of the phase of transit and their maximum length toward the beginning of support.  At the beginning of support they contract again, and toward
the end of the phase of support are once more somewhat elongated, for the
second time in the cycle.  In a cycle of walking this muscle twice has bursts
of activity, and both times they occur before the muscle has finished elon-
gating; in other words, each time the muscle starts to work in a yielding

regime.   At the end of support the extended caudal portion is actively
strained, which leads to flexion of the knee joint at the moment of pushing
away.

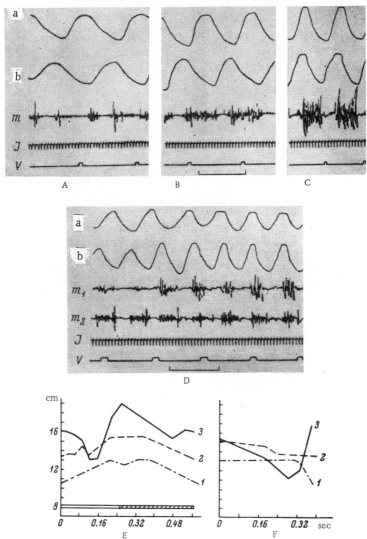

FIGURE 160. Electromyograms and changes in the length of the hind limb muscles in the cat:

A, B, C) m. biceps femoris pars posterior: A) walk, B) trot, C) gallop; D) m. biceps femoris
pars anterior ($m_1$) and m. biceps femoris pars posterior ($m_2$), switch from trot to gallop;
E, F) length change of m. biceps femoris: E) during walking (whole cycle), F) in the phase
of support before a jump. 1) anterior margin, 2) middle, 3) posterior margin. Other notation
as in Figure 156.

With the transition from trot to gallop the amplitude of the EMG of the cranial portion of m. biceps femoris increases several times (Figure 160, D, $m_1$). With it rises that of the caudal portion, and this diminishes deflection of the posterior tendon of the cranial portion. When the animal switches from a walk to a gallop, m. biceps femoris pars posterior changes also its phase of activity: during the gallop it works throughout the phase of support and is completely inactive in the phase of transit.

On the anterior margin of the cranial portion of this muscle ends m. tensor fasciae latae, which is active at the end of support and beginning of transit no matter what the intensity of locomotion (Figure 161, A, D). Its peak activity coincides with the lengthening of its fibers (Figure 161, D), in other words, this muscle works in a yielding regime. The amplitude of the EMG of m. tensor fasciae latae at the end of support increases especially sharply when the animal switches from a walk to a gallop (more than 20 times, whereas m. biceps femoris does not increase by more than 5 times). The angle of plumosity of the cranial portion of m. biceps femoris is larger in the cheetah than in other Felidae, and the strength of m. tensor fasciae latae is also considerably greater, its relative weight accounting for 3.1% of that of all the hind limb muscles while it constitutes only from 2.0 to 2.9% in other species (Table 18). The strengthening of this muscle prevents the anterior tendon of the cranial portion of m. biceps femoris from being deflected. There is in the cheetah in addition a separate well developed tendon from the caudal margin of m. biceps femoris to the calcaneal tubercle. As a result, in the phase of support an extending moment arises in the hip and talocrural joints simultaneously. These appreciable morphological changes of m. biceps femoris with cursorial specialization are not accompanied in the cats by a marked increase in the strength (relative mass) of this muscle (Table 18).

Two other muscles from the group of long postfemorals, m. semitendinosus and m. gracilis (Figure 162), work almost in cophase, and the nature of their activity depends on the intensity of locomotion (Figure 158, D). During the walk they work at the beginning of support and at the end of support—beginning of transit, and are not active in the middle of support. With the gallop the activity of these muscles is stepped up at the end of transit and it continues during the whole of the phase of support. We may therefore consider them to take active part during the walk in flexing the knee joint at the beginning of transit, while with the gallop they act as extensors of the hip joint. The length of these muscles increases toward the end of transit and smoothly decreases during support. The tendinous connections of these muscles with the extensors of the talocrural joint are strengthened in the cheetah, this being expressed in the formation of a powerful tendinous band from the distal margin of m. semitendinosus and gracilis which becomes fused with the Achilles' tendon. Thus, in the cheetah, specialization for running is manifested in the enhancement of the links with the talocrural joint extensors rather than with the reinforcement of these actual muscles (Table 18).

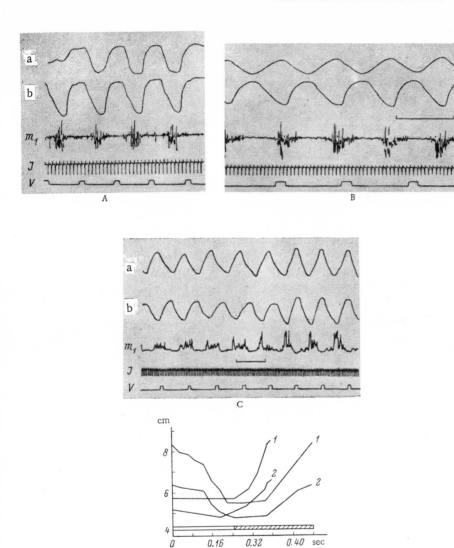

FIGURE 161. Electromyograms and changes in the length of the hind limb muscles in the cat:

A, B) m. tensor fasciae latae (intensification of EMG 16 times less on A than on B): A) walk, B) gallop;
C) m. iliacus, switch from trot to gallop (in this recording the EMG is integrated with the time constant
0.03 sec); D) length change of muscles (short lines: in the phase of support before a jump, long lines:
during a walk, full cycle). 1) m. tensor fasciae latae; 2) m. iliacus. Other notation as in Figure 156.

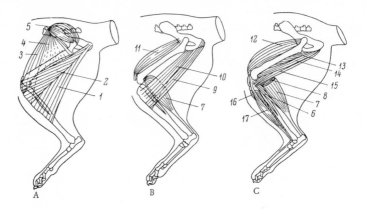

FIGURE 162. Hind limb muscles of the cat:

A) superficial, B) middle, C) deep layer. 1, 2) m. biceps femoris: 1) pars posterior,
2) pars anterior; 3) m. sartorius; 4) m. tensor fasciae latae; 5) m. gluteus medius;
6) m. soleus; 7, 8) m. gastrocnemius: 7) lateralis, 8) medialis; 9) m. semitendinosus;
10) m. gracilis; 11) m. rectus femoris; 12) m. vastus lateralis; 13) m. adductor;
14, 15) m. semimembranosus: 14) anterior, 15) posterior; 16) m. tibialis anterior;
17) m. ext. digitorum longus.

*Hip joint flexors*

The typical single-jointed flexor of the hip joint may be considered to
be only m. iliacus, which is fused into a single whole with m. psoas major,
the result being the formation of m. iliopsoas. However, its last part, apart
from flexing the hip joint, acts also to flex the spine. The great importance
of flexor-extensor movements of the spine for increasing the speed of motion of carnivores was noted by Hildebrand (1961, 1962), who mentioned that
these movements speeded up the cheetah's running by 10—15%. In connection with cursorial specialization the strength of this muscle was increased
in this animal (Table 18). In the phase of support m. iliacus becomes
34—36% longer during walking and jumping, and becomes actively effective
toward the very end of the phase of support in a yielding regime, which automatically leads to flexion of the hip joint the moment the foot pushes away.

The other flexors of the hip joint are double-jointed muscles. M. rectus
femoris works at the end of support—beginning of transit during walking and
only in the second half of support during a gallop (Figure 157, B, $m_2$). Usually
(Sinel'nikov, 1952; Klimov, 1955; and others) this muscle is included in the
complex of m. quadriceps femoris, whose capita are considered to be the
main extensors of the knee joint. But the activity of m. rectus femoris
coincides with that of m. iliopsoas rather than with that of the other capita
of m. quadriceps femoris, which are the knee joint extensors proper.

M. rectus femoris becomes about 27% longer during support and shorter in the adjustment period of the phase of transit (Figure 157, C, D). M. sartorius is active at the end of support—first half of transit during a walk and almost throughout the cycle (except for the beginning of support and the end of transit) during a gallop. During the walk this muscle is actively effective in the period of fiber elongation; during support the fibers become 75% longer (Figure 157, C, D). During a jump, elongation of this muscle proceeds at a still faster rate than during walking.

*Knee joint extensors*

Of the three femoral capita of m. quadriceps femoris which are the knee joint extensors proper, the EMG of only m. vastus lateralis was studied. This muscle is elongated by 15% in the first half of the phase of transit and is shortened by 14% in the second half (Figure 157, C, D). In the phase of support too it becomes longer in the first half and shorter in the second. Its active contraction begins in the last third of the phase of transit (Figure 157, A, $m_1$) and continues almost to the end of the phase of support. Hence, the single-jointed extensors of the knee joint contribute to straighten the limb before it lands and work in a yielding regime in the first part of support, actively contracting in the starting period. Comparing the length changes of m. rectus femoris and m. vastus lateralis, we see that they do not coincide either in intensity or in time. M. vastus lateralis changes in length in accordance with the change of the angle in the knee joint, while m. rectus femoralis changes its length in accordance with the change of the angle in the hip and knee joints.

Although specialization for running incurs the need for stronger extension of the knee joint, the cheetah's knee joint extensors proper do not show an increased relative strength (Table 18). The function of extension is reinforced by the short postfemoral muscles, which have an indirect influence on this movement (the mechanism of this influence was analyzed on p. 80).

*Talocrural joint extensors*

In the second half of the phase of transit the talocrural joint is extended due to active contraction of m. soleus. All its capita begin to be active in the second half of transit and continue to work to the end of support (Figure 156, A, $m_1$, B, $m_1$, D). The electrical activity of m. soleus ends a little before that of both capita of m. gastrocnemius (Figures 155 and 156, D). These muscles probably cease to contract almost at the same time, since the mechanical cycle of the slow m. soleus lasts longer than that of m. gastrocnemius (Denny-Brown, 1929).*

---

* The burst of electrical activity of fast muscles more or less coincides with their mechanical effect, but in slow muscles it ends before the mechanical effect disappears.

The length of m. soleus as a single-jointed muscle changes under the influence of the changing angle in the talocrural joint (Figures 154 and 156, E). The influence of the work of the knee joint on the length of the capita of m. gastrocnemius leads to smoother changes of their length in the walking cycle as compared with those of m. soleus. The amplitude of length change in the cycle is 25% for m. soleus and 12% for the gastrocnemial muscles. This influence of movements in the knee joint on the length of the gastrocnemial muscles is necessarily greatly enhanced in the cheetah, in which the origin of the gastrocnemial capita is shifted proximally along the lateral and medial lip of the femur. In connection with this, cursorial specialization in this animal goes along with a weakening rather than a strengthening of m. soleus (Table 18).

Flexion of the talocrural joint in the first part of the phase of support (like the simultaneously proceeding flexion of the knee joint) is due not to active work of flexor muscles but to the effect of the force arising in relation to forward movement of the body by inertia and the force being developed at this time by the extensors of the hip joint. The talocrural joint extensors are now working in a yielding regime. In the next part of the phase of support (probably close to the moment when the foot intersects the vertical dropped from the hip joint) the effect of this same force and the elastic strength of the elongated extensors bring about extension of the talocrural joint (the same applies to the knee joint).

M. soleus ceases to be active slightly before extension of the talocrural joint is ended still in the phase of support. The moment the foot leaves the ground the resistance to extension of the talocrural joint abruptly weakens, and if m. soleus were active to the end of the phase of support, the talocrural joint would begin to be overextended.

Apart from m. soleus, m. flexor digitorum longus and also the weak muscles m. fl. hallucis longus and m. tibialis posterior may take part in extending the talocrural joint. A comparison of the EMG of m. soleus with the EMG of m. fl. digitorum longus (Engberg and Lundberg, 1962; Engberg, 1964) suggests that they are active in the same part of the cycle.

*Talocrural joint flexors*

The flexors of the talocrural joint include the single-jointed m. tibialis anterior and the double-jointed m. ext. digitorum longus. M. tibialis anterior changes very little in length (about 12%) owing to the fact that its distal tendon is passed through a "block" (the tendinous ring of the transverse ligament of the tarsus) situated near the axis of movement of the talocrural joint. The fibers of this muscle enter into action (Figure 156, A, C, $m_2$) at the end of the phase of support, inhibiting extension of the joint, and they continue to be active after the foot has left the ground up to the end of flexion of the joint in the phase of transit. The muscle is completely passive during the rest of the cycle. The double-jointed m. ext. digitorum longus (Figure 156, C, $m_1$), which also participates in flexing the talocrural joint, works in the same regime as m. tibialis anterior. The effect is redoubled in the phase of transit by the fact that flexion of the knee joint causes the muscle to be drawn up proximally, and its flexing effect on the talocrural joint is promoted not only by actual contraction but also by the force of flexion of the knee joint.   The phase of the cycle in which the talocrural joint flexors work actively remains constant with the transition from walk to trot (Figure 156, B) and from walk to gallop (Figure 156, C). The relative weight of these muscles hardly changes in different representatives of the Felidae.

We see from the above data that a change of gait is reflected in the intensity of limb muscle activity and at the same time leads to a change of the moment at which a number of the hind limb muscles enter into activity. With high-speed gaits, which require a large amount of energy, all the muscles which in accordance with the topography of their attachment can play a dual role, working both in the phase of support and in the phase of transit, are switched over to the supporting phase (the most load-bearing in the cycle).   In addition, some of the muscles that are morphologically related to the main working components act as auxiliary mechanisms in the cycle of motion, improving the working conditions for the principal muscles.   Here specialization for running can lead to their being strengthened or, with a change of gaits to more intensive ones, to a rise in the amplitude of the EMG (as occurs, for instance, with m. tensor fasciae latae, which considerably enhances the EMG amplitude in the cat during the switch from the walk to the gallop and in the cheetah in comparison with other Felidae). Examination of the work of the cat's hind limb muscles convinced us of the possibility of determining the true function of the muscles according to the biomechanical usefulness of their work.

FIGURE 163. End of the phase of support before the cheetah springs

FIGURE 164. Movement of the hind limb skeleton of the cat in the phase before a leap of 170 cm

This study also showed that it is justifiable to divide the hind limb muscles into a number of functional groups, which we did while investigating the structural features of ricocheting animals (Gambaryan, 1955). For example, m. rectus femoris was included (Gambaryan, 1955, 1960, 1964) in the category of hip joint flexors, not knee joint extensors, and this was borne out by the experiments described above.

It is very hard to follow up the movement of the cheetah's skeleton in the phase of support according to the data of Hildebrand (1959, 1960, 1962), since he gives only three frames which are connected with the phase of hind support. It is therefore difficult to compare the details of movement in the domestic cat and cheetah. Still, extension of the hip, knee and talocrural joints in the galloping cheetah is similar to that in the leaping cat (Figures 163 and 164). We may infer that in the cheetah as opposed to other cats, cursorial specialization is expressed in the strengthening of a number of the most advantageously placed muscles, in a certain change of the ratios of the skeletal levers and in structural reorganizations leading to closer interaction of the joints.

We noted when analyzing the gallop of the domestic cat (p.209) that a stage of free crossed flight is practically absent, that is, the forelimbs work mainly to cushion impacts (at the end of the stage of free flight), while active flexion of the spine proves inadequate to allow for crossed flight. In other Felidae less specialized for fast running the gallop probably differs little from that of the domestic cat. In the cheetah, on the other hand, we observe a well-defined stage of crossed flight which is hardly smaller than the stage of extended flight. This stage arises not only due to active thrusting of the forelimbs but also in connection with flexion of the spine. Thus, the flexors of the cheetah's spine are better developed than those of other cats. Thus, m. psoas minor and m. iliopsoas in all constitute 7.6% of the weight of the hind limb muscles in the cheetah but only 4.7—6.0% in other cats (Table 18). The cheetah also has stronger extensors of the spine, which together with extension of the hip and knee joints and flexion of the talocrural joint promote extended flight. The spinal extensors include the complex muscle m. sacrospinalis, consisting of m. iliocostalis and m. m. longissimus dorsi, multifidus, semispinalis and interspinalis. In the cheetah m. sacrospinalis accounts for 23.8% of the weight of the forelimb and hind limb muscles, while in other species of Felidae it constitutes from 12.0 to 22.0% (Table 18).

The phase of front support in the Felidae (Figure 165) differs substantially from that in the Ungulata in the ratio of movement of the humerus with respect to the trunk. At the moment of landing the cat's humerus is positioned almost vertically (Figure 166). During the preparatory period it passes into a horizontal position and in the starting period the proximal end of the shoulder is again somewhat raised. At the beginning of the preparatory period the anterior end of the sternum is situated a little above and behind the elbow joint, toward the end of this period slightly below and in front of both the elbow and the shoulder joints, while toward the end of the starting period it is raised above the shoulder and passes beyond its anterior margin by almost half of the length of the humerus itself. Hence, the Felidae possess much greater mobility of the trunk between the forelimbs than the Ungulata. In the presence of this mobility the importance of m. serratus ventralis, whose main function is to keep the trunk between the legs, is diminished, whereas the importance of the muscles which originate caudally on the trunk and end proximally on the humerus is greatly increased. Of these muscles the most favorably placed are m. latissimus dorsi and m. endopectoralis. M. latissimus dorsi begins on the apexes of the spinous processes from the fifth to the last thoracic vertebra and on the lumbar fascia. The Felidae have in addition from two to four costal teeth the muscular bundles of which stretch to the humerus almost parallel to the axis of the body. M. latissimus dorsi ends on the crest of the tuberculum minus humeri together with m. teres major and on the crest of the tuberculum majus together with m. endopectoralis. M. endopectoralis originates on the caudal half of the sternum and its ensiform process and ends on the crest of the tuberculum majus humeri, on the tuberculi majus and minus (Figure 167) and on the coracoid process. Of these two muscles m. latissimus dorsi ends further distally on the humerus, and therefore when the leg moves out and forward, prior to the phase of front support, its bundles are especially strongly extended, so that its strength can be used most effectively in the phase of support. From the above it is clear why in the cheetah of the two muscles most favorably placed for drawing the trunk between the legs it is m. latissimus dorsi which is the better developed: its weight is 13.5% of that of the forelimb muscles, while in other Felidae it is from 7.2 to 11.6% (Table 19).

The weight of m. endopectoralis constitutes 10.4% in the cheetah and from 6.8 to 9.7% in other Felidae (in the snow leopard even 10.6%) (Table 19).

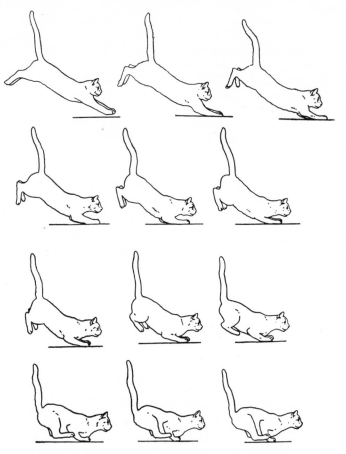

FIGURE 165. Successive frames showing a cat landing after a 170-cm leap

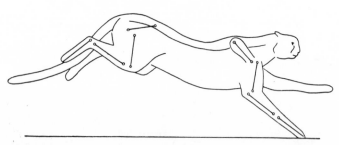

FIGURE 166. Position of the cheetah's skeleton at the moment of landing on the forefeet (from Hildebrand, 1962)

It is interesting to note that with an increased weight of m. endopectoralis in the snow leopard the relative weight of the second main muscle drawing the trunk between the legs (m. latissimus dorsi) is even lower than in other cats. If we take the total weights of these two muscles in the Felidae we obtain the following order: in the cheetah 23.9% of the weight of the fore-limb muscles, in the leopard 21.1%, in the puma 20.3%, in the jaguar 19.8%, in the snow leopard 19.4%, in the lion 19.1%, in the panther 18.2% and in the jungle cat 14.0%.

FIGURE 167. Ending of m. endopectoralis (1) and m. latissimus dorsi (2) on the humerus in the leopard (Felis pardus)

In the preparatory period of the phase of front support flexion of the shoulder and elbow joints slows down, contributing toward absorbing the shock on landing on the forefeet. In the starting period, extension of these joints along with active flexion of the spine promotes crossed flight. The yielding strain of the shoulder and elbow joint extensors in the preparatory period and their active contraction in the starting period renders the heavy load on them constant during the phase of front support. These muscles are naturally strongly developed in the cheetah in connection with cursorial specialization.

The shoulder joint extensors proper comprise m. supraspinatus and m. infraspinatus, while the elbow joint extensors include the capita of m. triceps brachii. The almost simultaneous extension of the shoulder and elbow joints has the result that the most advantageously placed of the elbow joint extensors is m. anconeus longus. This caput is therefore markedly strengthened in the cheetah, constituting 10.5% of the weight of the forelimb muscles, whereas in other Felidae it constitutes only 6.6—9.2%. However, m. anconeus longus may act as an extensor of the elbow joint only

TABLE 19. Relative weight of the forelimb muscles in the Felidae and Canidae (% of the total weight of the forelimb muscles)

| Muscle | Acinonyx jubatus | Felis pardus | Felis pantera | Felis uncia | Felis onza | Felis concolor | Felis leo | Felis lynx | Felis chaus | Nyctereutes procyonoides | Vulpes vulpes | Canis lupus | Lycaon pictus |
|---|---|---|---|---|---|---|---|---|---|---|---|---|---|
| m. brachiocephalicus[1] | 4.3 | 9.4 | 7.3 | 5.6 | 8.3 | 7.8 | 8.5 | 6.8 | 5.7 | 6.6 | 6.3 | 5.2 | 4.7 |
| m. acromiotrapezius | 1.2 | 1.5 | 1.6 | 1.2 | 1.7 | 2.0 | 1.4 | 1.3 | 0.7 | 1.9 | 1.1 | 1.3 | 1.1 |
| m. spinotrapezius | 1.3 | 1.4 | 1.7 | 1.7 | 1.7 | 1.3 | 1.3 | 1.4 | 1.3 | 2.1 | 1.2 | 1.5 | 1.6 |
| m. rhomboideus[2] | 3.0 | 2.9 | 2.0 | 2.9 | 2.7 | 3.0 | 2.7 | 2.6 | 2.9 | 3.4 | 2.9 | 2.8 | 2.7 |
| m. rhomboideus capitis | 0.4 | 0.7 | 0.6 | 0.6 | 0.9 | 0.5 | 0.8 | 0.4 | 0.4 | 0.5 | 0.3 | 0.3 | 0.4 |
| m. latissimus dorsi | 13.5 | 11.6 | 11.1 | 8.8 | 10.2 | 10.4 | 9.9 | 10.1 | 7.2 | 9.2 | 8.6 | 10.5 | 10.8 |
| m. sternomastoideus | 1.2 | 1.2 | 2.8 | 2.1 | 1.2 | 1.7 | 0.9 | 1.3 | 2.7 | 2.4 | 1.2 | 2.9 | 2.4 |
| m. ectopectoralis | 5.0 | 6.5 | 7.2 | 4.2 | 6.4 | 4.4 | 6.0 |  | 4.4 | 3.4 | 3.8 | 4.4 | 3.4 |
| m. endopectoralis | 10.4 | 9.5 | 7.1 | 10.6 | 9.6 | 9.7 | 8.2 | 13.5 | 6.8 | 8.2 | 8.6 | 10.4 | 10.9 |
| m. serratus ventralis | 8.5 | 7.8 | 5.7 | 9.0 | 7.9 | 7.4 | 7.5 | 7.2 | 8.7 | 11.4 | 9.0 | 9.4 | 8.4 |
| m. omotransversarius | 1.2 | 0.9 | 1.1 | 0.8 | 1.3 | 0.8 | 1.0 | 1.0 | 1.5 | 1.9 | 1.3 | 2.0 | 1.8 |
| m. deltoideus[3] | 2.5 | 1.8 | 2.6 | 1.6 | 1.8 | 2.1 | 2.0 | 2.3 | 1.7 | 1.8 | 2.0 | 2.5 | 2.2 |
| m. supraspinatus | 6.2 | 5.3 | 5.2 | 5.9 | 4.5 | 5.6 | 6.0 | 6.1 | 7.3 | 6.9 | 6.6 | 6.3 | 8.7 |
| m. infraspinatus | 5.1 | 3.4 | 5.9 | 4.7 | 3.3 | 4.3 | 4.3 | 4.8 | 3.0 | 3.9 | 4.8 | 4.4 | 5.1 |
| m. subscapularis | 4.9 | 4.0 | 4.1 | 4.3 | 4.4 | 5.0 | 4.5 | 5.1 | 6.4 | 3.9 | 3.8 | 4.1 | 4.4 |

| | | | | | | | | | | | | | |
|---|---|---|---|---|---|---|---|---|---|---|---|---|---|
| m. teres major | 3.5 | 2.5 | 2.7 | 2.4 | 2.7 | 3.5 | 3.2 | 4.0 | 2.3 | 1.7 | 2.5 | 1.8 | 2.6 |
| m. teres minor | 0.3 | 0.2 | 0.2 | 0.2 | 0.2 | 0.2 | 0.3 | 0.2 | 0.3 | 0.2 | 0.1 | 0.1 | 0.3 |
| m. coracobrachialis | 0.1 | 0.1 | 0.1 | 0.1 | 0.1 | 0.1 | 0.1 | 0.1 | 0.1 | 0.2 | 0.1 | 0.3 | 0.3 |
| m. anconeus longus | 10.5 | 7.0 | 7.0 | 7.4 | 6.6 | 9.0 | 9.1 | 9.2 | 8.8 | 9.1 | 10.9 | 10.3 | 9.9 |
| m. anconeus lateralis | 3.5 | 3.1 | 3.6 | 4.0 | 2.3 | 2.9 | 3.4 | 3.6 | 3.7 | 3.0 | } 7.4 | 3.4 | 2.7 |
| m. anconeus medialis[4] | 1.8 | 3.2 | 3.5 | 3.3 | 3.1 | 2.6 | 2.2 | 3.4 | 4.1 | 4.0 | | 3.5 | 4.8 |
| m. biceps brachii | 2.8 | 2.6 | 3.0 | 2.8 | 3.4 | 2.6 | 2.9 | 2.5 | 3.8 | 1.9 | 2.1 | 2.1 | 2.0 |
| m. brachialis | 0.8 | 1.2 | 1.3 | 1.1 | 1.7 | 1.4 | 1.4 | 1.4 | 1.3 | 1.4 | 1.0 | 1.5 | 1.3 |
| m. brachioradialis | — | 0.6 | 0.9 | 0.6 | 0.7 | 0.5 | 0.3 | 0.2 | — | 0.1 | — | — | — |
| m. pronator teres | 0.4 | 0.7 | 0.8 | 0.8 | 1.1 | 1.0 | 0.9 | 0.8 | 0.7 | 0.4 | 0.3 | 0.3 | 0.2 |
| m. supinator | 0.1 | 0.2 | 0.3 | 0.1 | 0.5 | 0.3 | 0.4 | 0.2 | 0.2 | 0.2 | 0.1 | 0.1 | 0.1 |
| m. ext. carpi radialis[5] | 0.9 | 1.4 | 1.7 | 1.3 | 2.0 | 1.6 | 1.4 | 1.4 | 2.4 | 1.6 | 1.5 | 1.4 | 1.4 |
| m. ext. digitorum[6] | 0.7 | 1.1 | 1.2 | 1.5 | 1.6 | 1.1 | 1.3 | 1.7 | 1.9 | 0.9 | 1.0 | 1.1 | 0.8 |
| m. abd. pollicis longus | 0.2 | 0.5 | 0.7 | 1.0 | 0.7 | 0.7 | 0.7 | 0.3 | 0.7 | 0.5 | 0.5 | 0.2 | 0.2 |
| m. ext. carpi ulnaris | 0.4 | 0.7 | 0.7 | 0.9 | 0.7 | 0.7 | 0.9 | 0.6 | 1.0 | 1.0 | — | 0.5 | 0.7 |
| m. fl. carpi ulnaris | 0.7 | 1.2 | 1.2 | 1.5 | 1.6 | 1.0 | 1.4 | 1.2 | 1.2 | 1.2 | 0.9 | 0.2 | 0.2 |
| m. fl. carpi radialis | 0.2 | 0.3 | 0.5 | 0.5 | 0.7 | 0.3 | 0.5 | 0.4 | 0.6 | 0.3 | 0.3 | 0.4 | 0.5 |
| m. fl. digitorum sublimis | 0.8 | 1.0 | 1.0 | 1.4 | 1.3 | — | 1.2 | 1.1 | 0.9 | 1.0 | 1.1 | 0.9 | 0.8 |
| m. fl. digitorum profundus | 2.6 | 3.2 | 3.2 | 3.9 | 3.0 | 3.5 | 3.5 | 3.4 | 4.9 | 2.8 | 2.8 | 2.6 | 2.3 |

[1]  m.  m.  clavotrapezius + clavomastoideus + clavodeltoideus.
[2]  m.  m.  rhomboideus thoracalis + rhomboideus cervicalis.
[3]  m.  m.  spinodeltoideus + acromiodeltoideus.
[4]  m.  m.  anconeus medialis + dorsoepitrochlearis + epitrochleoanconeus.
[5]  m.  m.  ext. carpi radialis longus + brevis.
[6]  m.  m.  ext. digitorum communis + ext. digitorum lateralis + ext. pollicis longus.

when there is simultaneous tension of the shoulder joint extensors.   Therefore, m. supraspinatus and m. infraspinatus are as a whole stronger in the
cheetah than in other cats, 11.3% of the weight of the forelimb muscles as
against 7.8—11.1% in other Felidae.

FIGURE 168.   A snow leopard pouncing on a wild goat

Apart from these features of the structure and development of the forelimb muscles in the cheetah which are connected with specialization for
swift running, we must take note of some other characteristics which are
related to the means of catching prey.   Having caught up with its victim,
a cheetah usually fells it and then tears it to pieces.   All the other Felidae
seize the prey in their claws, and if it is a large animal, pull themselves up
on it (Figure 168) and demolish it, but if it is small, they simply kill it with
the claws.   In seizing prey with the claws a heavy load is placed on m. fl.
digitorum profundus, the terminal tendons of which end on the distal phalanges.   The weight of m. fl. digitorum profundus amounts to only 2.6% of
the forelimb muscles in the cheetah as against 3.2—4.9% in other Felidae
(Table 19).   These animals pull themselves up on their prey mainly on
account of flexion of the elbow joint.   These flexors, m. biceps brachii,
m. brachialis, m. pronator teres, m. supinator and m. brachioradialis,
make up 4.1% of the weight of the forelimb muscles in the cheetah, and
from 5.1 to 7.4% in the other species.

**Cursorial adaptation in the Canidae**

The limb muscles were studied in only four species of the family
(Tables 18 and 19): the raccoon-dog (Nyctereutes procyonoides),
fox (Vulpes vulpes), wolf (Canis lupus) and Cape hunting dog

TABLE 20. Changes in the relative size of the limb segments (% of the total length of the lumbar and thoracic regions of the spine) in the Canidae in connection with cursorial specialization

| Species | Relative size of segments | | | | | | | | Degree of cursorial specialization |
|---|---|---|---|---|---|---|---|---|---|
| | femur (a) | tibia (b) | foot (c) | total (a+b+c) | shoulder (d) | forearm (e) | hand (f) | total (d+e+f) | |
| Lycaon pictus | 45.0 | 46.5 | 43.5 | 135 | — | — | — | — | Highly specialized |
| Canis lupus ... | 44.0 | 46.0 | 44.0 | 134 | 41 | 49 | 32 | 122 | Highly specialized |
| C. aureus ...... | 42.0 | 43.0 | 42.0 | 127 | 38 | 43 | 30 | 101 | Less specialized |
| Vulpes vulpes | 43.0 | 47.0 | 42.0 | 132 | 40 | 46 | 32 | 118 | Specialized |
| V. corsac ...... | 42.0 | 47.0 | 51.0 | 140 | 40 | 46 | 32 | 118 | Specialized |
| Nyctereutes procyonoides | 41.0 | 41.0 | 38.0 | 120 | 36 | 38 | 27 | 101 | Little-specialized |

(Lycaon pictus). The skeleton was studied in a large number of forms (Table 20).

Although we will not use the data yielded by an analysis of the muscles in domestic dogs,* the following is to be emphasized: their topography and weight ratios are very similar to those in wild dogs. The explanation for this is that we analyzed the motion mechanics of the skeleton in the domestic dog,** not the wolf, which was unfortunately filmed at a poor angle. The shots were taken from above at an angle of 20—25° and proved of little use in reconstructing the details of the skeleton's true position.

Of the species studied the one most highly specialized for running is the raccoon-dog, which feeds on small vertebrates, a number of invertebrates and also fruits and berries. For this animal successful hunting depends less on speed than on a talent for smelling out a young, wounded, resting or small animal concealed in a thicket. In trying to escape from enemies, the raccoon-dog has to confuse them, hiding in shrubs, reed thickets and other densely overgrown places. The speed of this animal is such that even man can overtake him (observations of I. M. Fokin, oral communication). Foxes feed mainly on small rodents; however, where enemies are concerned they need their capacity for swift flight, and this often serves them well also when capturing prey which is fleet of foot (hares, etc.). The staying power and swiftness of wolves and Cape hunting dogs may be considered to be one of the main trends of specialization (for more details see p. 204). Thus, of the Canidae investigated the raccoon-dog is the least specialized for fast running, while the level of cursorial specialization in the wolf, Cape hunting dog and to a slightly lesser extent the fox is more or less the same.

* This is because in working with these animals one may come across changes due to domestication which would be hard to explain correctly.
** The analysis of the dog's movements and of the Cape hunting dog's morphology was performed in collaboration with graduate student M.F. Zhukova.

FIGURE 169. Movement of the hind limb skeleton in the raccoon-dog (Nyctereutes procyonoides) in the phase of support before a leap

Comparing the relative size of the limb segments in the Canidae (Table 20), we see that in the jackal as compared with the wolf and the raccoon-dog as compared with other Canidae the legs are relatively shorter. The distal segments are especially reduced. The table shows that in the Canidae, adaptation to swift running leads to elongation of the limbs, particularly their distal segments.

Motion pictures of a domestic dog (short-haired pointer) were taken at a speed of 120 frames/sec. For an analysis of muscle activity and the change in the angles of the hind limb joints we drew in the skeleton on 10 frames of the phase of hind support (every other frame) when the entire phase lasted $1/6$ sec. On the skeleton we plotted the extreme points of insertion of all the double-jointed muscles and also m. semimembranosus anterior, which ends on the femur in the Canidae. The raccoon-dog was filmed at 230 frames/sec. At this speed the phase of support extended over 43 frames (Figure 169). The skeleton was sketched in on every fourth frame. The wave and intensity of extension of the hind limb joints differ markedly in the dog and raccoon-dog. In the dog in the phase of support (Figure 170) the hip joint is extended and the knee and talocrural joints at first flexed and then extended. In the raccoon-dog flexion with subsequent extension is observed in all these joints. Extension is especially intensive toward the end of support, and its intensity increases from hip to knee and from knee to talocrural joint.

On the whole, the sequence in which the joints begin to be extended and the relative intensity of extension are similar in galloping dogs and jumping cats (Figures 170 and 154,B). The amplitude of movement in the joints is also similar in cats and dogs, and at the same time very different from that in ungulates.

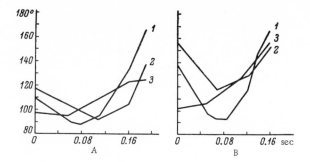

FIGURE 170. Changes of the angles in the hind limb joints during the phase of support of a gallop in the Canidae:

A) raccoon-dog; B) pointer. 1) talocrural, 2) knee, 3) hip joints. The abscissa gives the time in seconds, the ordinate the angles in degrees.

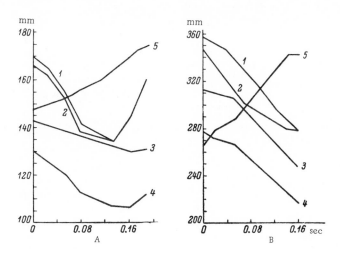

FIGURE 171. Changes in the length of the hind limb muscles in the phase of support during a gallop of the raccoon-dog (A) and pointer (B):

1, 2) posterior margin of m. biceps femoris posterior: 1) with tendinous band, 2) without tendinous band; 3) anterior margin of m. biceps femoris posterior; 4) posterior margin of m. gracilis; 5) m. sartorius. The abscissa gives the time in seconds, the ordinate the length in millimeters.

Despite the similarity of the flexor-extensor movements in the hind limb joints during the phase of support in the cat and dog, the work of their muscles differs distinctly, due to appreciable differences of structure. In the cat, m. biceps femoris pars posterior is elongated toward the end of the phase of support. It improves the working conditions of the anterior part of the muscle by acting on it as a brace. In dogs, m. biceps femoris pars posterior is not stretched at the end of the phase of support. Thus, if we examine the graph (Figure 171) without taking into account the influence of the brace, we see that the caudal bundles of m. biceps femoris pars posterior contract right up to the end of the phase of support. Hence, whereas in cats the stretching of the posterior bundles is related to their being inserted distally along the tibia, in dogs they are not stretched because they are inserted proximally on this bone (Figure 171).

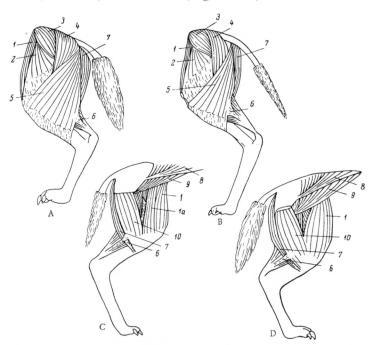

FIGURE 172. Hind limb muscles of dogs:

A, D) raccoon-dog (Nyctereutes procyonoides); B, C) Cape hunting dog (Lycaon pictus); A, B) from the lateral side; C, D) from the medial side. 1) m. sartorius; 2) m. tensor fasciae latae; 3, 4) m. gluteus: 3) medius, 4) superficialis; 5) m. biceps femoris; 6) tendinous band to m. gastrocnemius; 7) m. semitendinosus; 8) m. psoas minor; 9) m. iliopsoas; 10) m. gracilis.

The very proximal ending of m. gracilis results in dogs in that in the phase of support the anterior and posterior bundles work in opposite directions, whereas in cats their work is very similar (cf. Figures 171 and 158, E, F). The anterior bundles of m. gracilis in fact act on the posterior

bundles like a brace.   This function is especially clearly expressed in
m. sartorius.   The stretching of its bundles throughout the phase of support serves to pull up m. gracilis, on whose tendon these bundles end
(Figures 171 and 172).

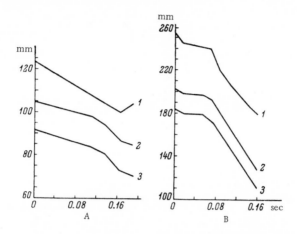

FIGURE 173.  Changes in the length of the hind limb muscles during
the phase of support in the raccoon-dog (A) and pointer (B):

1, 2) m. semimembranosus: 1) posterior, 2) anterior; 3) m. adductor.
Abscissa and ordinate as in Figure 171.

    The other long postfemoral muscles, like m. semimembranosus anterior
(Figure 173) and m. adductor in dogs, contract during the entire phase of
support.   Despite the difference we have noted in the work of m. biceps
femoris and m. gracilis in the cat and dog, the chief functional interactions
of all the hind limb muscles have much in common, due to the similar type
of change of the angles in the joints during the phase of support.  Because
of this similarity, the short postfemoral muscles turn out to be advantageously placed, an analogous relation between the work of the knee and talocrural joint muscles is observed, and so on.   In connection with this, the
changes of the musculature in the Canidae and Felidae also show many
common features during cursorial specialization.   As in the cheetah  in
comparison with other cats, so in the wolf, Cape hunting dog and fox in
comparison with the raccoon-dog is there a strengthening of the hip joint
extensors.   Of these the best developed are the short postfemoral muscles,
especially the most favorably placed group of adductors.   Thus the weight
of all the hip joint extensors constitutes 49.8% of the weight of the hind limb
muscles in the raccoon-dog, the adductors making up 8.9%, while in the fox
the respective values are 52.0% and 11.3%, in the wolf 52.5 and 12.1% and
in the Cape hunting dog 50.6 and 11.7%.   The development of the adductors,
apart from acting on the hip joint, indirectly serves to extend the knee joint,

and this extension is automatically transmitted to the talocrural joint.
Therefore, in dogs, as in cats, cursorial specialization leads not to a
strengthening but possibly even to a weakening of these two groups of
muscles.   Thus, in the raccoon-dog the knee joint extensors constitute
14.2% of the weight of the hind limb muscles, in the wolf 13.3% and in the
fox 13.9% (Table 18).   The gastrocnemial muscles represent 7.1% in the
raccoon-dog, 3.1% in the Cape hunting dog, 3.3% in the wolf and 4.4% in
the fox.   Specialization for running in both the Canidae and Felidae involves a strengthening of the extensors of the spine (m. sacrospinalis),
the total weight of which is 19.0% of the weight of the forelimb and hind
limb muscles in the wolf, 17.3% in the Cape hunting dog, but only 15.0%
in the raccoon-dog (Table 18).

In the phase of front support in dogs, as in cats, the greatest load is
taken by the extensors of the elbow and shoulder joints and the muscles
drawing the trunk between the legs.   However, while we can see a similarity in the development of the elbow joint extensors in the two families, in
relation to cursorial specialization, there is no such analogy in the development of the shoulder joint extensors and the muscles drawing the body between the legs.   As in the cheetah in comparison with other cats, so in the
wolf and fox in comparison with the raccoon-dog, m. anconeus is strengthened due to cursorial specialization (Table 19).   Whereas in the cheetah,
of the muscles drawing the trunk between the legs it is m. latissimus dorsi,
as the one most favorably placed, which is especially strengthened, in the
Canidae, the absence of costal bundles causes the most advantageously
placed muscle to be m. endopectoralis, which is also better developed in the
wolf and fox in comparison with the raccoon-dog (Table 19).   In contrast
to the cheetah, in the wolf and fox the shoulder joint extensors are not more
developed than in the raccoon-dog.


## Cursorial adaptation in the Ursidae

We mentioned above (p. 207) that bears are characterized by a strengthening of the forelimbs that is not primarily associated with running but did
serve as a basis for their subsequent use in running as well.   The use of
the forelimbs for climbing (the Malayan sun bear and to a lesser extent the
Asiatic black bear), for overturning stones (the European brown bear and to
a lesser extent the Asiatic black bear) and for digging up roots and tubers
(the black and brown bears) was the primary reason for their strengthening.
It was only later that their strength began to be used also in the stage of
front support during the gallop.   This special path of specialization led to
the development of a specific mode of running, during which the thrust of
the forelimbs proved no less important than that of the hind limbs.

Of the three species studied (the Malayan sun bear, Ursus malaja-
nus, the Asiatic black bear, Ursus tibetanus and the European brown
bear, Ursus arctos) the sun bear is the least specialized and the brown
bear the most highly specialized for running.   This is reflected in the
relative development of the limb muscles in the different species.   Thus,

TABLE 21. Relative weight of the hind limb muscles in the Mustelidae and Ursidae (% of the total weight of the fore and hind limb muscles)

| Muscles | Mustela putorius | Mustela eversmanni | Mustela nivalis | Vormela peregusna | Martes foina | Gulo gulo | Mellivora indica | Lutra lutra | Enhydra lutris | Ursus malajanus | Ursus tibetanus | Ursus arctos |
|---|---|---|---|---|---|---|---|---|---|---|---|---|
| m. gluteus medius | 2.1 | 1.9 | 1.6 | 0.9 | 1.9 | 3.1 | 3.3 | 2.3 | 4.0 | 2.5 | 2.5 | 2.4 |
| m. gluteus minimus [1] | 0.2 | 0.4 | 0.2 | 0.1 | 0.5 | 0.3 | 0.2 | 0.5 | 0.3 | 0.5 | 0.4 | 0.6 |
| Gluteal | 2.3 | 2.3 | 1.8 | 1.0 | 2.4 | 3.4 | 3.5 | 2.8 | 4.3 | 3.0 | 2.9 | 3.0 |
| m. quadratus femoris | 0.2 | 0.2 | 0.1 | 0.1 | 0.4 | 0.3 | 0.2 | 0.3 | 0.4 | 0.2 | 0.3 | 0.3 |
| m. pectineus | 0.3 | 0.1 | 0.3 | 0.2 | 0.4 | 0.5 | 0.7 | 0.4 | 0.5 | 0.6 | 0.5 | 0.5 |
| m. adductor | 2.9 | 2.9 | 2.2 | 3.0 | 4.4 | 4.8 | 3.4 | 2.9 | 2.8 | 2.9 | 4.1 | 4.5 |
| m. semimembranosus anticus | 3.0 | 2.3 | 3.2 | 3.4 | 4.0 | 2.0 | 2.2 | 3.2 | 3.1 | 1.5 | 3.7 | 3.2 |
| Short postfemoral | 6.4 | 5.5 | 5.8 | 6.7 | 9.2 | 7.6 | 6.5 | 6.8 | 6.8 | 5.2 | 8.6 | 8.5 |
| m. biceps femoris | 4.7 | 5.6 | 6.3 | 4.1 | 4.7 | 4.0 | 3.7 | 2.9 | 5.3 | 4.4 | 5.1 | 6.2 |
| m. semitendinosus | 3.0 | 3.0 | 1.1 | 1.4 | 2.8 | 2.2 | 1.1 | 1.5 | 2.0 | 1.4 | 1.0 | 0.7 |
| m. semimembranosus posticus | 1.2 | 1.9 | 1.9 | 1.5 | 2.0 | 1.1 | 1.1 | 2.0 | 1.1 | 1.5 | 3.4 | 1.5 |
| m. gracilis | 1.6 | 1.7 | 2.2 | 2.0 | 2.1 | 2.1 | 0.8 | 1.5 | 1.3 | 2.5 | 2.4 | 2.4 |
| Long postfemoral | 10.5 | 12.2 | 11.5 | 9.0 | 11.6 | 9.4 | 6.7 | 7.9 | 9.7 | 9.8 | 11.9 | 10.8 |
| Hip joint extensors | 19.2 | 20.0 | 19.1 | 16.7 | 23.2 | 20.4 | 16.7 | 17.5 | 20.8 | 18.0 | 23.4 | 22.3 |
| m. vastus lateralis | 2.7 | 1.9 | 2.2 | 2.4 | | 3.4 | 3.8 | 2.3 | 2.3 | | | 4.5 |
| m. vastus medialis [2] | 1.2 | 0.8 | 0.7 | 0.7 | | 1.3 | 1.2 | 1.0 | 0.9 | | | 1.5 |
| Knee joint extensors | 3.9 | 2.7 | 2.9 | 3.1 | 4.8 | 4.7 | 5.0 | 3.3 | 3.2 | 2.9 | 3.8 | 6.0 |
| m. gastrocnemius | 3.5 | 3.9 | 4.0 | 3.1 | 4.6 | 2.1 | 2.6 | 3.5 | 4.2 | 1.7 | 1.8 | 2.0 |
| m. psoas minor | 0.5 | 0.4 | 0.2 | 0.3 | 0.5 | 0.7 | 0.4 | 0.6 | 1.3 | 0.4 | 0.8 | 1.0 |
| m. iliopsoas | 1.6 | 2.3 | 1.8 | 1.1 | 1.2 | 1.1 | 1.2 | 1.0 | 3.0 | 2.0 | 1.5 | 1.7 |

TABLE 21 (continued)

| Muscles | Mustela putorius | Mustela evers-manni | Mustela nivalis | Vormela pere-gusna | Martes foina | Gulo gulo | Melli-vora indica | Lutra lutra | Enhydra lutris | Ursus mala-janus | Ursus tibeta-nus | Ursus arctos |
|---|---|---|---|---|---|---|---|---|---|---|---|---|
| m. gluteus superficialis | 0.9 | 1.1 | 1.6 | 0.5 | 1.2 | 0.8 | 1.2 | 1.0 | 4.0 | 1.4 | 2.2 | 1.5 |
| m. tensor fascia latae | 0.5 | 0.5 | — | 0.4 | 0.9 | 0.8 | 0.5 | 0.5 | — | 0.5 | — | 1.2 |
| m. sartorius | 2.1 | 2.3 | 1.9 | 1.8 | 2.2 | 2.7 | 2.1 | 1.4 | 2.9 | 2.8 | 2.9 | 2.3 |
| m. obturator | 0.7 | 0.6 | 0.7 | 0.4 | 1.1 | 1.5 | 0.9 | 0.7 | 1.1 | 1.0 | 1.3 | 1.0 |
| m. gemelli | 0.3 | 0.1 | 0.3 | 0.2 | 0.1 | 0.6 | 0.1 | 0.1 | 0.2 | 0.1 | 0.1 | 0.1 |
| m. popliteus | 0.3 | 0.3 | 0.4 | 0.2 | 0.4 | 0.3 | 0.2 | 0.3 | 0.4 | 0.3 | 0.3 | 0.2 |
| m. rectus femoris | 2.1 | 2.2 | 2.0 | 2.0 | 3.1 | 2.2 | 2.0 | 1.7 | 1.4 | 1.4 | 1.9 | 2.0 |
| m. plantaris | 1.1 | 1.2 | 1.0 | 1.3 | 1.5 | 1.5 | 0.7 | 1.2 | 1.3 | 0.8 | 0.8 | 1.4 |
| m. tibialis anterior | 1.4 | 1.3 | 1.4 | 1.0 | 2.1 | 1.2 | 1.0 | 1.4 | 1.3 | 0.5 | 0.5 | 0.4 |
| m. ext. digitorum [3] | 0.7 | 0.7 | 1.6 | 0.7 | 0.7 | 0.6 | 0.5 | 0.6 | 1.0 | 0.6 | 0.5 | 0.7 |
| m. fl. digitorum [4] | 1.9 | 1.6 | 1.5 | 1.5 | 1.8 | 1.5 | 1.2 | 1.4 | 1.3 | 1.1 | 1.1 | 1.4 |
| m. peroneus [5] | 0.7 | 0.7 | 0.5 | 0.7 | 0.9 | 0.6 | 0.6 | 0.7 | 1.1 | 0.6 | 0.5 | 0.4 |
| Hind limb | 41.5 | 42.1 | 39.4 | 35.4 | 50.8 | 43.0 | 37.6 | 36.7 | 48.6 | 35.4 | 43.2 | 45.7 |
| m. sacrospinalis | 25.0 | 23.6 | — | 19.5 | — | 13.8 | — | 21.0 | 36.2 | — | — | 10.1 |

[1] m. gluteus minimus + piriformis.
[2] m. vastus medialis + intermedius.
[3] m. ext. digitorum longus + ext. hallucis longus.
[4] m. fl. digitorum longus + fl. hallucis longus + tibialis posterior.
[5] m. peroneus longus + brevis + digiti quinti.

TABLE 22. Relative weight of the forelimb muscles in the Mustelidae and Ursidae (% of the total weight of the fore and hind limb muscles)

| Muscles | Mustela putorius | Mustela evers-manni | Mustela nivalis | Vormela pere-gusna | Martes foina | Gulo gulo | Melli-vora indica | Lutra lutra | Enhydra lutris | Ursus mala-janus | Ursus tibeta-nus | Ursus arctos |
|---|---|---|---|---|---|---|---|---|---|---|---|---|
| m. brachiocephalicus [1] | 9.9 | 11.1 | 7.5 | 7.1 | 3.9 | 4.8 | 5.4 | 8.3 | 4.0 | 5.2 | 4.9 | 2.9 |
| m. spinotrapezius | 1.1 | 1.7 | 1.0 | 2.3 | 0.8 | 1.2 | 1.9 | 1.2 | 1.3 | 1.0 | 1.1 | 1.0 |
| m. acromiotrapezius | 0.9 | 2.2 | 1.6 | 1.5 | 1.0 | 1.0 | 1.2 | 1.5 | 1.3 | 1.3 | 1.1 | 0.9 |
| m. rhomboideus [2] | 1.4 | 1.6 | 1.4 | 1.8 | 1.2 | 1.3 | — | 1.5 | 1.2 | 1.5 | 1.6 | 1.8 |

| | | | | | | | | | | | | |
|---|---|---|---|---|---|---|---|---|---|---|---|---|
| m. rhomboideus capitis | 0.6 | 0.8 | 0.8 | 1.0 | 0.4 | 0.6 | 1.1 | 1.1 | 0.5 | 0.5 | — | 0.3 |
| m. latissimus dorsi | 4.5 | 3.7 | 5.8 | 6.8 | 4.0 | 5.5 | 8.0 | 6.5 | 7.0 | 6.1 | 6.2 | 5.0 |
| m. sternomastoideus | 0.7 | 0.8 | 2.4 | 3.5 | 1.1 | 1.2 | 1.7 | 1.8 | 1.8 | 2.6 | 2.1 | 1.2 |
| m. endopectoralis | 5.6 | 4.4 | } 9.8 | 6.0 | 4.1 | 2.9 | 2.9 | 5.8 | 3.1 | 3.0 | 2.0 | 1.6 |
| m. ectopectoralis | 2.7 | 3.3 | | 3.7 | 2.4 | 2.2 | 2.8 | 3.7 | 3.4 | 4.4 | 4.7 | 5.4 |
| m. serratus ventralis | 5.1 | 5.4 | 4.4 | 6.4 | 4.4 | 4.0 | 4.0 | 5.7 | 5.5 | 4.6 | 4.2 | 5.0 |
| m. omotransversarius | 2.4 | 3.1 | 2.0 | 2.5 | 1.1 | 1.3 | 1.4 | 2.3 | 1.4 | 1.0 | 1.3 | 1.2 |
| m. deltoideus | 0.5 | 0.3 | 0.4 | 0.4 | 0.8 | 1.2 | 1.2 | 0.9 | 0.6 | 2.0 | 2.0 | 1.9 |
| m. supraspinatus | 2.5 | 1.9 | 2.1 | 2.0 | 1.9 | 2.4 | 2.5 | 2.4 | 2.9 | 2.3 | 1.7 | 1.8 |
| m. infraspinatus | 1.0 | 0.9 | 1.1 | 1.1 | 1.1 | 1.8 | 2.1 | 1.0 | 1.4 | 1.9 | 1.5 | 1.6 |
| m. subscapularis | 2.1 | 1.6 | 2.1 | 4.7 | 2.6 | 2.8 | 2.3 | 3.1 | 3.6 | 3.1 | 2.9 | 2.4 |
| m. teres major | 0.6 | 0.6 | 0.8 | 0.5 | 0.6 | 0.7 | 1.0 | 0.8 | 0.8 | 1.2 | 1.1 | 0.8 |
| m. teres minor | 0.1 | 0.1 | 0.1 | 0.1 | 0.1 | 0.1 | 0.1 | 0.1 | 0.1 | 0.2 | 0.1 | 0.2 |
| m. coracobrachialis | — | — | 0.1 | 2.6+1.7 | 0.1 | 0.2 | 0.1 | 0.6 | — | 0.5 | 0.4 | 0.4 |
| m. anconeus longus[3] | 3.0 | 3.0 | 2.9 | 1.8 | 3.0 | 3.7 | 3.4 \| 2.2 | 3.5 | 2.1 | 3.8 | 3.2 | 4.6 |
| m. anconeus lateralis | 2.5 | 1.5 | 2.0 | 1.0 | 1.8 | 2.8 | 3.2 | 1.6 | 1.1 | 1.6 | 1.5 | 2.3 |
| m. anconeus medialis[4] | 2.6 | 2.5 | 2.8 | 1.0 | 3.3 | 2.9 | 1.3 | 2.6 | 1.6 | 2.4 | 1.7 | 1.6 |
| m. biceps brachii | 0.9 | 0.8 | 0.9 | 0.7 | 1.1 | 1.4 | 0.9 | 0.8 | 0.8 | 1.9 | 1.6 | 1.4 |
| m. brachialis | 0.8 | 0.7 | 1.0 | 0.6 | 0.8 | 0.8 | 0.9 | 0.6 | 0.6 | 1.6 | 1.1 | 1.0 |
| m. brachioradialis | 0.3 | 0.4 | 1.1 | 0.6 | 0.3 | 0.5 | 0.8 | 1.1 | 0.5 | 0.7 | 1.2 | 0.2 |
| m. pronator teres | 0.4 | 0.2 | 0.7 | 0.4 | 0.5 | 0.7 | 0.7 | 0.5 | 0.3 | 0.8 | 0.7 | 0.6 |
| m. supinator | 0.2 | 0.2 | 0.2 | 0.2 | 0.1 | 0.2 | 0.2 | 0.1 | 0.1 | 0.5 | 0.3 | 0.2 |
| m. ext. carpi radialis[5] | 0.8 | 0.8 | 1.2 | 1.0 | 0.8 | 0.9 | 0.7 | 0.9 | 0.6 | 0.8 | 0.4 | 0.5 |
| m. ext. digitorum[6] | 0.6 | 0.6 | 0.5 | 0.7 | 0.6 | 0.8 | 1.0 | 0.4 | 0.7 | 1.1 | 0.7 | 0.8 |
| m. abd. pollicis longus | 0.3 | 0.3 | 0.2 | 0.3 | 0.3 | 0.3 | 0.4 | 0.3 | 0.2 | 0.6 | 0.4 | 0.4 |
| m. ext. carpi ulnaris | 0.4 | 0.4 | 0.4 | 0.4 | 0.5 | 0.7 | 0.5 | 0.4 | 0.2 | 0.6 | 0.4 | 0.5 |
| m. fl. carpi ulnaris | 1.0 | 1.1 | 0.7 | 0.8 | 0.5 | 1.7 | 0.9 | 0.8 | 0.5 | 0.9 | 0.9 | 0.9 |
| m. fl. carpi radialis | 0.2 | 0.2 | 0.1 | 0.2 | 0.3 | 0.3 | 0.1 | 0.2 | 0.2 | 0.4 | 0.3 | 0.4 |
| m. palmaris longus | 0.7 | 1.0 | 0.5 | 0.7 | 0.6 | 1.1 | 0.8 | 1.0 | 0.2 | — | 0.5 | 0.7 |
| m. fl. digitorum sublimis | — | 0.1 | 1.0 | 0.1 | 0.6 | 0.1 | 0.1 | 0.1 | 0.6 | 0.7 | 0.4 | 0.1 |
| m. fl. digitorum profundus | 2.0 | 2.4 | 1.3 | 2.3 | 1.7 | 2.6 | 3.6 | 1.6 | 1.0 | 3.5 | 2.6 | 2.5 |
| Forelimb | 58.5 | 60.7 | 60.6 | 64.9 | 49.5 | 57.0 | 62.1 | 63.6 | 51.4 | 62.6 | 56.7 | 54.1 |

[1] m. clavotrapezius + clavomastoideus + clavodeltoideus.
[2] m. rhomboideus cervicalis + rhomboideus thoracalis.
[3] m. anconeus longus + dorsoepitrochlearis pars scapularis.
[4] m. anconeus medialis + epitrochleoanconeus + dorsoepitrochlearis.
[5] m. ext. carpi radialis longus + brevis.
[6] m. ext. digitorum communis + lateralis.

the weight of the forelimb muscles in the brown bear constitutes only 54% of the weight of the fore and hind limb muscles, while in the black bear it accounts for 57% and in the sun bear 63% (Table 21). Hence, in the Ursidae adaptation to running leads to an ever increasing development of the importance of the hind limbs. The fact that the brown bear has retained more powerful forelimbs than hind limbs shows that it is in general little specialized cursorially in comparison with the other species.

The habits typical for the different forms can explain some structural features of their organs of movement. The sun bear is very slow-moving, but a good tree-climber; it often hangs down on its back by its strong hook-shaped claws and in this position picks off leaves and fruit. It suspends itself on half-bent limbs, clinging to the branch with its claws. In this position the weight of the head draws the neck dorsally, whereas normally the head draws it in the opposite direction. This habit of hanging down on its back has left a characteristic mark on the animal's whole appearance. The neck is very short, and appears to be hunched, with the result that the head is positioned almost between the legs. Measuring the distance between the center of gravity of the head and neck and the first thoracic vertebra, we see that it is smallest in the sun bear, thus preventing the head from being thrown back when the animal is suspended.

Overturning stones in the search for invertebrates and the food stores of rodents (Figure 174, A) and digging up roots and tubers impose a heavy load on the muscles flexing the hand. Digging and climbing also involve an increased load on the flexors of the digits, especially m. fl. digitorum profundus, which terminates on the distal phalanges. Turning over stones and also breaking off branches (Figure 174, B) goes along with strengthening of all the flexors of the hand, including those of the digits. The strength of the brown bear may be judged from a number of observations. For example, when gathering apricots, brown bears, which find climbing trees difficult, break off branches 20—25 cm thick; in so doing they generally hold onto the trunk with three paws and pull the branch toward it with the other paw (Figure 174, B). Lifting stones, breaking branches and climbing trees lead to a significant increase of the loads on the muscles drawing the body between the legs. All the characteristic habits of bears moreover result in a large load also on the elbow joint flexors. Hence, we may expect bears to show the following characteristics: strengthening of m. fl. digitorum profundus and the flexors of the hand, the first of these mainly in the sun bear and all of them in the brown and black bears; strengthened flexors of the elbow joint and, finally, a strengthening of the muscles drawing the shoulder in toward the trunk. In the sun bear we may additionally expect the muscles pulling the head in the ventral direction (i. e., upward, when the animal is hanging down) to be reinforced. In the Felidae, as mentioned above (p. 240), in connection with adaptation to seizing prey in the claws (Figure 168), m. fl. digitorum profundus is strengthened. It is thus hardly surprising that whereas in the cheetah and Canidae the weight of this muscle constitutes 1.2—1.3% of that of the fore and hind limb muscles, in the true cats it amounts to 1.6—2.6% (Table 19), in the brown bear 2.5%, the black bear 2.6% and in the sun bear even 3.5% (Table 22). The elbow joint flexors change accordingly: in the Canidae and cheetah they constitute 1.3—1.9%, in cats 2.0—2.9%, in the brown bear 2.6%, in the black bear 3.9% and in the sun bear 4.1% of the weight of the fore and hind limb muscles. Strengthening of all the flexors of the hand, with the exception of m. fl. digitorum profundus, is typical also for the Felidae and Ursidae. Thus, in the cheetah the weight of the hand flexors without m. fl. digitorum profundus is 0.7%, in the wolf 1.0%, in the other cats 1.3—2.0% (Table 19), in the sun bear 2.0% and in the brown and black bears 2.1% of the weight of the fore and hind limb muscles (Table 22). Strengthening of the muscles drawing the trunk between the legs is in dogs and cats connected with cursorial adaptation, and in cats also with pulling the body up toward the prey (Figure 168). Hanging from the claws, branch breaking, weight lifting when turning over snags and stones (Figure 174) require efforts in quite another direction, since in this case the legs have to be drawn in toward the body, not the body drawn between the legs. These two different types of interaction between forelimbs and trunk must be reflected primarily in the structure of the pectoral muscles. M. endopectoralis (Figure 175) begins in bears on the ensiform process of the sternum and on the last rib cartilages of the true ribs and ends on the tuberculi majus and minus humeri and on the coracoid process. It is placed very conveniently for drawing the trunk between the legs, but it can have hardly any effect on bringing the legs toward the body. M. ectopectoralis, which begins on the manubrium and the first few segments of the sternum and ends on the crest of the tuberculum majus humeri, can, on the contrary, scarcely participate in drawing the trunk between the legs but instead, owing to the transverse arrangement of its fibers with respect to the trunk, it functions as a muscle pulling the legs in toward the body. Naturally, m. ectopectoralis is specially strengthened in bears, while in cats and dogs, especially in those adapted for swift running, it is m. endopectoralis which is strengthened. In the Ursidae the weight of the superficial pectoral muscle constitutes from 4.4 to 5.4% of the weight of the fore and hind limb muscles and the deep pectoral muscle only 0.9—2.1%. In the Canidae the weight of m. ectopectoralis represents 1.7—2.3% and in the Felidae 1.5—3.6%, while that of m. endopectoralis represents 4.3—5.9% in the Canidae and 3.5—6.0% in the Felidae (Table 22).

The above data show that bears have a number of structural peculiarities in the organs of locomotion which did not derive from cursorial specialization. Strengthening of the forelimb muscles in connection with adaptation to a specific mode of life led to increased loads on them during running. Thus, while with the slow light lateral gallop in the dog (running speed of about 25 km/hr) the stage of hind support lasts $\frac{1}{5}$ sec, front support lasts $\frac{1}{8}$ sec and free flight (crossed and extended) lasts $\frac{1}{15}$ sec. When the light lateral gallop is accelerated to 35 km/hr, 7 frames correspond to hind support (filming speed 64 frames/sec), 6 frames to front support and 11 frames to free flight. A similar time distribution for support can be seen in the horse during the slow gallop, in which the hind limbs are also on the ground much longer than the forefeet. But with the slow gallop of the bear hind support corresponds to 36 frames (filming speed 120 frames/sec) and front support already 42 frames.

When a bear gallops, the thrust of the forelimbs, as in other carnivores, is accomplished on account of extension of the elbow and shoulder joints and the drawing of the body between the legs. Due to the strong protrusion of the humerus and its almost vertical position at the very beginning of the stage of support, m. ectopectoralis (Figure 176), which is strongly developed in the Ursidae in connection with their mode of life, can be involved in this act of drawing the trunk between the legs. Apart from this, a vertical position of the shoulder in the initial moment of support brings m. latissimus dorsi into a convenient position for drawing the trunk between the legs (see the analysis of the similar action of this muscle in the cheetah). This muscle is therefore markedly developed in bears, and we can understand why it is especially when we consider that its position is also quite favorable for drawing the humerus toward the body, since the muscle has a large force lever. The weight of m. latissimus dorsi constitutes 5.0—6.2% of the weight of the fore and hind limb muscles in bears as against 3.7—6.6% in cats and dogs. Finally, we should note that in bears, the elbow joint extensors also begin to be strengthened, whereby the relative weight of m. triceps brachii increases in proportion to the degree of cursorial specialization, in the brown bear representing 7.8%, in the black bear 6.9% and in the sun bear 6.1% of the weight of the fore and hind limb muscles (Table 22). The most favorably placed m. anconeus longus is very slightly weaker in the brown bear than in the fox, wolf and cheetah (Figure 177), the most highly specialized runners of the Felidae and Canidae investigated.

FIGURE 174. Some means of obtaining food employed by the European brown bear (Ursus arctos): A) overturning stones; B) breaking off branches.

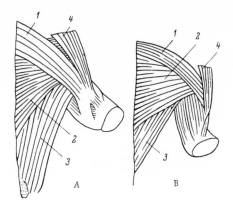

FIGURE 175. Pectoral muscles in the leopard (Felis pardus) (A) and the European brown bear (Ursus arctos) (B):

1, 2) m. ectopectoralis:   1) pars cranialis, 2) pars caudalis;  3) m. endopectoralis; 4) m. clavodeltoideus.

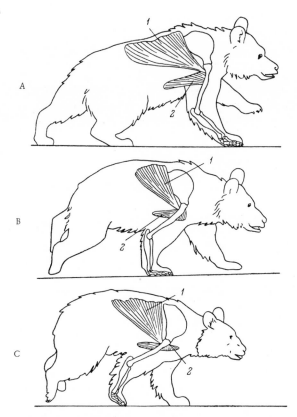

FIGURE 176. Diagram showing the activity of m. latissimus dorsi (1) and the pectoral muscles (2) in the bear during the phase of front support:

A) beginning, B) middle, C) end of the phase of support.

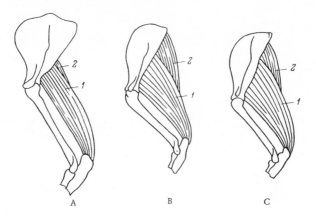

FIGURE 177. Diagram showing the insertion of m. anconeus longus on the scapula of the bear (Ursus arctos) (A), wolf (Canis lupus) (B) and cheetah (Acinonyx jubatus) (C):

1, 2) m. anconeus longus: 1) main part, 2) proximal bundles.

Finally, we must remember one more structural feature of the shoulder girdle muscles in the sun bear.  When the animal is hanging down on its back the head is held up partly on account of tension in m. sterno-mastoideus, the weight of which in the sun bear amounts to 2.6% of the weight of the fore and hind limb muscles, in the black bear 2.3%, and in cats, dogs and the brown bear from 0.6 to 1.6%.

### Cursorial adaptation in the Mustelidae

The family Mustelidae is of special interest for functional morphology, as this small group of mammals, which are closely related systematically, shows very different trends of specialization (hunting in burrows, digging, climbing, swimming, pursuit of large animals over land).  Future studies will discover the patterns of change of the locomotory organs in this family; our aim in this section is to determine the influence that the primary mode of life of this group (hunting in burrows) has had on the mechanics of their movement during running and the corresponding adaptive changes in the skeleton and muscles.

We have already pointed out (p. 206) that adaptation to hunting in burrows leads to increased mobility of the spine, the flexion and extension of which impart acceleration to the trunk also during locomotion above ground. Flexion and extension of the spine is best achieved with simultaneous push-ing away or suspension of the limbs, and therefore the Mustelidae developed the bound, in which the hind feet push away together and the forefeet land together.  The locomotory characteristics of the Mustelidae are  easily

traced from footprints.   In a galloping cat or dog the hind footprints are
placed well in front of the fore footprints (Figure 178), but in the marten
they are grouped such that the print from the forefoot lying in front usually
falls between the hind footprints (Figure 179).   The tracks of martens are
sometimes in pairs, but in this case they are grouped in fours with a good
distance between them.   Hence, unlike dogs and cats, in which the extended
and crossed stages of flight are more or less the same, in martens the extended stage always lasts much longer than the crossed stage. Only when
the gallop slows down considerably does the extended stage of flight become
shortened so that the difference between the two stages is canceled out. It
is interesting to observe that dogs and cats may also display prints of all
four feet close together.   This is observed in cats during jumps in pursuit
of prey and in dogs clearing obstacles.   Thus, in these animals the prints
are close together if their trajectory of flight is steep.   In martens, on the
other hand, the reason for the footprints' being close together is that the
body is elongated; their flight trajectory, however, is not steep. Thus, whereas a cat jumps 170 cm at an angle of 40—42°, the marten does the same at
an angle of at most 25° (Figures 153 and 180).

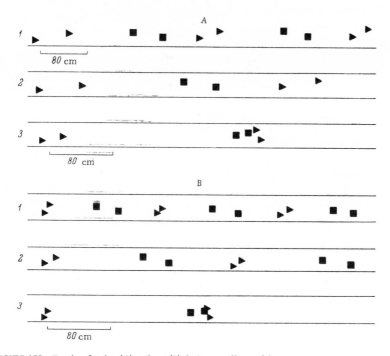

FIGURE 178.  Tracks of a dog (A) and cat (B) during a gallop and jumps:

1) slow, 2) fast gallop; 3) 170-cm jump over an obstacle. Triangles denote hind footfalls, squares
front footfalls.

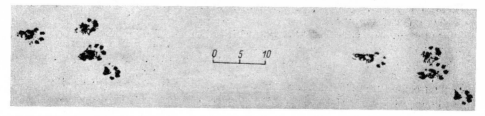

FIGURE 179. Footprints of the stone marten (Martes foina) running at a speed of 18 km/hr. Photo by R.G. Rukhkyan.

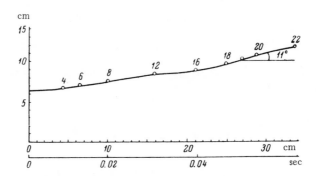

FIGURE 180. Kinematics of the center of gravity of the stone marten in the phase of support before a jump of 160 cm.

Figures on the curve represent the serial Nos. of frames in which the center of gravity was determined. The angle of departure of the center of gravity (11°) is shown. The abscissa gives the time in seconds and the length in centimeters, the ordinate the height in centimeters.

Forceful extension of the spine (Figures 148 and 149) during the phase of hind support imparts acceleration to the trunk, and this promotes extended flight. The fact that the hind feet land close to the forefeet allows us to assume that flexion of the spine in the phase of front support is induced by inertia of motion and not by the active work of muscles, as a result of which they cannot transmit propulsive force to the body. Proceeding from this assumption, it is natural to expect that in the Mustelidae the spinal extensors will be strengthened and the strength of its flexors will be little changed. The increased importance of vertical flexures of the vertebral column when the animal moves forward is observed in the sea otter (Enhydra lutris), which is specialized for swimming (Gambaryan and Karapetyan, 1961). In swimming, both upward movements (extension of the spine) and downward movements (flexion of the spine) are propulsive. And therefore in the sea otter both the extensors and the flexors of the spine are additionally strengthened. In the Mustelidae examined, the weight of the spinal extensors (m. sacrospinalis) is greater than in other carnivores.

Thus, in Mustela putorius, m. sacrospinalis constitutes 25.0% of the weight of the fore and hind limb muscles, in Martes zibellina 25.3% and in Enhydra lutris even 36.2%, whereas in the Canidae its relative weight varies between 15.0 and 19.0% and in the Felidae between 10.2 and 20.8% (Table 18). Only in the spurt champion, the cheetah, does the relative weight of m. sacrospinalis approach that in the Mustelidae, being 23.8% of the weight of the limb muscles. Along with this, the Muste-lidae differ little from other carnivores with respect to the degree of de-velopment of the spinal extensors (m. iliopsoas and m. psoas minor). The total weight of these muscles is 1.4—2.4% of the weight of the fore and hind limb muscles, while in cats and dogs it is from 2.2 to 2.6%, and in the cheetah even 4.3%. In the sea otter, however, their weight reaches 4.4%.

Apart from this feature in the structure of the spinal extensors which is primary for the Mustelidae, adaptation to hunting in burrows led to a marked reduction in the length of the limbs. For instance, the sum of lengths of the hind limb segments in the genera Vormela and Mustela constitutes 70.0—75.0% of the length of the lumbar and thoracic regions of the spine, while for the forelimbs the figures are 58.0—62.0%; the corre-sponding values in the genus Martes are 80.0—85.0% and 65.0—70.0%. Adaptation to pursuing large animals above ground brings about a length-ening of the limb segments. Thus, in Gulo gulo the length of the hind limb segments represents 120.0% and that of the forelimb segments 102.0% of the length of the lumbar and thoracic vertebrae.

## PATHS AND CAUSES OF REORGANIZATION OF THE LOCOMOTORY ORGANS IN CARNIVORES

In examining the main trends of adaptation to running in the Carnivora we have noted a number of specific mechanisms of land locomotion gov-erned by selectivity in feeding and the methods of obtaining food. The groups that became the most highly specialized for swift running are those which adapted to capturing large animals (Felidae, Canidae). But even these groups do not neglect any possible means of overcoming their victim without running fast (stalking, lying in wait).

In cheetahs, like other cats, hunting begins by stalking the prey, and the final jump is replaced by a sprinting burst over a short distance. Wolves and other dogs, whose prey include large, swift-running animals, generally conduct a long chase rather than perform spurts. Hence, cursorial speci-alization might be called "dashing" in the Felidae and "sustained running" in the Canidae.

Because they originally hunted in burrows, the Mustelidae came to pos-sess a very mobile spine. Thus, even when running over land, marked ex-tension of the spine contributes to the main thrust, leading to an extended stage of flight. In the phase of front support the spine is flexed almost fully owing to the forces of inertia.

Bears, which eat very different kinds of food, have adapted to climbing trees, to overturning stones and tree fragments, to breaking off thick branches and other techniques calling for great strength in the forelimbs. Strengthening of the forelimbs during running promotes a powerful thrust, increasing the crossed stage of flight.

In fact, these two groups of mammals are little adapted for running. Still, they do include some more or less cursorially specialized forms. We found that in the most specialized Mustelidae and Ursidae the load on the hind limbs progressively increases during the gallop.

Hence, cats and dogs develop speed when traveling over land by a strong thrust of the hind limbs and a somewhat lesser thrust of the forelimbs. In general speed is also promoted by the extensor movements of the spine, that is, these animals possess dorsomobile dilocomotory running with two forms of specialization: spurts and staying power. Forelimb thrust is of great importance for bears and spinal extension for the Mustelidae; however, in accordance with their degree of adaptation to running they gravitate increasingly to the form of running typical for the Felidae and Canidae.

The vertical mobility of the spine characteristic with this mode and these types of running leads to an appropriate reorganization of the spine. The spinous processes are narrowed toward the apex, and instead of a supraspinous ligament a system of interspinous tendinous formations appears which makes for freer movement between adjacent vertebrae. The articular processes do not form any hinges of the type characteristic for the Ungulata. The transverse processes are directed obliquely forward and down, and their apexes are not united by an intertransversal ligament. The special importance of spinal mobility for the Mustelidae results in a strong reduction of the spinous and transverse processes, the replacement of interspinous ligaments with interspinal muscles, etc.

The fact that flexion and extension of the spine play a substantial role in the locomotory cycle of cats and dogs leads to a strengthening of the main flexors (M. iliopsoas, m. psoas minor, m. quadratus lumborum) and extensors (m. sacrospinalis). For the Mustelidae the extensors of the spine assume special importance in running, and therefore these are more developed in them than in the cursorially specialized Felidae (cheetah).

Specialization for running in cats and dogs brings about not only a lengthening of the limb segments but also a reinforcement of the most important groups of muscles making for speed and firmness of front and hind support. The range of elongation and contraction of the long postfemoral muscles in the running cycle is very large in these animals due to the considerable amplitude of flexor-extensor movements in the knee and hip joints. They therefore have no use for a plumose structure of m. biceps femoris, as is characteristic for the ungulates and which can appear only in the presence of small ranges of muscle contraction. The strong development of the short postfemoral muscles with cursorial specialization in the Felidae and Canidae is probably also related to the specific flexor-extensor movements in the hip and knee joints. The indirect influence of the short postfemorals in extending the knee joint, and the influence of this extension on that of the talocrural joint, lead to a certain weakening of the femoral capita of the quadriceps and gastrocnemial muscles.

The large amplitude of vertical movements of the trunk with respect to the forelimbs would be somewhat impeded by the development of m. serratus ventralis.  This is probably why cursorial specialization in cats and dogs, unlike in ungulates, leads to a weakening of this muscle.  In a cat landing on the ground the vertical position of the humerus makes the development of m. latissimus dorsi and m. endopectoralis, which draw the trunk between the legs, especially favorable.  This is why these two muscles are strongly developed in cursorially specialized cats and dogs.  Another feature of such animals are strongly developed shoulder and elbow joint extensors.  Particularly strengthened is m. anconeus longus in the presence of a relatively reduced olecranon, which promotes intensive and at the same time swift movement in the phase of support.

In the Ursidae, forelimb activity with a large load, not connected with running, leads to strong development of m. ectopectoralis, which is the muscle that is weakened in the Felidae and Canidae. The force of contraction of this muscle is used by a bear also in the phase of front support during the gallop.

On the whole, our study of the paths of specialization for land locomotion in some carnivores enabled us to work out the typical hunting strategies which were inherent in the ancestors of individual groups; in other words, we were able to ascertain the primary divergence of the Carnivora.  In the subsequent phylogeny of these groups the types of locomotion which developed in connection with this primary divergence were reflected in the trends of changes of their skeleton and muscles even when the hunting techniques of some groups converged secondarily.

*Chapter 7*

## ADAPTATION TO RUNNING IN THE LAGOMORPHA

MECHANICS OF RUNNING

The Lagomorpha* include both weakly specialized and highly specialized
runners, so that it is of special interest to analyze the distinctive path of
cursorial adaptation in this group. Yet no biomechanical analysis has yet
been made concerning running in these animals. The data to be found in
the literature (Camp and Borell, 1937; Bohmann, 1939; Dondogin, 1950;
Gureev, 1964; and others) are either very meager or simply inaccurate.
In studying the morphology of the Lagomorpha, the first four authors
(Dondogin's study is particularly detailed) related it to the depth of specialization for running attained by certain representatives, but they paid
little attention to the actual techniques of running. Gureev made an attempt
to describe the running of these animals and even presented successive
poses in the running cycle (Gureev, 1964, fig. 5). However, both his description and the frames shown in the figure unfortunately bear very little relation to the reality.

Thus, there was clearly a need for an in-depth study of the cursorial
mechanics of the Lagomorpha. An examination of motion pictures showing
the gallop of the Altai pika (Ochotona alpina, Figure 181) and the wild
rabbit (Oryctolagus cuniculus, Figure 182) revealed on the one hand
some common features of locomotion in these forms and on the other a number of specific characteristics distinguishing the running of pikas from that
of hares and rabbits. When a pika jumps from stone to stone, it generally
lands on the forefeet, at the very edge of the stone. After landing the whole
body sinks down at first, and then, together with marked flexion of the spine,
is raised so that the hind feet are planted on this edge of the stone
(Figure 181). Hence, in this animal we can distinguish a stage of extended
flight, beginning after the hind feet have pushed away, and a stage of landing
on the forefeet. A second stage of flight (crossed) is virtually absent. On
slow running trips a pika may go over to a trot or some other symmetrical
gait. In hares and rabbits we observe the typical gallop, with phases of
support of the hind limbs and forelimbs and stages of extended and crossed
flight (Figure 182). In the phase of front support the spine is strongly flexed

---

* In view of the fact that the adaptive changes of the locomotory organs of the Lagomorpha were the subject
of a special investigation performed by Klebanova, Polyakova and Sokolov, we will just give here some
very general considerations on cursorial adaptation in this group.

to be able to promote the acceleration necessary for crossed flight.* And in the phase of hind support the spine is markedly extended so as to increase the speed of motion. Thus, strong flexion of the spine is uniformly characteristic for the Leporidae and Lagomyidae. In the latter this movement helps raise the body to the level of the next stone, while in the Leporidae it serves to implement the crossed stage of flight.

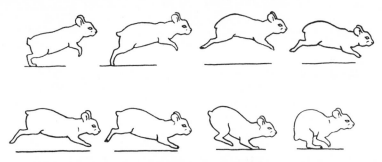

FIGURE 181. Gallop of the Altai pika (Ochotona alpina)

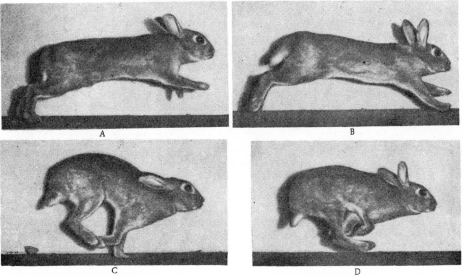

FIGURE 182. Different stages of the gallop in the wild rabbit (Oryctolagus cuniculus). Photo by G.P. Gambaryan:

A) support on hind limbs; B) extended flight; C) support on forelimbs; D) crossed flight.

* During the work of the flexors of the spine, the force of the thrust is transmitted by the forelimbs.

By all accounts the common ancestors of all the Lagomorpha were structurally similar to the Upper Paleocene E u r y m y l u s. Like the recent pikas, these animals probably lived in stony deposits of alluvium. The likelihood of this being the original mode of life of lagomorphs is based on the following.

All the recent representatives are characterized by the typical half-bound. This gait, we will remember, is also typical for cats, in which it was worked out as an adaptation to one or several jumps when capturing prey. A precisely calculated jump is in this case best achieved by the simultaneous thrusting off of both hind limbs. More economical for prolonged running is the alternating gallop, which came to characterize all the Canidae, the Ungulata, and even the cheetah, whose precursors, like all the other cats, used the half-bound. The fact that the half-bound has been preserved even in such highly specialized runners as the Lagomorpha shows that their ancestors needed precisely calculated jumps. The high-speed asymmetrical gait of the ancestors of all groups of mammals was the primitive ricocheting jump. If any branch of the mammals adapted to primitive ricocheting saltation comes to adopt stony deposits as its main habitat, the original gait proves unsuitable. With jumping from stone to stone, the forward deflection of the hind limbs after the phase of hind support would, with the primitive ricocheting jump, be proportional to the distance between the stones. Under these conditions the animal would land on the hind feet during large jumps and on the forefeet during small ones. These different methods of landing would naturally undermine the accuracy of distance-judging. Therefore, life among stones must have gone along with inhibition of hind limb progression right up to the point where the animal landed on the forefeet. Unlike tree branches, on which animals can fasten, pulling the body up by means of the forelimbs, stones are often flat, and, when wet, also slippery. It is therefore better for an animal which has landed on its forepaws on the edge of a stone to raise the whole body straight up and to pull the trunk and hind limbs toward the stone not by active pulling with the forelimbs but by flexion of the spine. With the secondary transition to running in open spaces, such a type of limb activity and work of the spine leads to the gallop, during which crossed flight is promoted mainly by strong flexion of the spine instead of active front propulsion. The active forelimb thrusting characteristic for ungulates and carnivores had the result that the forelimbs and hind limbs participated almost to an equal extent in the cycle of motion. In the Lagomorpha, on the other hand, the work of the fore and hind limbs came to be clearly differentiated: the forelimbs act mainly to absorb the shock on landing and the hind limbs and spine work to impart the necessary acceleration to the center of gravity. The characteristic half-bound might be one more argument in the theory that stone deposits were the primary habitat of the Lagomorpha.* In jumping from stone to stone the animal comes to a halt before each jump, which it makes from one spot.

---

* The gallop of hares is not an absolute half-bound. Tracks often show that the anterior margin of the hind paw is placed at the level of the middle, less often the posterior margin of the print of the other paw. In carnivores and ungulates moving at a lateral and diagonal gallop all the footprints fall in one longitudinal line, and their tracks differ so markedly from those of hares (cf. Figures 76, 178 and 183) that there is no doubt that the hares in fact prefer the half-bound.

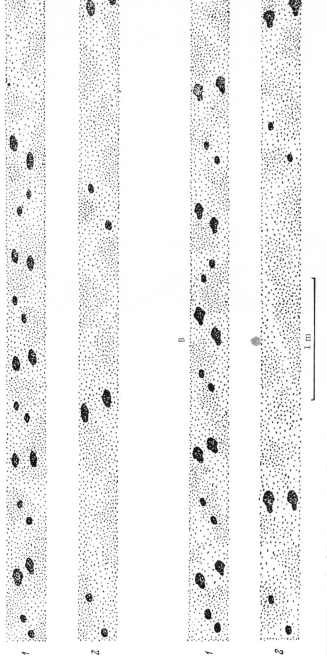

FIGURE 183.  Tracks of the European hare (Lepus europaeus) (A) and the blue hare (Lepus timidus) (B):

1) slow, 2) fast gallop.

The thrust is produced simultaneously by the two hind limbs, that is, the half-bound is adopted.  The preservation of this type of movements with the secondary transition to running in open spaces proves that the ancestors of these animals dwelt in stone deposits for a long period.  Hence, the differentiation of the role of the forelimbs and hind limbs, the active part played by flexion of the spine during running, and also the adoption of the half-bound find their explanation in the primary habitat of the Lagomorpha among stones.

The common mechanics of motion that we have noted for the Lagomorpha enables us to call their type of running *dorsomobile metalocomotory*, a term which embraces the whole group, even though some representatives may show specific features related to the mechanics of running.  The difference between the Lagomyidae and the Leporidae is expressed in the type of activity of the forelimbs, which in the hares act as shock absorbers and in the pikas actively pull the body up after the animal has landed on a stone. Within the Leporidae we also observe differences in the mechanics of motion which are connected with deep specialization for running and the adoption of various habitats.  For instance, the blue and Manchurian hares prefer to live in forests, while the European and Tolai hares favor more open spaces.  Rabbits dig complicated burrows and instead of running long distances usually just scuttle from one refuge to another.  The difference between the running of the European and the blue hare can be detected by examining their footprints.  Measurement of different phases of the gallop cycle on tracks of L. e u r o p a e u s showed that the stage of extended flight (measured from the hind to the front footprints) is practically equal in length to the stage of crossed flight (measured from the front to the hind footprints). When running is accelerated, both these stages of flight increase almost to the same extent, and often the stage of crossed flight may even be longer than that of extended flight.  In L. t i m i d u s, extended flight always exceeds crossed flight.  Furthermore, when running is speeded up, the stage of extended flight increases more than that of crossed flight, which brings out the difference between the two species still more clearly (Figure 183).  Thus, in the European hare the stages of extended and crossed flight equal about 0.9—0.95 m when the animal is running slowly, while upon acceleration, extended flight is increased to 1.4—1.6 m but crossed flight to 1.8—2.0 m. In the blue hare extended flight averages 0.8—0.9 m and crossed flight 0.6—0.7 m during slow running; upon acceleration, extended flight increases to 1.8—2.1 m but crossed flight only to 0.8—0.9 m.  This difference in the type of tracks left by these two species is similar to the difference observed in footprints of ungulates (see Chapter 4).  Forest-dwelling ungulates, like the blue hare, have a greater stage of extended flight, while those which inhabit open spaces, like the European hare, have equal stages of extended and crossed flight.  As in the Ungulata, these differences in the length of the flight stages are probably related to the steepness of the jump trajectory during running.  As we have already worked out for the ungulates, the angle of departure of the center of gravity is smaller in animals living in open spaces than in forest dwellers.  If in calculating the angle of departure of the center of gravity in L. e u r o p a e u s and L. t i m i d u s from the formula

$$\sin 2a = \frac{gl}{v^2},$$

we take the length of the jump $l$ to be equal to the distance from the hind footprints to the front footprints, and $v$ as equal to the maximum running speeds registered during our observations in nature, the other values are easily computed from the formula. For L. t i m i d u s:

$$\sin 2\alpha = \frac{9.81 \cdot 2.1}{196} = \frac{20.6}{196} = 0.105, \quad 2\alpha = 6°, \quad \alpha = 3°.$$

For L. e u r o p a e u s:

$$\sin 2\alpha = \frac{9.81 \cdot 1.8}{289} = \frac{17.6}{289} = 0.061, \quad 2\alpha = 3°30', \quad \alpha = 1°45'.$$

We see from these calculations that in the Leporidae, as in the Ungulata, the differences in the ratios of extended and crossed flight are determined by the angle of departure of the center of gravity. The decrease in the relative size of the stage of crossed flight in the blue hare is also due to the fact that most of the propulsive force arising on account of spinal flexion and forelimb activity is used for cushioning impacts, the force of which increases in proportion to sin $\alpha$.

## STRUCTURAL FEATURES OF THE SPINE

The need for active flexion and extension of the spine during running leads in the Lagomorpha to a strengthening of the muscles responsible for these movements and also to corresponding reconstructions in the vertebral column. These features are most clearly expressed in L e p u s e u r o p a e u s. In the thoracic region of this species the spinous processes slope backward up to anticlinal vertebra X, and the following thoracic vertebrae XI and XII (if these are present) and also all the lumbar vertebrae bear forward-sloped spinous processes. On the apexes of the spinous processes are well defined anteroposterior widenings, to which the weakly expressed supraspinous ligament is attached. The spinous processes gradually become higher as far as the fourth thoracic vertebra, and then again lower to the end of the thoracic region. The slope of the spinous processes progressively diminishes in the thoracic region, so that the apexes are brought closer together and the interspinous parts of the supraspinous ligament are reduced. Due to all these transformations the anterior section of the thoracic region assumes in the Leporidae the structure of a typical organ of spinal rigidity. The forward inclination of the very small spinous processes in the lumbar region gradually increases in the cranial direction, whereby there is an increase in the interspinous spaces and in the length of the supraspinous ligament in this region. As a result, unlike the anterior section of the thoracic region, the lumbar region of the spine possesses great vertical mobility. The spread of the articular processes of the lumbar vertebrae is of interest (Figure 184). In hares these processes serve as powerful levers improving conditions for extension of the lumbar region. The articular processes are strongly developed right up to the thoracic region, and on the thoracic vertebrae they are gradually reduced to nothing. At the same time costal tubercles spread out in the thoracic region which play a role similar to that of the articular processes of the lumbar vertebrae.

The transverse processes of the lumbar vertebrae are in the form of fairly wide, flat plates (except for the transverse process of the last lumbar vertebra), directed slightly downward and forward. Their ends bear caudal and cranial outgrowths distributed in a plane at an angle to the body axis (Figure 184). From the ventral side on the transverse processes are inserted the bundles of m. psoas major and m. quadratus lumborum, and from the dorsal side m. iliocostalis, m. semispinalis and m. longissimus dorsi. It may be assumed that these processes in fact serve to increase the area of attachment of the flexors and extensors of the lumbar region and at the same time prevent lateral flexures of the spine. On the ventral surface of the first three lumbar and last two or three thoracic vertebrae the ventral spinous processes grow out in L. europaeus. The most strongly developed is the spinous process of lumbar vertebra III, which is strongly sloped forward. The slope gradually diminishes, and it disappears from the first lumbar vertebra. The bundles of m. psoas major and m. quadratus lumborum are inserted on the whole surface of these processes. According to Brovar (1935, 1940), the appearance and spread of the spinous processes are in general related to the development of a rigid spine. It would seem that the ventral spinous processes in the Leporidae, similarly to the dorsal processes in the anterior section of the thoracic region preventing flexion of the spine, must serve to diminish extension of the spine. However, the absence of a supraspinous ligament and the fact that the bundles of the flexor muscles of the spine are inserted on these processes indicate that these processes act as levers in L. europaeus and L. timidus, facilitating flexion of the spine. The structure of the spine is similar in other species of Leporidae. In rabbits the ventral spinous processes are weakly developed. In pikas they are in the form of a weakly expressed narrow ridge on the ventral surface of the lumbar vertebrae. As for the dorsal spinous, articular and transverse processes, they are in general quite similar in structure in pikas to those of the other Lagomorpha, but somewhat more weakly developed.

Besides the above-described adaptive features of the spine, active flexion and extension of the spine are in the Lagomorpha associated with the strengthening of the flexor and extensor muscles. In all these animals the spine is flexed by m. quadratus lumborum and m. psoas major, and in the Leporidae additionally by m. psoas minor, which is absent in the Lagomyidae. M. quadratus lumborum begins on the ventral surface of the ilium, on the ventral surface of the transverse processes and on the ventral surface of the bodies of the lumbar vertebrae. The deep-lying bundles pass over from one segment to the next, while the more superficial ones are linked up to every other one or two segments. M. iliopsoas covers m. quadratus lumborum. It begins on the last one or two thoracic and on all the lumbar vertebrae, in the middle of their bodies, on the transverse processes, lateral to the point of insertion of m. quadratus lumborum. Its bundles are also attached to the ventral spinous processes. This muscle ends on the trochanter minus of the femur. M. psoas minor departs in the Leporidae from the bodies of the last two, sometimes last three, lumbar vertebrae and ends on the pubic tubercle of the innominate bones.   This complex of muscles works both to influence flexion of the spine and to flex the limb in the hip joint.   The total weight of the lumbar flexor muscles constitutes 5.3—9.5% of the weight of the fore and hind limb muscles in pikas and hares (Table 23), whereas it amounts to at most 5.0% in carnivores and ungulates.   This applies even to the highly specialized cursorial carnivores, in which active mobility of the spine is also observed.   Thus, for instance, in the cheetah the weight of the lumbar flexors is only 4.3% of the weight of the limb muscles.   Even in animals specialized for swimming, such as the sea otter, the relative weight of this group of muscles is only 4.4%.

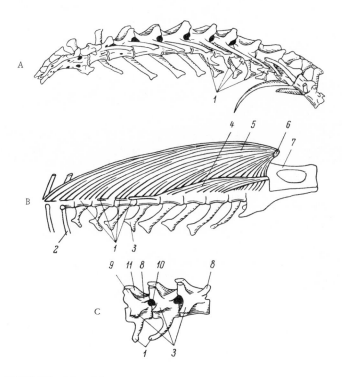

FIGURE 184.  Spine of Lepus europaeus:

A) dorsal; B) ventral, with muscles of the left side; C) lumbar vertebrae III—V, transverse processes broken off.  1) ventral processes;  2) ribs; 3) transverse processes;  4) m. psoas minor; 5) m. iliopsoas; 6) trochanter minus of femur; 7) pelvis; 8, 9) articular processes: 8) caudal, 9) cranial; 10) intervertebral foramen; 11) dorsal spinous process.

TABLE 23. Relative weight of the hind limb muscles in the Lagomorpha (% of the total weight of the fore and hind limb muscles)

| Muscles | Ochotona alpina | Lepus timidus | Lepus europaeus | Lepus tolai |
|---|---|---|---|---|
| m. gluteus medius . . . . . . . | 1.9 | 1.8 | 4.4 | 2.0 |
| m. gluteus minimus[1] . . . . . . | 2.0 | 2.0 | 0.9 | 2.0 |
| Gluteal . . . . . . . . . | 3.9 | 3.8 | 5.3 | 4.0 |
| m. quadratus femoris . . . . . . | 0.1 | 0.4 | 0.7 | 0.3 |
| m. pectineus . . . . . . . . . | 0.2 | 0.2 | 0.2 | 0.5 |
| m. m. adductores[2] . . . . . . . | 1.8 | 6.2 | 7.9 | 6.2 |
| m. semimembranosus anticus[3] . . | 3.5 | 3.3 | 3.0 | 2.7 |
| Short postfemoral . . . | 5.6 | 10.1 | 11.8 | 9.7 |
| m. biceps femoris[4] . . . . . . | 6.3 | 11.6 | 14.4 | 15.7 |
| m. semitendinosus . . . . . . . | 1.5 | 3.6 | 3.6 | 3.7 |
| m. semimembranosus posticus . . | 1.8 | 2.8 | 2.8 | 4.2 |
| m. gracilis . . . . . . . . . | 1.4 | 1.9 | 2.3 | 1.9 |
| Long postfemoral · · · · · · · · | 11.0 | 19.9 | 23.1 | 25.5 |
| Hip joint extensors · · · · · · · | 20.5 | 33.8 | 40.2 | 39.2 |
| m. vastus lateralis . . . . . . . | 2.9 | 5.5 | 6.6 | 6.3 |
| m. vastus medialis[5] . . . . . . . | 1.8 | 2.5 | 2.1 | 2.2 |
| Knee joint extensors . . . . . | 4.7 | 8.0 | 8.7 | 8.5 |
| m. gastrocnemius[6] . . . . . . . | 2.9 | 3.7 | 3.8 | 3.9 |
| m. psoas minor . . . . . . . . | — | 0.1 | 0.1 | 0.1 |
| m. iliopsoas . . . . . . . . . | 4.0 | 4.7 | 6.6 | 4.8 |
| m. gluteus superficialis . . . . | 0.8 | 1.2 | 1.2 | 1.2 |
| m. tensor fasciae latae . . . . . | 1.9 | 1.8 | 3.7 | 3.3 |
| m. sartorius . . . . . . . . . | 0.2 | 1.4 | — | — |
| m. obturator[7] . . . . . . . . | 0.5 | 0.8 | 0.9 | 0.7 |
| m. gemelli . . . . . . . . . | 0.1 | 0.1 | 0.1 | 0.1 |
| m. popliteus . . . . . . . . . | 0.2 | 0.3 | 0.3 | 0.3 |
| m. rectus femoris . . . . . . . | 3.0 | 3.0 | 2.6 | 2.7 |
| m. plantaris . . . . . . . . . | 0.9 | 1.6 | 1.6 | 1.1 |
| m. tibialis anterior . . . . . . | 0.9 | 0.7 | 0.8 | 1.2 |
| m. ext. digitorum[8] . . . . . . | 1.7 | 2.3 | 2.5 | 1.1 |
| m. peroneus[9] . . . . . . . . . | 0.6 | 0.8 | 0.6 | 0.5 |
| m. fl. digitorum[10] . . . . . . | 1.3 | 1.3 | 1.6 | 1.1 |
| Hind limb muscles . . . . . . | 45.0 | 64.6 | 67.1 | 68.5 |
| m. sacrospinalis . . . . . . . . | 15.6 | 25.7 | 32.0 | 29.5 |
| m. quadratus lumborum . . . . | 1.3 | 1.7 | 1.8 | 1.8 |

[1] m .m. gluteus minimus + piriformis.
[2] m. m. adductor longus + brevis + maximus.
[3] m. m. semimembranosus anticus + praesemimembranosus.
[4] m. m. biceps femoris anticus + posticus.
[5] m. m. vastus medialis + intermedius.
[6] m. m. gastrocnemius medialis + lateralis + soleus.
[7] m. m. obturator externus + internus.
[8] m. m. ext. digitorum longus + ext. hallucis longus.
[9] m. m. peroneus longus + brevis + digiti quinti.
[10] m. m. fl. digitorum longus + fl. hallucis longus + tibialis posterior.

The strengthening of the flexors of the spine, the spread of the ventral spinous processes and the mechanically favorable position of the lumbar transverse processes promote sharp active flexion of the spine in the region of the loin and the lumbar-thoracic section, allowing for crossed flight in the Leporidae and for bringing the hind legs to rest on a stone in the pikas.  Active extension of the spine in the Lagomorpha contributes to hind propulsion, the result being extended flight.  In these animals, therefore, the extensors of the spine are also strengthened.  The complex extensor m. sacrospinalis is very well developed, constituting 29.0% of the weight of the limb muscles in the blue hare, 33.0% in the European hare and even 36.0% in the Tolai hare.  According to the level of relative development of the spinal extensors, the Lagomorpha can be compared only with the Mustelidae, in which good spinal mobility leads to a marked increase in the size of this complex muscle.  Among the Mustelidae, however, only the highly specialized sea otter attains such a relative weight of the extensors of the spine (36.0% of the weight of all the limb muscles, see Chapter 6, Table 21). Such a degree of development of these muscles in the Lagomorpha can be explained only by their important role during running.  Their strong development in the Leporidae even in comparison with most of the Mustelidae indicates the influence of deep cursorial specialization.  For example, in Ochotona alpina, which is the lagomorph least specialized for swift running, m. sacrospinalis constitutes only 14.0% of the weight of the limb muscles.

## STRUCTURAL FEATURES OF THE SKELETON AND LIMB MUSCLES

As pointed out above, the Lagomorpha show two extreme forms of adaptation to swift running.  The Lagomyidae are little-specialized runners, adapted to short scuttles from one hiding place to another.  The Leporidae are highly specialized for sustained, high-speed running.  A comparison of the structural characteristics of the limbs in these two families may therefore be instructive in clarifying the patterns of change of the skeleton and muscles of the limbs in connection with cursorial specialization.

Specialization for running involves a change in the relative length of the limb segments (Table 24).  We see from the table that the most notable feature in the hind limb is the lengthening of the foot.  Thus, in Caprolagus brachyurus the foot is 1.45 times larger than in Ochotona daurica.  The foot increases in length almost to the same extent in Lepus europaeus.  In the more highly specialized L. tolai, the foot is 1.53 times bigger, and in L. timidus even 1.54 times. In C. brachyurus the femur is 1.12 times longer than in O. daurica and 1.15 times longer than in the other hares.  An analogous lengthening takes place in the forelimbs, in which the forearm increases in size most of all: about 1.18–1.44 times in hares as compared with the pikas, while the hand is 1.1–1.31 times and the shoulder 1.1–1.25 times larger (Table 24).

TABLE 24. Changes in the relative size of the limb segments (% of the length of the lumbar and thoracic regions of the spine) in the Lagomorpha in relation to cursorial specialization

| Species | Femur (a) | Tibia (b) | Foot (c) | Total (a + b + c) | Shoulder (d) | Fore-arm (e) | Hand (f) | Total (d + e + f) | Degree of cursorial specialization |
|---|---|---|---|---|---|---|---|---|---|
| Lepus timidus | 40.2 | 48.2 | 46.1 | 134.5 | 34.3 | 37.6 | 24.6 | 96.5 | Highly specialized |
| L. europaeus | 40.5 | 47.5 | 43.3 | 131.3 | 33.7 | 37.9 | 20.3 | 91.9 | Highly specialized |
| L. tolai | 40.2 | 48.5 | 45.9 | 134.6 | 31.4 | 37.1 | 24.0 | 92.5 | Highly specialized |
| Caprolagus brachyurus | 39.1 | 46.0 | 43.5 | 128.6 | 30.6 | 30.9 | 24.5 | 86.0 | Little specialized |
| Ochotona daurica | 35.0 | 46.3 | 30.0 | 111.3 | 27.5 | 26.3 | 18.7 | 72.5 | Nonspecialized |

Note. Data received from R. S. Polyakova.

It was mentioned on p. 133 that in the forest-dwelling ungulates adapted to the saltatorial-cursorial form of running and also in animals living in shrubby habitats adapted to the cursorial form with elements of the saltatorial-cursorial form (goitered gazelle), the length of the limb segments was relatively greater than in ungulates which live in open spaces and have the cursorial form of running. The difference between them was especially noticeable in the relative length of the foot. Interestingly, a similar phenomenon is observed in the Lagomorpha. The foot of L. timidus is slightly longer than that of L. europaeus and L. tolai, although when running fast, the two latter species develop greater speed than the blue hare. In the Manchurian hare the length index of the foot is not lower than in the European hare, despite the fact that this species shows a relatively poor degree of cursorial specialization (Table 24).

We stated in the first section of this chapter that the differences in the mechanics of running between L. timidus and L. europaeus is expressed in the different steepness of their jump trajectories. The steeper flight trajectory of L. timidus brings about the need to use relatively more force when jumping (formula (9)), thereby requiring relatively longer legs. On the whole, the small difference in the relative length of the foot is due to the fact that despite the steeper trajectory, which causes an increase in the angle of departure, the blue hare runs more slowly, and this lesser speed also brings about an increase of the relative force (formula (9)).

As in other mammals, in the Leporidae and Lagomyidae, in the phase of hind support the hip, knee and talocrural joints are flexed in the preparatory period and extended in the starting period. The yielding stretch of the hip, knee and talocrural joint extensors in the preparatory period and also their active contraction in the starting period cause considerable loads to be placed on these muscles.

For the Lagomorpha, as for other mammals, the heaviest load in the phase of support is borne by the extensors of the hip joint, which comprise three groups of muscles: gluteal and long and short postfemoral.  In connection with cursorial specialization, therefore, the long postfemoral muscles come to be inserted more proximally along the tibia and there is a strengthening of the two groups of postfemoral extensors of the hip joint and, to a lesser extent, of the gluteal muscles.  This is clearly seen in comparing the pika with the blue hare, and particularly in comparing the pika with the more specialized European and Tolai hares (Table 23).  The weight of the long postfemorals is 17.0% of that of the fore and hind limb muscles in L. t i m i d u s, 20.1% in L. e u r o p a e u s, 22.2% in L. t o l a i, and only 9.1% in O. a l p i n a, in other words, in comparison with the pika the relative weight of these muscles is 1.87 times greater in the blue hare, 2.21 times greater in the European hare and even 2.54 times greater in the Tolai hare. The weight of the short postfemorals represents 7.2% of the weight of the limb muscles in the pika, 12.5% in L. t i m i d u s, 14.5% in L. e u r o p a e u s and 14.0% in L. t o l a i, that is, in comparison with the pika the relative weight of this group is 1.74 times greater in the blue hare, 1.94 times greater in the Tolai hare and 2.02 times greater in the European hare.  No less strengthened in the hares are the knee joint extensors, namely the three femoral capita of m. quadriceps femoris, which in pikas constitute 4.7% of the weight of the limb muscles, and in hares from 8.0 to 8.7%, i. e., the relative weight of the knee joint extensors proper is 1.70—1.85 times greater in hares than in pikas.  The relative weight of the talocrural joint extensors is less appreciably greater in the Leporidae in comparison with the Lagomyidae; they constitute 2.9% of the weight of the limb muscles in the pika and 3.7—8.9% in the hare, that is, about 1.28—1.35 times more.

When a pika has landed on its forefeet, the trunk and head are raised, the trunk is drawn between the legs and the shoulder and elbow joints are extended.  The scapulae are now advanced with the vertebral margin forward.  A whole series of muscles move the scapulae forward:  m. m. rhomboideus, acromiotrapezius, omotransversarius, serratus ventralis pars cervicalis, and sternoscapularis (shoulder girdle muscles) and m. supraspinatus and m. infraspinatus (shoulder joint muscles).  As we have noted, the movement of the forelimbs in the preparatory period of the phase of front support in hares is achieved mainly on account of inertia of motion of the body.  It is only when the preparatory period ends, and the forearm is already directed slightly backward, that an active load is placed on the forelimbs in order to promote support when the spine is flexed.  The shoulder and elbow joints are extended at this time.  As in the Lagomyidae, the shoulder joint extensors are a number of the shoulder girdle and shoulder joint muscles listed above.  However, whereas in the pika these muscles act to raise the whole body, draw the trunk between the legs and markedly extend the shoulder joint, in hares the load on these muscles is chiefly reduced to resistance to flexion, making for adequate support when the spine is flexed.  The most advantageously placed of the above muscles for drawing the trunk forward and for extending the shoulder joint is m. subclavius seu sternoscapularis, which originates on the first two rib cartilages, the manubrium and the first segment of the sternum.  It ends on the anterior

margin of the scapula and the fascia of m. supraspinatus. The muscle is
inserted onto the clavicle along its path. The attachment of bundles on the
clavicle leads to a more horizontal direction of the fibers with respect to
the thorax and to a more oblique (anteroposterior) attachment of the fibers
on the scapula (Figure 185). A horizontal position of the fibers relative to
the thorax is best for drawing the trunk between the legs, while the oblique
arrangement relative to the scapula facilitates extension of the shoulder
joint. In a number of mammals, m. subclavius gradually widens out, and
its bundles pass over from the clavicle onto the fascia of m. supraspinatus
and then onto the scapula (several fossorial rodents: Gambaryan, 1960;
Perissodactyla: Gambaryan, 1964; and others). A similar widening of
m. subclavius is observed also in the Lagomorpha, for which it becomes
especially profitable in connection with the position of the bundles described
above. This can probably explain the strong development of this muscle in
the pikas, in which (Table 25) its relative weight (4.0% of the weight of the
limb muscles) is 3.6 times greater than in hares (1.1%). However, even in
hares this muscle is more important than in rodents, in which the relative
weight of m. subclavius is not more than 0.5% of that of the limb muscles.

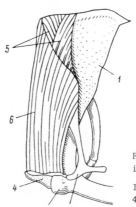

FIGURE 185. M. sternoscapularis
in Lepus europaeus:

1) scapula; 2) acromion; 3) humerus;
4) manubrium; 5) m. supraspinatus;
6) m. sternoscapularis.

    In the hares as compared with the pikas there is also a weakening of a
number of muscles which take part in pushing the trunk between the legs
and at the same time in extending the shoulder joint (Table 25). These
muscles include on the one hand, those that begin on the dorsal side of the
neck and on the head (m. acromiotrapezius and m. rhomboideus capitis)
and on the other, the muscle which originates on the ventral surface of the
neck and the anterior part of the thorax (m. serratus ventralis pars cervicalis). When the body is pulled up between the legs the distance between
the vertebral margin of the scapula and the ventral line of the neck and
thorax diminishes, whereby there is a decrease in the possible amplitude
of contraction of m. serratus ventralis pars cervicalis. The reason for this is

TABLE 25. Relative weight of the forelimb muscles in the Lagomorpha (% of the total weight of the fore and hind limb muscles)

| Muscle | Ochoto-na alpina | Lepus timidus | Lepus euro-paeus | Lepus tolai |
|---|---|---|---|---|
| m. brachiocephalicus [1] | 1.9 | 0.8 | 0.9 | 0.7 |
| m. acromiotrapezius | 1.2 | 0.5 | 0.5 | 0.4 |
| m. spinotrapezius . . | 1.4 | 0.7 | 0.7 | 0.6 |
| m. rhomboideus [2] . . | 1.7 | 1.6 | 1.2 | 1.2 |
| m. rhomboideus capi-tis . . . . . . | 1.1 | 0.2 | 0.1 | 0.1 |
| m. latissimus dorsi . . | 2.9 | 2.6 | 2.2 | 1.9 |
| m. sterno-mastoideus | 2.0 | 0.3 | 0.6 | 0.3 |
| m. ectopectoralis . . . | 4.5 | 3.6 | 3.0 | 3.1 |
| m. endopectoralis . . | 3.8 | 1.2 | } 2.4 | 1.0 |
| m. sternoscapularis . . | 4.0 | 1.4 | | 1.1 |
| m. serratus ventralis | 4.5 | 3.6 | 3.1 | 3.4 |
| m. atlantoscapularis | 1.4 | 0.7 | 0.5 | 0.6 |
| m. deltoideus . . . | 1.0 | 0.4 | 0.5 | 0.2 |
| m. supraspinatus . . | 2.6 | 2.9 | 2.9 | 2.5 |
| m. infraspinatus . . | 1.9 | 2.0 | 2.1 | 2.1 |
| m. subscapularis . . | 2.7 | 1.7 | 1.5 | 1.7 |
| m. teres major . . | 2.2 | 1.6 | 1.8 | 1.6 |
| m. teres minor . . | — | 0.1 | 0.1 | 0.1 |
| m. coracobrachialis . . | 0.1 | 0.1 | 0.2 | 0.1 |
| m. anconeus longus . . | 5.4 | 3.4 | 3.5 | 3.7 |
| m. anconeus lateralis | 1.6 | 1.8 | 1.7 | 1.6 |
| m. anconeus medialis [3] | 0.9 | 0.7 | 0.6 | 0.2 |
| m. biceps brachii . . | 1.1 | 0.9 | 0.8 | 0.9 |
| m. brachialis . . . . | 0.7 | 0.4 | 0.3 | 0.3 |
| m. pronator teres . . | 0.1 | 0.1 | — | — |
| m. supinator . . . | 0.1 | — | — | — |
| m. ext. carpi radia-lis [4] . . . . . . . | 0.7 | 0.3 | 0.3 | 0.2 |
| m. ext. digitorum [5] . . | 0.5 | 0.3 | 0.2 | 0.3 |
| m. abd. pollicis lon-gus . . . . . . | 0.2 | 0.1 | 0.1 | 0.2 |
| m. ext. carpi ulnaris | 0.2 | 0.1 | 0.2 | 0.1 |
| m. fl. carpi ulnaris . . | 0.6 | 0.3 | 0.2 | 0.2 |
| m. fl. carpi radialis | 0.2 | 0.1 | 0.1 | 0.1 |
| m. palmaris longus | 0.1 | 0.1 | — | — |
| m. fl. digitorum subli-mis . . . . . . | 0.4 | 0.4 | 0.4 | 0.5 |
| m. fl. digitorum pro-fundus . . . . . | 1.3 | 0.8 | 0.6 | 0.6 |
| Total | 55.0 | 35.8 | 33.3 | 31.6 |

[1] m. m. clavotrapezius + clavomastoideus + clavodeltoideus.
[2] m. m. rhomboideus thoracalis + rhomboideus cervicalis.
[3] m. m. anconeus medialis + epitrochleoanconeus + dorsoepitrochlearis.
[4] m. m. ext. carpi radialis longus + brevis.
[5] m. m. ext. digitorum communis + ext. digitorum lateralis + ext. pollicis longus.

that the weakening of m. serratus ventralis in hares as compared with
pikas is not so well expressed as the weakening of the first two muscles.
Thus, the relative weight of m. serratus ventralis constitutes 2.3% of the
weight of the limb muscles in the pika and 1.2—1.8% in the hare, that is,
1.3—1.9 times less.   M. acromiotrapezius accounts for 1.2% and m. rhomboideus capitis 1.07% in the pika, whereas the respective values for the
hare are 0.2—0.5 and 0.16%, i.e., these muscles are weakened by a factor
of 2.4—6.7 in the Leporidae (Table 25).   The specific differences in the
work of the forelimbs in the Lagomyidae and Leporidae are also manifested
in the function of shoulder joint extension.   In a pika pulling up the trunk
between the legs, the shoulder girdle muscles simultaneously extend the
shoulder joint.   But in the hares, the forelimbs hold the anterior part of the
trunk in check, lending support for strong flexion of the spine, as a result of
which a stage of crossed flight appears.   Extension of the shoulder joint in
hares, therefore, must be effected mainly on account of the extensors proper
of this joint, and not by the muscles of the shoulder girdle.   And whereas
above we noted a greater strength of the shoulder girdle muscles in the
pikas, where the hares are concerned, the strengthening of m. supraspinatus
constitutes 14.0—15.9% of the weight of the muscles of the free forelimbs*
and m. infraspinatus 10.0—11.7% in the Leporidae, while the respective
figures for the Lagomyidae are 10.5 and 7.6%.

   Extension of the elbow joint is very important in pikas and hares: in the
former for raising the trunk and in the latter for creating adequate support.
Consequently their relative weight differs little, being 18.1—20.6% of the
weight of the muscles in the free limbs in the hares and 21.7% in the pikas.**

## PATHS AND CAUSES OF REORGANIZATION OF THE LOCOMOTORY ORGANS IN LAGOMORPHS

   The ancestors of the Lagomorpha apparently lived in stone deposits.
Leaping from stone to stone caused them to develop the half-bound with
active flexion of the spine after the animal landed on the forefeet.  Maintenance of this type of movement after the transition to open spaces led to
the appearance of dorsomobile metalocomotory running, which is characterized by a stage of crossed flight, arising due to sharp active flexion of the
spine in the lumbar and lumbar-thoracic regions.

* The index of m. supraspinatus and m. infraspinatus is calculated in this case in relation to the sum of weights of the muscles only of the free forelimb.  This is done because the muscles of the shoulder girdle are relatively much larger in pikas than in hares.
** In calculating the index in relation to the weight of the forelimb muscles together with the shoulder girdle muscles, similar, but inverse, ratios would be obtained.

The outcome of this cursorial specialization was that the segments of
the limbs became markedly longer in the hares in comparison with the
pikas, and the hip, knee and talocrural joint extensors were strengthened.
All these changes are in principle similar to the transformations arising
upon adaptation to other modes and forms of running (ungulates, carnivores).
But the characteristic feature of the lagomorphs is the strengthening of the
spinal flexors (m. iliopsoas, m. quadratus lumborum), which are developed
to a very high degree even in comparison with those in highly cursorially
specialized Felidae (which also show active flexion of the spine). Moreover,
in the Leporidae ventral spinous processes are present on the last thoracic
and first lumbar vertebrae.  Bundles of the flexor muscles of the spine are
inserted on these, additionally strengthening the function of active flexion
of the vertebral column.

An analysis of the gait, mode of life and type of running of the Lagomorpha
enables us to visualize the original mode of life of the group and to consider
the recent pikas, in which the features of life characteristic for the ancestors
of the group are preserved very fully, as "living fossils."

*Chapter 8*

## ADAPTATION TO RUNNING IN THE RODENTIA
## AND MARSUPIALIA

### MECHANICS OF RUNNING

The enormous diversity of rodents, very many of which adapted to swift running quite independently, lends special interest to a study of cursorial specialization in this group of mammals. The characteristic asymmetrical gait of most rodents is the primitive ricocheting jump, which indicates that the Rodentia departed from the common stem of the Mammalia at a stage when the gallop had not yet appeared. The development of the gallop in some rodents no doubt took place independently.

In a number of groups, Dipodidae, Dipodomyidae, Pedetidae, etc., highly specialized forms appeared whose main gait became the bipedal ricochet. The marsupials investigated here also proved to be adapted to the bipedal ricochet, and in general we can find much in common, where the mechanics of running is concerned, between the various species of rodents adapted to this gait and the kangaroo.* The cursorial mechanics of rodents which move by the gallop and those which use the bipedal ricochet is different. And naturally, also the forms of adaptation to swift running are manifold in the rodents.

The ancestors of the whole order Rodentia are still unknown. The most ancient of the known fossils representing this order, the Ischyromyidae, probably already possessed the principal features of the suborder Sciuromorpha. They may also have been climbing animals, in which case their terrestrial descendants could not have secondarily adopted the primitive ricocheting jump, which generally disappears when animals take to climbing. The Ischyromyidae can therefore hardly be considered to be the source of all the later branches of Rodentia. In view of this, in order to explain the phylogenetic pathways within the order, apart from studying the morphology of the most primitive extinct forms, we must investigate the trends of ecological specialization of the recent forms so as to reconstruct the mode of life of the ancestors. A precondition for such a reconstruction is, of course, the possibility of deriving all the recent trends of specialization from the mode of life of the ancestral forms.**

---

* It would be more useful to compare the kangaroo with other marsupials in order to study the adaptive changes of the locomotory organs in the Marsupialia in relation to cursorial specialization. However, due to the lack of relevant material we were obliged to compare the kangaroo with the rodents adapted to the bipedal ricochet.

** The majority of the recent rodents and all the known extinct forms were small animals, and therefore the ancestors could hardly have been larger than the Norway rat.

The numerous trends of specialization followed by different groups of the order may serve as an argument that their precursors did not display a sharply expressed specialization for one particular mode of locomotion. They could not have been specialized for running, since digging involves heavy work, the economy of which requires thorough reorganization of the organs of movement (Gambaryan, 1960). For a similar reason it is hard to imagine that these ancient animals were highly specialized for swimming. Nor could climbing have been typical, for in this case they would have had to be adapted for jumping from branch to branch. And such leaps would have demanded a change in the rhythm of activity of the neuromuscular chains typical for the primitive ricocheting jump, in which the degree of forward deflection of the hind limbs is proportional to the size of the jump. Maintenance of the working rhythm of the neuromuscular mechanisms characteristic for the primitive ricocheting jump, as mentioned above (p. 56), would have led to the result that sometimes the hind feet and sometimes the forefeet would have landed on the next branch, and this would have interfered with precise calculation of the size of jumps. With the arboreal mode of life, therefore, forward deflection of the hind limbs is inhibited after the phase of hind support, whereby the primitive ricocheting jump is transformed into a gallop. In view of the great economy of the gallop in comparison with the primitive ricocheting jump (more details in Chapter 2), we cannot acknowledge climbing to be the original mode of life of the ancient rodents. Nevertheless, these forms probably did live in forests. This assumption is based on the fact that life in open spaces calls either for well developed cursorial specialization or for burrow-digging, and neither of these trends of specialization was feasible for the primitive rodents.

From the above we may infer that the precursors of the rodents were small animals which probably lived above ground, not under the forest cover. Their favorite temporary refuges (fallen tree trunks, debris, crevices in bark and in the earth, dense thickets in the undergrowth, etc.) were their means of escape from enemies. Their diet was probably quite varied: seeds, leaves, grass, mushrooms and, to a lesser extent, animal food. In running from one refuge to the next, the main asymmetrical gait was the primitive ricocheting jump. The descendants of these forms came to adopt very different habitats. Adaptation to feeding on fruits, seeds and leaves of trees promoted the transition to climbing, which was also another way of fleeing from predators. Life at the water's edge led to adaptation to swimming and diving, both to obtain food and as a means of escape. Penetration into the steppe resulted in further specialization for feeding on the green parts of plants and at the same time in adaptation to digging burrows or in specialization for feeding on grains (more caloric food) and also in the development of the ability to run fast. Once the open spaces were conquered, primitive ricocheting saltation was perfected by means of gradual transition to the bipedal ricochet. Animals which were originally tree climbers were also able to secondarily adapt to running swiftly in the open. But in this case they adapted not to the primitive ricochet but to the specialized bipedal ricochet and to the gallop.

In order to judge the depth and course of specialization for running in rodents we made observations in nature and also analyzed motion pictures of various species. Although the data are not as full as they might be, they do characterize the main features of running.

A study of motion pictures and the data in the literature (Howell, 1932, 1944; Böker, 1935; Bartholomew and Caswell, 1951; Bartholomew and Reynolds, 1954; Eble, 1955; Krapperstück, 1955, and others) show that many species are characterized by the primitive ricocheting jump which upon specialization becomes the bipedal ricochet. This gait is probably typical for all the Myomorpha, many of the Hystricomorpha and, possibly, some of the Sciuromorpha.

In Chapter 2 we discussed the conditions under which relative force changes in relation to a change in the length of the limbs, the angle of departure, the size of the jump, the body mass, etc. (see formula (6)). The formula shows that, other conditions being equal, the relative force developed by a limb diminishes in proportion to its elongation (h) and increases in proportion to the increase in body mass (m). However, whereas the length of the leg increases linearly with the increment of body weight, the body mass increases as the cube. We therefore introduced the efficiency index for various builds of animals, calculated as the product of the jump index times the cube root of the body weight (Table 26).* Apart from the morphological indexes, which will be discussed below, the depth of specialization of the rodents presented in Table 26 can be characterized both according to the relative size (index) of the jump and according to the efficiency index of build. The jump index is suitable when comparing animals of more or less the same weight. For instance, a comparison of the mode of life of the hamsters P h o d o p u s   s u n g o r u s ,   C r i c e t u l u s   m i g r a t o r i u s  and C a l o m y s c u s   b a i l w a r d i  reveals that the last species shows the highest degree of specialization. The mouselike hamster C . b a i l w a r d i  lives among rocks and stone deposits of the semidesert regions in the southern USSR; it is a very mobile and adept creature. It apparently does not have burrows as such; it reproduces in selected gaps under stones. Because of their nimbleness, these animals are quite successful in escaping the clutches of predatory birds and beasts, even in places where the hamster population is higher than that of other rodents (Gambaryan and Martirosyan, 1960). The migratory hamster C r i c e t u l u s   m i g r a t o r i u s  has a much wider distribution than the mouselike hamster; it occurs from the semideserts to alpine meadows and lives in steppes, forest-steppe, among rocks and also around human habitations. It digs quite complicated burrows, but it can also do without them, collecting food stores and raising their young in the most diverse kinds of shelters. It is much less mobile than C . b a i l w a r d i , and if it occurs in the same places as this species, it falls prey far more often to predators, even if its population is smaller in the area (Sosnikhina, 1950; Bashenina, 1951; Gambaryan and Martirosyan, 1960). The most mobile of these three species is the striped hairy-footed hamster, P h o d o p u s s u n g o r u s . It inhabits steppes and deserts, choosing areas with a cover of grass; it escapes into burrows or else shakes off enemies in the grass (Flint and Golovkin, 1961). A study of footprints gives us a clearer idea of the depth of specialization for the ricochet in these animals. The differences in the length of their jumps may be seen in Table 26. But there

---

* The necessary force is proportional to the mass (cube), whereas the force developed by a muscle depends on its cross section (square); division of these quantities gives us first-order quantities. Therefore, for the efficiency index we take the product of the length of the jump times the linear body weight (the cube root of the body weight).

are also other interesting features of running, which tracks readily reveal. For example, in Phodopus sungorus the width of the tracks is even greater than the length of the body. The foot is placed at an angle of 30—45° to the sagittal plane of the trunk (Figure 186, A). In Cricetulus migratorius the tracks are narrower, even during slow bounds, when the jumps are the same size as those of P. sungorus. Their width is about equal to the width, not the length, of the body, and the paws are placed almost parallel to the trunk (the angle rarely exceeds 10—15°) (Figure 186, B). When this hamster makes a bigger jump, the prints of the forepaws are brought close to those of the hind paws (Figure 186, B, 2—3). The process of change of the tracks in relation to the size of the jump is traced particularly well in Calomyscus bailwardi, in which during small hops the distance between the front footfalls and hind footfalls is almost equal to the length of the body, while with larger jumps the front and hind footfalls come closer and closer together, and with maximum jumps the prints of the forepaws sometimes disappear altogether. This pattern of change is due to the fact that as the size of the jump increases, the hind limbs are deflected increasingly farther forward in the air, so that the landing of the hind feet and forefeet comes to coincide more and more in time. With maximum jumps landing is already accomplished secondarily on the hind paws.

TABLE 26. Relative size of jumps of rodents and the kangaroo and efficiency indexes of build during running (the jump index is the ratio of the size of the jump to the length of the body)

| Species | Weight of body, g | Jump index | Efficiency index of build |
|---|---|---|---|
| Phodopus sungorus | 20.0 | 1.5 | 4.1 |
| Cricetulus migratorius | 30.0 | 2.5 | 7.8 |
| Calomyscus bailwardi | 20.0 | 10.0 | 27.0 |
| Rhombomys opimus | 200.0 | 3.0 | 17.6 |
| Meriones vinogradovi | 200.0 | 3.5 | 20.4 |
| M.persicus | 180.0 | 4.5 | 25.0 |
| M.blackleri | 180.0 | 4.0 | 22.6 |
| M.meridianus | 90.0 | 6.5 | 28.0 |
| Rattus norvegicus | 350.0 | 2.2 | 15.5 |
| Mus musculus | 15.0 | 6.0 | 15.0 |
| Apodemus sylvaticus | 20.0 | 6.5 | 18.0 |
| Myocastor coypus | 4,000.0 | 1.0 | 16.0 |
| Allactaga severtzovi | 400.0 | 12.0 | 88.0 |
| A.elater | 100.0 | 15.5 | 72.0 |
| A.bobrinskii | 100.0 | 16.0 | 74.0 |
| Alactagulus acontion | 80.0 | 8.0 | 35.0 |
| Pygerethmus platyurus | 40.0 | 5.5 | 19.0 |
| Dipus sagitta | 150.0 | 14.5 | 77.0 |
| Paradipus ctenodactylus | 200.0 | 19.5 | 114.0 |
| Jaculus turcmenicus | 120.0 | 21.0 | 103.0 |
| Macropus rufus | 70,000.0 | 8.5 | 350.0 |

Note: For the jerboas data are presented from an analysis of material submitted by I.M.Fokin.

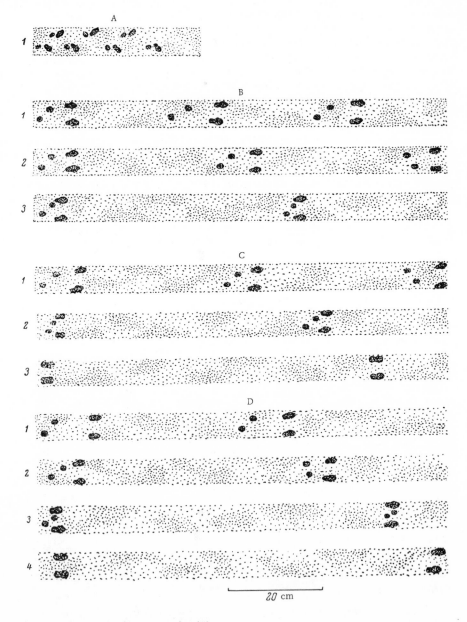

FIGURE 186.  Tracks of hamsters and gerbils:

A) Phodopus sungorus; B) Cricetulus migratorius; C) Calomyscus bailwardi; D) Meriones meridianus. 1) slow running; 2–4) progressive increase in the speed of running and the size of jumps.

Various species of Gerbillinae also show markedly different jump
indexes.   As in the hamsters, this can serve as one of the many criteria of
the depth of cursorial specialization.   Thus, large gerbils, which have a
small jump index, are very much confined to their burrows (Kambulin, 1941;
our own observations) and are undoubtedly less specialized for running
than the midday gerbil, M e r i o n e s   m e r i d i a n u s ,   which possesses
high overall mobility (Gambaryan et al., 1960).   Interestingly, according to
the nature of the changes in the tracks in relation to the size of jumps the
same changes are observed in the gerbils as in the mouselike hamster.
For example, in M e r i o n e s   m e r i d i a n u s ,   the distance between the front
and hind footfalls is 5—6 cm during a jump of 40 cm;  this decreases to
2—3 cm when the jump is of 55—60 cm, while during a jump of 65 cm the front
footfalls are found to be placed between the prints of the hind paws, and with
a jump of 70—75 cm forepaw prints are absent altogether and the creature
jumps only on the hind paws (Figure 187).   As in C a l o m y s c u s   b a i l -
w a r d i ,   in gerbils the foot is placed almost parallel to the trunk during a
jump.   An analogous position of the foot is observed also in all the jerboas,
although this is rather hard to establish from tracks, as only the distal parts
of the foot rest on the ground.   Still, the prints of the digits give an idea of
the general direction of the thrust characteristic for all the Dipodidae.
Hence, we can determine from tracks the direction of the thrust imparted to
the body by the limbs.   If this direction is strongly deflected from the body
axis (P h o d o p u s   s u n g o r u s), part of the thrust force is used unproduct-
ively for running speed.   Therefore, with cursorial specialization, even as
little expressed as it is in the migratory hamster, Norway rat and some
other rodents, the paws are placed parallel to the trunk during running.

FIGURE 187.  Tracks of the midday gerbil (M e r i o n e s   m e r i d i a n u s).  Photo by I.M.Fokin.

Although the jump index for the hamsters and gerbils presented in Table 26 can alone serve as quite an objective criterion of the degree of their specialization for running, when comparing these groups of rodents we are compelled by the difference in their size to introduce an additional index, that of the efficiency of build. It cannot be said that this makes for a fully objective assessment of the depth of cursorial specialization. We have to take into account that a small animal running fast is subjected to considerable drag, which scarcely has any importance for large animals running at the same speed or small ones running less fast (more details in Chapter 2). This can probably explain, for instance, the low efficiency index of build in the small five-toed jerboa, Allactaga elater, which may in fact be more specialized for running than the other five-toed and three-toed jerboas. The arbitrariness of all these numerical indexes becomes particularly apparent when comparing very different animals, where there is evidently no point in using such indexes. For instance, kangaroos are highly specialized ricocheters, but it is very doubtful that they are three times as highly specialized for running as the jerboas. But while fully aware of the arbitrariness of the efficiency index of build, we still consider it useful, especially when comparing closely related groups of different-sized animals.

Common to all animals adapted to the primitive ricocheting jump are the nature of the change in the amplitude of movement in the limb joints and the type of movement of the body during an increase in the size of jumps. We see from the motion pictures showing the primitive ricocheting jump in Rattus norvegicus (Figure 33), Cricetulus migratorius (Figure 188), Calomyscus bailwardi (Figures 189 and 190) and Meriones blackleri (Figure 191) that during a jump of 15—20 cm the phase of hind support ends when the feet are placed almost in the vertical plane. At this time the hip and knee joints are extended hardly more than by 90° and the body is horizontal. In the stage of free flight the feet are drawn forward, so that toward the end of this stage they are positioned almost in the horizontal plane. The anterior end of the body is lowered, whereby the shock is absorbed wholly by the forelimbs and the anterior section of the spine. When the jump is larger, at first only a slightly greater extension of the knee and hip joints and inclination in the phase of support is observed. The body hardly alters its position before it lands; only the hind limbs are brought markedly farther forward, so that the ends of the hind paw digits are placed almost at the level of the elbow joint, stretched out in front of the forepaws, as is seen in the Persian jird, Meriones persicus (Figure 192). In this position the impact is cushioned almost fully by the forepaws and the thoracic region of the spine. During such a jump, the center of gravity of the anterior half of the body, as with small jumps, is set lower than the center of gravity of the posterior half. With a further increase in the size of the jump toward the end of free flight (Figure 193) the vertebral column is strongly bent, with the result that the center of gravity of the posterior half of the body drops below that of the anterior half, the hind limbs reach the ground before the forelimbs, and it is they that cushion the jolt, even though the forelimbs then land too (Figures 35 and 194). With a still larger jump, landing takes place with the body almost vertical (Figure 36), and in this case the forelimbs

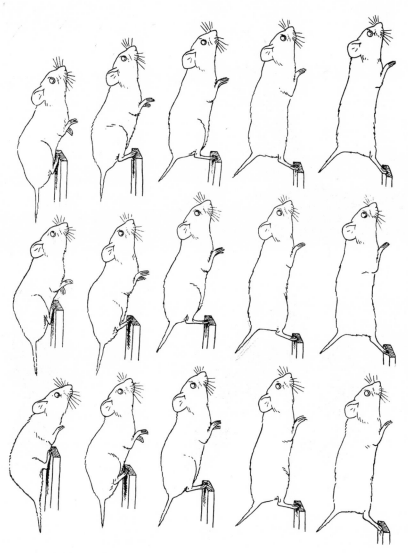

FIGURE 188.  Phase of support before a jump in the migratory hamster (Cricetulus migratorius)

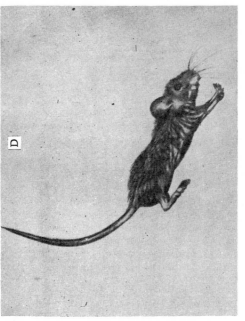

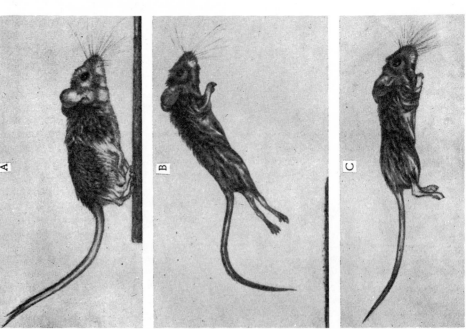

FIGURE 189.  Stages of the primitive ricocheting jump in the mouselike hamster (Calomyscus bailwardi):

A) stage of support;   B, C, D) stage of flight:   B) beginning, C) middle, D) end.

FIGURE 190.   Phase of support before a jump in  C. bailwardi

are excluded from the locomotory cycle.  It is interesting that changes of this type in the movement of the limbs and body are almost identical in the Gerbillinae and Calomyscus bailwardi with an increased size of jumps.  Judging from published data, a similar pattern is found in Peromyscus, Conilurus and some other rodents (Bartholomew and Caswell, 1951).

Examining motion pictures of a jerboa's ricochet (Figure 38), we notice an interesting feature in the position of the body.  In the Dipodidae the trunk is kept nearly horizontal practically throughout the cycle of the ricocheting jump (Figure 39), that is, the trunk is maintained rigid and immobile as far as its structure allows, and all propulsive movements are effected due to the work of the legs.  In the kangaroos, in contrast to the Dipodidae, the body alters its position during the jump from almost vertical to horizontal (Badoux, 1965).  Due to the close resemblance of the type of movement in animals using the primitive ricochet, the immobility of the trunk in the jerboas and its mobility in the kangaroo, we can distinguish three forms of running in the ricocheting animals: primitive, rigid and supple ricochet.  However, there is a common feature for all these three forms, the fact that the size of the jump depends entirely on the work of the hind limbs. We therefore classify together mammals adapted to any form of the ricochet into one group of animals employing the metalocomotory mode of running.

These types of movement do not cover all the paths of cursorial specialization in the large group Rodentia; whereas, as mentioned, the most primitive rodents were probably small animals, adapted to the primitive ricocheting jump, their descendants, which adopted the most different biotopes, were able to take up other gaits.  Some took to climbing trees and adapted to jumping from branch to branch. We have pointed out that the working rhythms of the neuromuscular chains typical for the primitive ricocheting jump proved unsuitable when it became

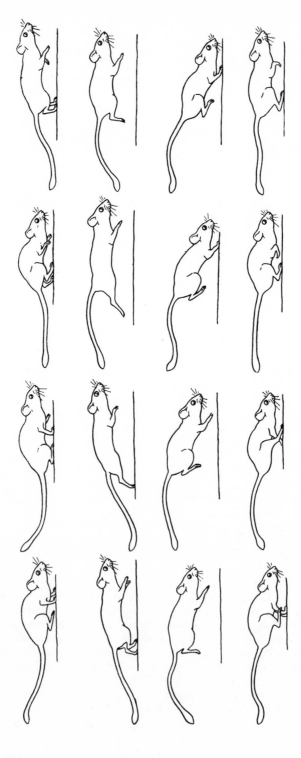

FIGURE 191.  Primitive ricocheting jump of the Asia Minor gerbil (Meriones blackleri)

necessary to calculate the size of jumps precisely, and so when the climbers went over secondarily to a terrestrial mode of life they retained the newly acquired rhythm of leg work and adopted a new asymmetrical gait, the gallop.   The transition of these secondarily terrestrial forms to life in the open brought about the appearance of adaptation to running and to digging burrows.   The prevalence of any of these trends of specialization depends on the choice of biotopes.   Specialization for fast running is more or less proportional to the need for scuttling from one burrow to another. Thus, with a continuous cover of vegetation, where temporary refuges could be dug close to one another, there was no need for deep specialization for swift running, but rather specialization for digging was required.   On the other hand, in semideserts and deserts, where large excursions had to be made in search of food, highly specialized runners appeared.   Because they retained the rhythm of leg work developed by the tree-climbing ancestors, the main asymmetrical gait of these animals became the gallop.   Adaptation to climbing is associated with active pulling up of the body by the forelimbs, and therefore the propulsive forelimb thrust is well expressed in these animals during the gallop.

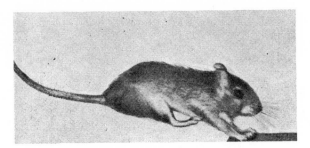

FIGURE 192.  Persian jird (M e r i o n e s   p e r s i c u s) about to land.
Photo by G.P.Gambaryan.

A study of the footprints of various representatives of the Sciuridae showed that the characteristic gait of these animals is the gallop.   But the tracks left by Sciuridae are very different from those of carnivores, ungulates and lagomorphs.   A characteristic feature of the gallop in these three groups is that with an increase in the speed of motion the stages of both extended and crossed flight are increased.   With a rise in the steepness of a jump (clearing obstacles), usually only the stage of extended flight is increased, while that of crossed flight is even diminished.   On the other hand, with the gallop of sousliks C i t e l l u s   c i t e l l u s   and   C i t e l l u s u n d u l a t u s   and two species of squirrels, the stage of crossed flight remains almost unchanged when there is a considerable change in the size of the stage of extended flight (Figure 195).   Thus in the red squirrel (S c i u r u s   v u l g a r i s) the stage of crossed flight does not change in size with an increase of extended flight from 50 to 110 cm.   Only with a further increase in the size of jumps does the stage of crossed flight also begin to be enhanced. An analogous pattern is seen in the sousliks. In the ground squirrel Citellus citellus the stage of crossed flight does not increase with an increment of extended flight from 20 to 60 cm. In the long-tailed Siberian souslik C. undulatus crossed flight does not change when extended flight increases from 20 to 90 cm, and only when the leap is further increased, if an obstacle is not being hurdled, does it increase slightly. It is only in the long-clawed ground squirrel Spermophilopsis leptodactylus that a full resemblance is observed in the type of tracks with those of the carnivores, ungulates and lagomorphs. With a 22-cm stage of extended flight crossed flight is equal to 7 cm, while with 40 cm of extended flight it increases to 19 cm (Figure 195).

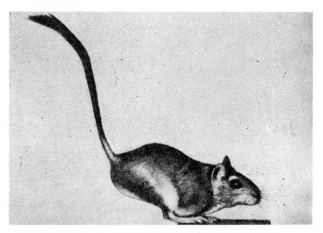

FIGURE 193. M.persicus landing. Photo by G.P.Gambaryan.

FIGURE 194. Calomyscus bailwardi about to land on four feet. Photo by R.G.Rukhkyan.

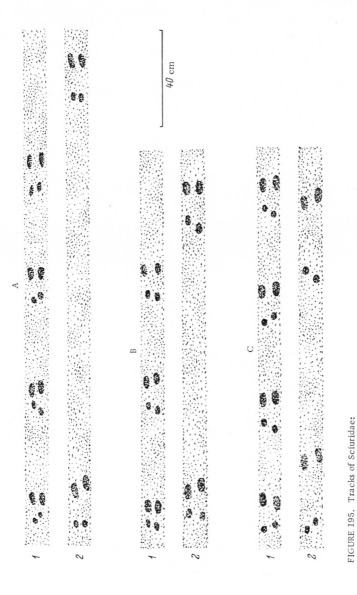

FIGURE 195.  Tracks of Sciuridae:

A) Sciurus vulgaris; B) Citellus citellus; C) Spermophilopsis leptodactylus. 1) slow, 2) fast gallop.

We studied motion pictures of the gallop in C. citellus (Figure 196) to ascertain the causes of this feature of the tracks of Sciuridae. After the end of the phase of hind support in the stage of free flight in this species and a squirrel (Figures 197 and 198), as in animals adapted to the primitive ricocheting jump, the hind limbs begin to be deflected forward. As a result, before the animal lands on the forepaws, the hind limbs are brought forward almost to the same extent during both small and large jumps. In S p e r m o - p h i l o p s i s   l e p t o d a c t y l u s (Figure 199) the stage of extended flight is better expressed. It is interesting to note that according to the relative force of the thrust of the forelimbs and hind limbs C. citellus is also more similar to the animals moving by means of the primitive ricocheting jump rather than to those using the gallop. Thus, the relative thrust in the ground squirrel during a jump of 41 cm was 2.14, while the impact on landing was 1.09, that is, according to these indexes the species stands closer to the gerbils than to the pika and weasel (cf. the data in Table 3).

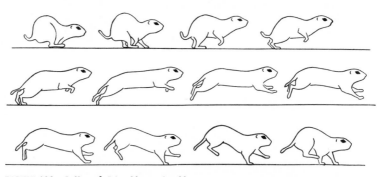

FIGURE 196. Gallop of C i t e l l u s   c i t e l l u s

A number of works have been written on the structural changes taking place in the locomotory organs in relation to different trends of specialization in the Sciuridae. Peterka (1937) attempted to link up the structural features of the organs of movement in these animals with characteristics of their mode of locomotion. Bohmann (1939) compared the movements of the squirrel, marmot and hare and described their skeleton and muscles. On the basis of this he tried to establish a connection between the changes in the locomotory organs of rodents and their adaptation to digging, climbing and running. Both these authors gave only a description of the organs of movement in a number of animals and data on their mode of life, believing these features to be mutually dependent. Sokolov (1964) studied the motion mechanics of ten species of Sciuridae which occur in the USSR. His findings seem rather unconvincing, and therefore we will discuss his material in comparison with our own.

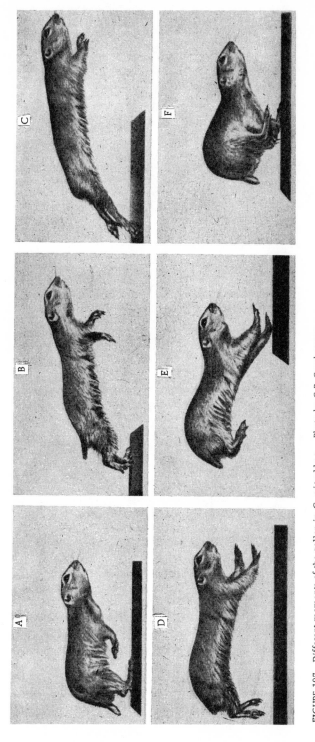

FIGURE 197.  Different moments of the gallop in C. citellus .  Photo by G.P. Gambaryan:

A,B) stage of hind support:  A) beginning,  B) end;  C—E) stage of free flight:  C) beginning,  D) middle,  E) end;  F) crossed flight, passing into the beginning of the stage of hind support.

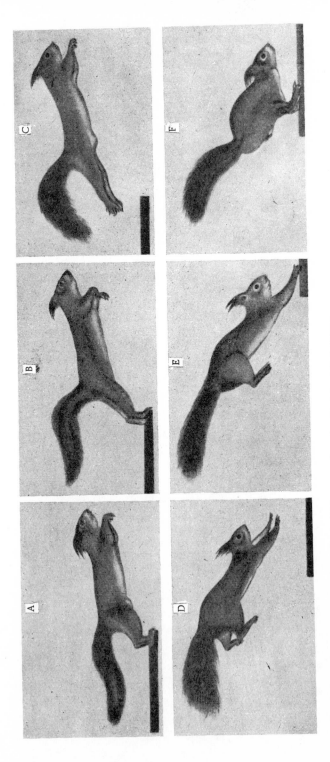

FIGURE 198.  Different moments of the gallop in S c i u r u s  v u l g a r i s .  Photo by R.G. Rukhkyan:

A, B) stage of hind support:  A) middle, B) end;  C—E) stage of free flight: C) beginning,  D) middle,  E) end;  F) crossed flight, passing into the stage of hind support.

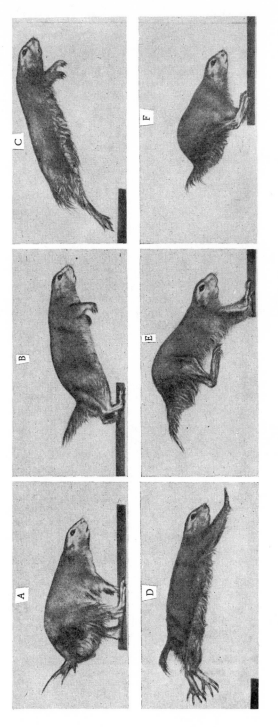

FIGURE 199.  Different moments of the gallop in Spermophilopsis leptodactylus.  Photo by R.G. Rukhkyan:

A, B) stage of hind support: A) beginning, B) end; C, D) stage of free flight: C) beginning, D) middle; E) stage of front support (beginning); F) crossed flight passing into the stage of hind support.

TABLE 27. The angle of departure as a factor of speed and the size of jumps in the Sciuridae

| Species | Extended stage of flight (cm) at a speed (m/sec) of: | | | | | Angle of departure | Maximum jumps (cm) |
|---|---|---|---|---|---|---|---|
| | 4.0 | 4.5 | 5.0 | 5.5 | 10.0 | | |
| Sciurus persicus ...... | 80 (3) | — | 110 (2) | 120 (3) | — | 13—15° | 285 (1) |
| S. vulgaris .......... | 70 (2) | 85 (4) | 90 (6) | 140 (11) | — | 12—15° | 290 (2) |
| Citellus citellus ...... | 50 (4) | 60 (7) | — | — | — | 8—9° | 80 (1) |
| C. undulatus ......... | 50 (3) | 60 (6) | 85 (1) | — | — | 8—10° | 120 (4) |
| Spermophilopsis leptodactylus ....... | — | 50 (1) | 55 (3) | 60 (4) | 80 (5) | 2°15'—7° | 100 (1) |

Note. The number of observations is given in parentheses.

Analyzing the running of Sciuridae, Sokolov filmed the Daurian souslik, long-clawed ground squirrel, squirrel and marmot on a treadmill with similar speeds of the belt (about 2 m/sec). The more or less equal and very low speeds of running which were obtained in connection with the experimental conditions were probably responsible for the inaccuracies in the description of the mechanics of running. Sokolov (1974) writes: "Hence, a squirrel makes a jump along a much less sloping trajectory than a souslik or marmot." To prove this theory he presents the flight trajectory of the Daurian souslik and squirrel. The trajectory of a moving body is conventionally depicted in the form of the line of movement of the center of gravity if it is not stipulated that any other point is being shown. Sokolov makes no such stipulation, and therefore we have to assume that he is showing the trajectory of the center of gravity. In this case, besides the frequent errors in the curve (absence of the influence of front and hind support), there are far more substantial errors. The point is that with the equal speeds obtained in accordance with the experimental conditions the squirrel made larger jumps than the souslik. But the size of the stage of free flight is directly proportional to sin 2 $\alpha$ (formula 3), and therefore the flight trajectory of the squirrel should theoretically be not less steep but steeper than that of the souslik. This hypothesis was checked by studying the running speeds and tracks of Sciuridae in nature. We often managed to combine observations of running speeds with investigation of tracks (Table 27), and we found that in running along the ground the red squirrel travels by bounds 70—140 cm long; if the animal is looking for food on the ground, it runs much more slowly, and also makes shorter jumps. In all we analyzed more than 1,000 cycles of galloping on the ground in the red squirrel.* In C. citellus the jumps did not generally exceed 60 cm even during panic-stricken scuttles from one burrow to another. The maximum jumps recorded were of 80 cm (Table 27). In C. undulatus the jumps were appreciably bigger than in C. citellus. Even jumps of 120 cm were recorded during obstacle clearing. When we investigated numerous

* We did not once record that a symmetrical gait was used in traveling along the ground. However, the use of symmetrical gaits has been occasionally observed (trot, walk) during observations of animals in captivity.

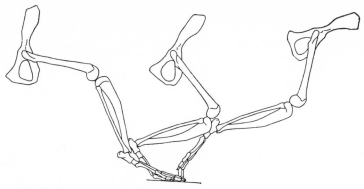

FIGURE 200. Diagram showing the movement of the hind limb skeleton in the phase of support in Citellus citellus

cycles (more than 1,000 for Spermophilopsis leptodactylus and over 700 for C. citellus) we noted the use of symmetrical gaits only occasionally, when the animals were right near the burrows. S. leptodactylus performs very small jumps when galloping. In Table 27 we give the results only of those track measurements which were done with simultaneous recording of the running speed as well. In calculating the angle of departure of the center of gravity from the measurements, it is assumed that the initial velocity of the center of gravity during each jump was equal to the average running speed. The least sloping trajectory was noted in S. leptodactylus and the steepest in the squirrels. Similar conclusions will be reached if we analyze the mode of life of the Sciuridae. For climbing squirrels, adapted to jumping from branch to branch, the possibility of making large leaps is very important, while speed on the ground is secondary. One-time large springs are best performed with a steep trajectory (at an initial speed of jumping of around 6 m/sec and an angle of departure of 45° the squirrel can jump 3.5 m along the horizontal; both the speed and the size of this jump come close to the maximum for squirrels). The long-clawed ground squirrel is adapted to sustained running on open sands. It encounters practically no obstacles along its path. A gently sloping trajectory of each jump is more favorable both for speed and for economy of running (see Chapter 3 for more details). The sousliks and marmots are inhabitants of steppes with well developed herbage. Bounding through the grass obliges them to make steeper jumps than S. leptodactylus. They jump 60—65 cm when running slowly (4.0 m/sec). Under these conditions the height of their jump is quite adequate for clearing the low herbage. However, running with the more economical gently inclined trajectory of the center of gravity during jumps is unsuitable for them. Hence, both a direct study of various Sciuridae in nature and an analysis of the demands made by their mode of life suggest that the least sloping trajectory during running is characteristic for the long-clawed ground squirrel and the steepest for the squirrels. A similar

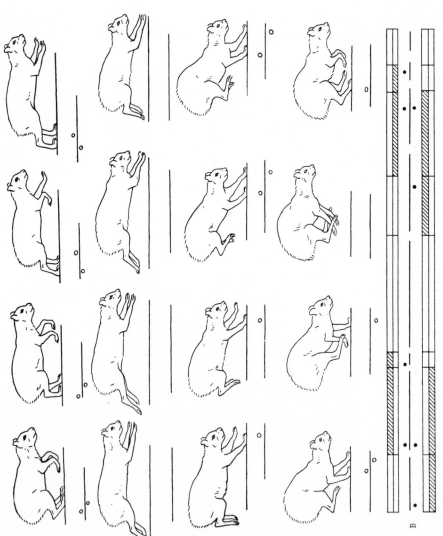

FIGURE 201.  Gallop of the agouti (Dasyprocta agouti) (A) and its support graph (B)

impression is gained if we examine the phases of the gallop in the Scuiridae (Sokolov, 1964, fig. 9).  Consequently, an analysis of the diagrams given by this author and the material obtained during a study of tracks and running speeds in nature contradict Sokolov's conclusions.

Looking at the work of the locomotory organs in the running cycle in the Sciuridae (Figure 200), we see that the length of each cycle depends on the activity of both the fore and the hind limbs.  Probably also of some significance for increasing the cycle are the flexor-extensor movements of the spine.  In fact, the mode of running of the Sciuridae is very similar to that of of the Felidae and Canidae, and may be termed the *dorsomobile dilocomotory* mode of running of rodents.

A study of motion pictures of the running (gallop) of the agouti shows that in the Dasyproctidae only the forelimbs and hind limbs work during the cycle, the spine remaining rigid (Figure 201).  This makes their running very similar to that of the ungulates.  We may therefore call their mode of running *dorsostable dilocomotory*.

STRUCTURAL FEATURES OF THE SPINE

We saw in the previous section that the distribution of the activity of the forelimbs and hind limbs and also that of the spine depends to a large degree in rodents on the mode and form of running.  During the primitive ricocheting jump, the forelimbs act exclusively to cushion the impact, the spine helping here too.  With the bipedal ricochet both the anterior section of the spine and the forelimbs are excluded from this function altogether.  In the running of Sciuridae and agoutis the forelimbs not only cushion impacts but also lend a second thrust, bringing about the crossed stage of flight.

The profound morphogenetic influence of the mechanics of movement of the three types of the metalocomotory mode of running is well illustrated by the morphofunctional features of the spine.  As already pointed out, during the primitive ricocheting jump a large shock-absorbing load is brought to bear on the anterior section of the spine.  The rigid ricochet is achieved through rigidity of the sacrolumbar region of the spine, which is subject to the largest bending moments.  At the same time in this region during the rigid ricochet, the impact imparted to the anterior section of the body must also be cushioned.  With the supple ricochet the considerable overall mobility of the spine promotes the necessary shock absorption.

Descriptions of the shock-absorption chains reflected in the sequence of movements of the joints on landing, in the structure and arrangement of the bone cells and the hyaline cartilage, in the structure of the thoracic region, and also in other morphological characteristics related to shock absorption have been given in a number of works (Lesgaft, 1905; Gambaryan, 1951; Gambaryan and Dukel'skaya, 1955; Egorov, 1955; Kummer, 1959a; etc.). We will therefore discuss here data on the structure of the shock-absorption chains whose features can be derived from the specifics of running inherent in each of the three groups of mammals adapted to various types of meta-locomotory running.

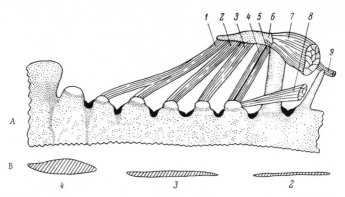

FIGURE 202. Diagram showing the structure of the shock-absorption organ of the spine in rodents:

A) organ of shock absorption; B) cross section of tendinous bands. 1—4) tendinous bands from triangular plate to spinous processes: 1) to cervical vertebrae IV and V, 2) to cervical vertebra VI, 3) to cervical vertebra VII, 4) to thoracic vertebra I; 5) supraspinous ligament; 6) triangular plate; 7) m.semispinalis; 8) m.multifidus; 9) supraspinous ligament from thoracic vertebrae III to II.

For animals adapted to the primitive ricocheting jump the most characteristic of the shock-absorption chains is the organ of shock absorption of the spine (Gambaryan, 1951), consisting of a series of elements united into a single functional whole. The components of this organ are (Figure 202): 1) the strongly developed spinous process of thoracic vertebra II; 2) a triangular plate, movably connected with the apex of this spinous process; 3) a number of tendinous fibers which stretch from the ventral surface of the triangular plate to the spinous processes of the first thoracic vertebra and the last cervical vertebrae; 4) the supraspinous ligament, which runs along the apexes of the spinous processes on the thoracic vertebrae; 5) the bundles of m. semispinalis dorsi, which end on the triangular plate.

For a better understanding of the function and origin of the shock-absorption organ of the spine in rodents it is useful to examine the more common principle of action of spinous processes. Brovar (1935, 1940) mentions that spinous processes are one of the components of the spine's organ of rigidity working to resist dorsal flexion. The supraspinous ligament is the second component of this organ of rigidity. Both parts work on the principle of a lattice girder. A girder is a structure in which a force tending to bend a body, i. e., acting to break it, is resolved into two components, one of which compresses the beams of compression and the other expands the beams of expansion (Figure 98). The work of the girder is illustrated by the following scheme: in beam AB force $F$ acts to break it, while in the girder ABCD the same force $F$ works to compress beams AC and BD and to expand beam CD (Figure 98). The spinous processes with the supraspinous ligament are constructed according to the girder principle, the spinous processes being the beams of compression and the supraspinous ligament the beam of expansion. This interpretation of the function of the spinous processes enabled Brovar (1935, 1940) to explain their strong development in the withers in mammals, to examine the part they play in fish, etc.

More recently a similar interpretation has been given on a mathematical basis (Kummer, 1959a, 1959c). The rigidity requirements of the spine are opposed to the shock-absorbing requirements, since in the first case all the components of the girder work to decrease possible changes of its configuration, while in the second case an arrangement which can smoothly change its parameters is needed. Despite this, the organ of shock-absorption of the spine probably arises from the typical withers, which act as the organ of rigidity of the spine. Proof of such an origin of the shock-absorbing apparatus in rodents adapted to the primitive ricocheting jump can be seen in a number of forms in which we can trace the full transition from typical withers to an organ of shock absorption of the spine. A good example of such a comparative anatomical series is the structure of the anterior section of the spine in various species of the subfamily Cricetinae. Thus, the hamster Mesocricetus brandti has typical withers consisting of a number of spinous processes of the thoracic vertebrae with a well developed supraspinous ligament (Figure 203). The best developed spinous process is that of thoracic vertebra II; the spinous processes in front of and behind this one are somewhat more weakly developed. That of the first thoracic vertebra is much weaker than that of the second, while those of vertebrae III and IV are just a little smaller. The spinous process of the last cervical vertebra is directed vertically, while the processes of the thoracic vertebrae are slightly sloped backward, the slope increasing from the first to the ninth or tenth vertebra. The supraspinous ligament is attached to the apexes of the thoracic spinous processes (Figure 203). It is best developed on the section from the last cervical vertebra to thoracic vertebrae V—VI. Such a structure of the anterior part of the thorax may serve as an example of typical withers, fulfilling the role of a spinal organ of rigidity which essentially does not differ from the analogous formations described for the Ungulata (Brovar, 1935, 1940). In contrast to Mesocricetus brandti, in Cricetulus migratorius the spinous process of thoracic vertebra II projects markedly above the level of the spinous processes of the third and following thoracic vertebrae and above the level of the process of the first vertebra (Figure 203). Another difference between these species is that in C. migratorius the backward slope of the spinous process of thoracic vertebra III is more pronounced than that of the process of the second vertebra. In connection with this, the supraspinous ligament, which runs between the spinous processes of thoracic vertebrae II and III, is markedly longer. Although the ligament, which consists of fibrous connective tissue, has a low coefficient of flexibility, its elongation still makes for a substantial degree of change of the parameters in the anterior part of the thorax. The reorganization described evidently diminishes rigidity in the region of the withers in C. migratorius as compared with M. brandti. Moreover, in the latter species, on the apex of the spinous process of the second thoracic vertebra a fibrous formation is attached which probably originates from the slightly thickened part of the supraspinous ligament that originally stretched between the spinous processes of thoracic vertebrae I and II but is now inserted in this species only on the spinous process of the second thoracic vertebra. This formation has the shape of a simple triangular plate which on the ventral side receives several weakly expressed tendinous bands stretching to the spinous processes of the first thoracic and the last two (more rarely three) cervical vertebrae.

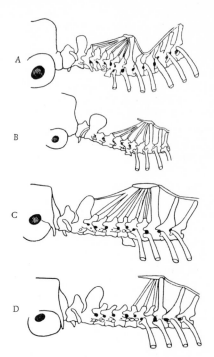

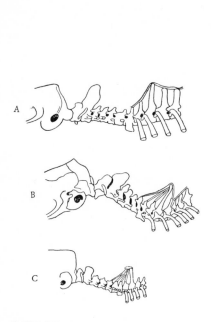

FIGURE 203.  Spine in the anterior part of the thoracic region and neck in the Cricetinae:

A) Mesocricetus brandti; B) Cricetulus migratorius; C) Calomyscus bailwardi.

FIGURE 204.  Spine in the anterior part of the thoracic region and neck in the Gerbillinae (A, B) and Muridae (C, D):

A) Meriones vinogradovi; B) Gerbillus gleadovi; C) Rattus norvegicus; D) Conilurus sp.

Thus, whereas in Mesocricetus brandti the spinous processes together with the supraspinous ligament make up a typical rigidity organ of the spine, in Cricetulus migratorius it is transformed into a shock-absorbing organ of the spine in the primitive form.  Subsequent perfection of the shock-absorption organ is manifested in a reorganization of the region both behind and in front of thoracic vertebra II.  Remembering that this organ probably appears quite independently in various groups of rodents, the difference in the paths of its perfection becomes understandable.  For instance, the changes in the region of the spine behind the second thoracic vertebra develop in at least two ways.  In the first case (Gerbillinae, Cricetinae, Figure 203), the spinous processes of thoracic vertebrae III and IV are greatly reduced in height, leading to an increase in the length of the supraspinous ligament and thereby to greater possibilities for the parameters of this region to be altered.  In the second case (Muridae, Figure 204 and the coypu) the reduced size of the spinous processes of the third thoracic vertebra onward is negligible, but the backward slope of the processes is markedly increased.  The different direction of the spinous processes of thoracic vertebrae II and III also lengthens the supraspinous ligament and enhances mobility in the anterior section of the thorax. In both cases the spinous process of the second thoracic vertebra is markedly enlarged and becomes vertically directed, and to its apex is attached the increasingly differentiated triangular plate from which the tendinous bands stretch to the spinous processes of the thoracic and cervical vertebrae. In Rattus norvegicus, apart from the bands of the triangular plate, the supraspinous ligament (which is quite undeveloped in hamsters and gerbils) runs to the spinous process of the first thoracic vertebra. As a result, this process is reached by two independent tendinous bands. In the most complete form, up to six tendinous bands stretch to the triangular plate in this species.

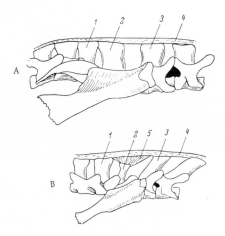

FIGURE 205. Sacrolumbar region in the Cricetinae:

a) Mesocricetus brandti; B) Calomyscus bailwardi. Spinous processes: 1, 2) of sacral vertebrae II and I; 3) of last lumbar vertebra; 4) supraspinous ligament; 5) brace from spinous process of sacral vertebra I to supraspinous ligament.

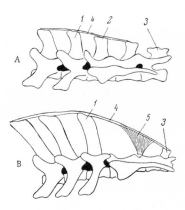

FIGURE 206. Sacrolumbar region in the Norway rat (Rattus norvegicus) (A) and William's jerboa (Allactaga williamsi) (B):

Spinous processes: 1) of last lumbar vertebra; 2) of first sacral; 3) of third sacral vertebra; 4) supraspinous ligament; 5) tendinous brace from the region of sacral vertebra II to the supraspinous ligament.

With specialization for the primitive ricocheting jump, the shock-absorption organ of the spine gradually becomes perfected, but subsequently this organ undergoes a secondary simplification. Several representatives of various groups of rodents which are specialized for the primitive ricocheting jump confirm the above. As we have already mentioned, the migratory hamster has a primitive shock-absorption organ of the spine, the structure of which is quite easily derived from the typical withers characteristic for Mesocricetus brandti. In Meriones vinogradovi we see a very perfect organ of shock absorption of the spine, while in the more specialized Gerbillus gleadovi (Figure 204) it is already secondarily simplified. The same can be found by studying representatives of the Muridae. Thus, in the short-tailed bandicoot rat, Nesokia indica, which is little specialized for the primitive ricocheting jump, the organ of shock absorption differs little from that in Cricetulus migratorius. In the Norway rat, Rattus norvegicus, which is more specialized for this type of locomotion, this organ has reached a high degree of perfection (Figure 202 and 204), and in the coypu Conilurus sp. (Figure 204), which is much more specialized than R. norvegicus for the primitive ricocheting jump, it is secondarily simplified. This perfection and then simplification can find their explanation in the mechanics of the primitive and specialized ricocheting jumps. Specialization for the primitive ricocheting jump goes along within an increase in the size of jumps. Naturally, the force of the impact which must be cushioned increases in proportion to the degree of specialization for the primitive ricochet. However, at the same time as the jump increases in size, there is an increase in the degree of forward deflection of the hind limbs. The forelimbs absorb the shock until the hind limbs have reached the ground. With increased deflection of the hind limbs the time interval between the landing of the fore and the hind limbs decreases, and as a result, shock absorption comes to be more and more evenly distributed between the front and hind legs. Furthermore, when hind limb deflection reaches a certain degree in the stage of free flight, deflection in the air occurs now on account of the bending of the spine, whereby the center of gravity of the posterior part of the body comes to be placed lower than that of the anterior part: the hind limbs reach the ground even sooner, at which point, of course, the shock-absorbing role of the forelimbs is gradually reduced to nothing. Whereas in the initial stages of specialization for the primitive ricochet only the jump itself is lengthened, subsequently less time is also taken for the hind limbs to be deflected forward. For example, animals of the genus Gerbillus can progress only on the hind limbs when the length of the jump is as small as 25 cm (Kirchshofer, 1958), whereas the Persian jird, Meriones persicus, would for that need a jump of at least 1 m and Meriones meridianus one of at least 70—75 cm. Naturally, at the beginning the organ of shock absorption is perfected according to the degree of specialization for the primitive ricochet, when the degree of specialization is proportional to the size of the jump, but later, when the role of the hind limbs as shock absorbers is increasingly enhanced, the organ is simplified.

Along with these changes in the thoracic region of the spine, transformations take place in the sacrolumbar region. Whereas in the thoracic section there is a gradual perfection of the organ of shock absorption and then a secondary simplification, in the sacrolumbar region reorganization proceeds only in the direction of progressive perfection. To trace this process in the shock-absorption apparatus of the sacrolumbar region it is easiest to compare different representatives of rodents which are variously specialized for the primitive ricocheting jump. In Mesocricetus brandti the spinous processes of the last lumbar vertebrae and first sacral vertebra are almost of the same size and parallel to each other (Figure 205); the same picture is observed in Cricetulus migratorius, which is also primitively adapted for the ricocheting jump. But in

Calomyscus bailwardi the spinous process of the second sacral vertebra grows out in the caudal direction while that of the last lumbar vertebra grows out in the cranial direction (Figure 205).  A sharp reduction is observed in the size of the spinous process of the first sacral vertebra; the supraspinous ligament passes directly from the spinous process of the second sacral vertebra to the last lumbar vertebra.  From the spinous process of the first sacral vertebra to the supraspinous ligament, on the other hand, there stretches a fanwise widening ligament.  A similar change in the structure of this region is seen in the Muridae which are specialized for the primitive ricocheting jump.  Thus, in Rattus norvegicus we see the very beginning of divergence in the direction of growth of the spinous processes of the last lumbar and first sacral vertebra (Figure 206).  However, in this species there is still a cranial slope of the spinous process of the first sacral vertebra, that is, a slope in the same direction as that of the spinous process of the last lumbar vertebra (Figure 206).  In Conilurus sp., as in Calomyscus bailwardi, there is a well defined divergence in the directions of growth of the spinous processes of the first sacral and last lumbar vertebrae.  There is also an analogous process of reduction in the length of the spinous process of the first sacral vertebra and a similar transfer of the supraspinous ligament from the spinous process of the last lumbar to the second sacral vertebra. This divergence in the direction of growth of the spinous processes and the gradual reduction of the spinous process of the first sacral vertebra have a double purpose.  On the one hand, any connective tissue, including the little-extensible fibrous tissue of the supraspinous ligament, possesses some degree of elasticity, and therefore with an increase in length of the supraspinous ligament the possibility for flexor-extensor movements in the sacro-lumbar region is also enhanced.  On the other hand, despite the better possibility for flexion and extension in this region, these movements remain quite limited, since the two spinous processes with the poorly elastic supraspinous ligament still represent a girder of rigidity.*

Specialization for the rigid ricochet demands adaptation that will maximally increase the rigidity of the sacrolumbar region of the spine, and yet the very large bending loads arising during headlong and large leaps call for improved shock-absorbing properties of this region.  Naturally, the above-described process of an improved shock-absorbing role of the sacro-lumbar region has to be further developed in animals adapted to the rigid ricochet.  As a result of this, in the jerboas (Figure 206) we observe a marked enlargement of the spinous processes of the last lumbar vertebrae. At the same time they show a pronounced forward slope, whereby the spinous process of the penultimate lumbar vertebra is sloped more forward than that on the last vertebra.  In view of this the supraspinous ligament is lengthened not only between the loin and the sacrum, but also in the caudal part of the actual loin.  In the sacrum the first two, sometimes three, spinous processes are reduced, and the supraspinous ligament stretches between the last lumbar and the third or fourth sacral vertebrae.  From the first two or three spinous processes of the sacral vertebrae tendinous bands extend

* Rigidity is associated with a greater abruptness of impact and shock absorption with a lesser abruptness, but nevertheless the progressive perfection of the organ of rigidity of the lumbar region is also associated with an improvement of its shock-absorbing properties.

fanwise to this ligament.    When the animal is at rest these bands promote partial give of the supraspinous ligament, while during jumps they act as a brace (the brace principle is discussed in detail in Chapter 3).    This improves conditions for shock absorption in the sacrolumbar region (the limits of elastic stretching are widened) and in addition, the conditions for rigidity are also enhanced, as tension of these fanlike bands leads to better resistance of the supraspinous ligament to straightening, this resistance being proportional to the tangent of the angle of sag at the point of attachment of the ligaments leading from the spinous processes of sacral vertebrae I and II to the supraspinous ligament.

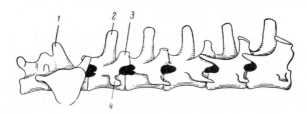

FIGURE 207.  Sacrolumbar region of the kangaroo (Macropus rufus):

1) spinous process of first sacral vertebra;  2) spinous process of last lumbar vertebra;  3) articular process;  4) transverse process.

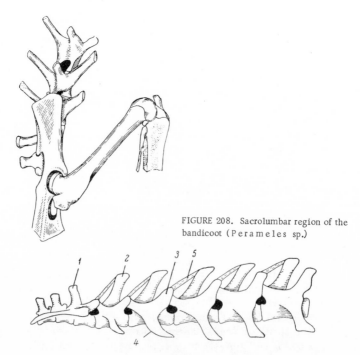

FIGURE 208.  Sacrolumbar region of the bandicoot (Perameles sp.)

FIGURE 209.  Sacrolumbar region of the agouti ( Dasyprocta agouti):

1) spinous process of first sacral vertebra;  2) spinous process of last lumbar vertebra;  3) cranial articular process;  4) transverse process;  5) tendinous attachment from articular to spinous process.

With specialization for the supple ricochet the structure of the sacro-lumbar region, unlike the changes associated with specialization for the rigid ricochet, develops by way of increasing mobility. Thus, in the kangaroo the spinous processes of the last lumbar and first sacral vertebrae are small, narrowed at the apex, and directed parallel to each other (Figure 207). The supraspinous ligament is absent in this region, and the interspinal muscle, which is almost fully reduced in the Dipodidae, is well developed here. The presence of a readily extensible muscle instead of a poorly extensible fibrous supraspinous ligament lends the possibility of making flexor-extensor movements of a broader amplitude in the sacro-lumbar region.

It is interesting to observe that in P e r a m e l e s, another marsupial highly specialized for the bipedal ricochet, the skeleton of the sacrolumbar region (Figure 208) resembles in structure and form that of the jerboas rather than that of the kangaroo. It may therefore be assumed that the Marsupialia became specialized not only for the supple but also for the rigid ricochet.

The thoracic section of the spine in the Sciuridae and Dasyproctidae forms typical withers, in which the spinous processes increase in size from the first to the third or fourth vertebra and then diminish again in the lumbar direction. The apexes of the spinous processes of the lumbar vertebrae are markedly widened anteroposteriorly, especially in some species of sousliks. Nevertheless, due to the relatively small size of the spinous processes, vertical flexures of the spine in this region are not rigidly limited in the Sciuridae. The rigidity of the spine in the agouti is apparently promoted by both the widening of the spinous processes and the activity of the interspinal muscles. Of the special formations improving spinal rigidity in this animal we should mention the strongly developed articular processes, the apexes of which almost reach the level of the apexes of the spinous processes. The ligaments between the articular and the spinous processes which connect two adjacent lumbar vertebrae greatly restrict both vertical and lateral flexures of the spine (Figure 209). This distinctive arrangement apparently makes for adequate spinal rigidity in the agouti, like the similar arrangement in the Ungulata.


TRANSFORMATIONS OF THE APPENDICULAR SKELETON IN
CONNECTION WITH ADAPTATION OF RODENTS AND
KANGAROOS TO DIFFERENT MODES AND
FORMS OF RUNNING


### Adaptation to the metalocomotory mode

As we noted above, the hind limbs fulfill the main locomotory function in all three forms of the metalocomotory mode of running. Therefore, as cursorial specialization is enhanced they should promote an increase in speed and in the length of jumps. Elongation of the limbs is useful both for speed and for large jumps (Table 28). The influence of cursorial specialization on the lengthening of the limb segments is clearly traced in many groups of rodents. Analyzing the data in Table 26 and the data on the biology of rodents presented in the first section of this chapter, and also drawing on published data on the biology of forms not studied here, we can make an approximate assessment of the degree of specialization for different types of ricochet in rodents and kangaroos in a six-point system (Table 28). From all four groups of animals investigated we took both little-specialized forms, classified by point 1, and the most highly specialized. Only in the group of gerbils are the differences in the degree of cursorial specialization of the extreme forms markedly smaller than in the other groups. Nevertheless, a comparison of representatives of all these forms making up anatomical series enables us to judge the trend of change in the length ratios of the limb segments in connection with this specialization. These changes are quite clearly manifested even in the Gerbillinae.

Thus, if we compare the indexes of the segments in different Cricetinae, we see that in Calomyscus bailwardi in comparison with Phodopus sungorus the foot index is increased 1.80, the tibia index 1.31, and the

TABLE 28. Changes in the length ratios of the limb segments depending on specialization for different forms of the ricocheting jump in mammals (% of the total length of the lumbar and thoracic regions of the spine)

| Species | Relative size | | | | | | | | Degree of specialization (six-point scale) |
|---|---|---|---|---|---|---|---|---|---|
| | femur (a) | tibia (b) | foot (c) | total (a+b+c) | shoulder (d) | fore-arm (e) | hand (f) | total (d+e+f) | |
| Mesocricetus raddei | 40.2 | 36.5 | 26.2 | 102.9 | 32.9 | 28.1 | 18.0 | 79.0 | 1 |
| M. brandti | 39.0 | 34.6 | 23.6 | 97.2 | 32.0 | 30.0 | 15.6 | 77.6 | 1 |
| Phodopus sungorus | 38.2 | 45.7 | 31.6 | 115.5 | 33.6 | 36.6 | 20.7 | 90.9 | 1 |
| Cricetulus migratorius | 39.6 | 44.0 | 35.3 | 118.9 | 34.4 | 33.6 | 18.7 | 86.7 | 2 |
| Calomyscus bailwardi | 46.6 | 59.7 | 56.9 | 163.2 | 33.6 | 36.0 | 23.7 | 93.3 | 3—4 |
| Nesokia indica | 33.7 | 40.2 | 37.8 | 111.7 | 30.0 | 27.3 | 21.4 | 78.7 | 1 |
| Rattus norvegicus | 41.0 | 44.3 | 40.2 | 125.5 | 29.0 | 26.2 | 18.0 | 73.2 | 2 |
| Conilurus sp. | 50.2 | 81.3 | 78.8 | 210.3 | 27.4 | 33.2 | 21.3 | 81.9 | 4—5 |
| Meriones blackleri | 41.0 | 51.0 | 46.6 | 138.6 | — | — | — | — | 3—4 |
| M. vinogradovi | 42.0 | 51.0 | 45.2 | 138.2 | — | — | — | — | 3 |
| M. persicus | 41.8 | 51.8 | 46.9 | 140.5 | — | — | — | — | 3—4 |
| M. meridianus | 41.0 | 52.0 | 48.6 | 141.6 | 29.0 | 30.0 | 18.0 | 77.0 | 3—4 |
| Gerbillus gleadovi | 41.6 | 68.4 | 76.8 | 186.8 | 29.2 | 36.0 | 27.0 | 92.2 | 4—5 |
| G. pyramidum | 43.0 | 71.0 | 79.6 | 193.6 | 30.2 | 37.3 | 28.0 | 95.5 | 4—5 |
| Sicista betulina | 34.2 | 45.8 | 52.3 | 132.3 | 27.1 | 31.0 | 26.5 | 84.6 | 1 |
| Pygerethmus platyurus | 53.8 | 78.3 | 85.7 | 217.8 | 25.0 | 31.2 | 15.6 | 71.8 | 5 |
| Alactagulus acontion | 56.2 | 81.3 | 97.6 | 235.1 | 23.1 | 25.9 | 15.6 | 64.6 | 6 |
| Allactaga elater | 64.0 | 95.0 | 112.5 | 271.5 | 25.0 | 30.5 | 19.5 | 75.0 | 6 |
| A. jaculus | 56.2 | 81.3 | 97.6 | 235.1 | 23.1 | 25.9 | 15.6 | 64.6 | 6 |
| Eremodipus lichtenschteini | 57.0 | 84.5 | 98.0 | 239.5 | 23.8 | 31.4 | 15.7 | 70.9 | 6 |
| Macropus bennetti | 44.0 | 68.2 | 52.5 | 164.7 | 28.2 | 37.0 | 13.4 | 78.6 | 6 |
| M. rufus | 47.5 | 77.5 | 56.8 | 181.8 | 29.4 | 35.8 | 14.2 | 79.4 | 6 |
| M. agilis | 52.2 | 83.5 | 72.0 | 207.7 | 29.6 | 37.5 | — | — | 6 |

femur index 1.22 times, so that the overall length of the limb is 1.41 times greater. Mesocricetus raddei and M. brandti are very similar to P. sungorus in depth of specialization. Their running speed attains 2.0, maximum 2.5 m/sec. In connection with the larger overall size of the two species of Mesocricetus, the absolute size of their limbs is also greater, and this leads to a relative decrease in the average force developed in these animals in comparison with P. sungorus. Hence, according to the indexes of both tibia and foot, Phodopus undergoes more change than Mesocricetus. Gerbils show considerably smaller differences in depth of cur-

sorial specialization than hamsters. Various species of Meriones are
actually not much less changed than Gerbillus pyramidum with
regard to their specialization for the ricocheting jump. Nevertheless,
even in comparison with the Persian jird (Meriones persicus) in
Gerbillus pyramidum the foot is 1.82 times larger, the tibia
1.46 times, the femur 1.12 times, and the relative length of the whole limb
1.59 times. A much more pronounced difference in depth of specialization
for the ricocheting jump is observed in different species of the family
Muridae, among which Nesokia indica moves by means of the primitive
ricocheting jump with the smallest jumps and Conilurus sp. virtually
goes over to the bipedal ricochet. Therefore, also the degree of difference
between Conilurus sp. and Nesokia indica is much greater than
between the Cricetinae and Gerbillinae. In Conilurus sp. as compared
with Nesokia indica the foot is 2.08 times larger, the tibia 2.03 times,
the femur 1.49 times and the overall length of the limb 1.89 times greater.

The most substantial differences in the depth of specialization for the
ricocheting jump among the animals in Table 28 are observed among the
Dipodidae: the Sicistinae are characterized by a little-differentiated primitive ricocheting jump, while the five-toed and three-toed jerboas are some
of the most highly specialized rodents where the bipedal ricochet is concerned. It is therefore not surprising that in Allactaga elater the
foot is 2.15 times, the tibia 2.07 times, the femur 1.54 times and the overall relative length of the limb more than 2.04 times larger than in Sicista betulina.

It is noteworthy that in the kangaroos, highly specialized for the supple
bipedal ricochet, the relative size of the foot is much smaller than in the
Dipodidae, while the indexes of the tibia are sometimes, on the contrary, larger.
The femoral indexes are also smaller in the kangaroo than in the jerboas.

The general tendency toward a lengthening of the hind limb segments
with specialization for all types of the ricocheting jump is manifested almost uniformly in all the systematic groups of mammals listed in Table 28.
However, in the small animals (Cricetinae, Gerbillinae, Muridae, Dipodidae)
it is the foot which is primarily lengthened, whereas in the kangaroo it is
the tibia. Since elongation of the hind limbs is in general useful (formula (9)),
it would seem not to matter on account of which segment it is achieved.
However, the concentration of the force muscle groups on the proximal
segment of the limb in the region of the femur leads upon elongation of the
distal segments, to more effective utilization of the limb levers for increasing the speed of motion, since in these conditions the limb acts on the
principle of a speed lever. In the phase of support the limb rotates about
the point of support, and if we draw the appendicular skeleton of a hamster,
jerboa and kangaroo (Figure 210) making the femur of equal length, we see
that the points of application of force of the main muscles, the extensors of
the hip and knee joints, are distributed near the point of support in the hamster and far from it in the jerboa. It thus follows from the figure that in
the jerboa the hind limb skeleton above all fulfills the role of levers gain-
ing in speed. In the kangaroo, in connection with the smaller relative size
of the foot, the hind limb works to a lesser degree than in the Dipodidae on
the principle of levers of speed.

As pointed out earlier, an increase in the body mass causes a marked
increase in the loads tending to break the foot. An increase in these loads
takes place in proportion to the increase in the length of the foot and the
body mass, and so in the great gray kangaroo (Macropus giganteus),

which has a huge body mass, the foot had to be secondarily shortened.   An
analogous change in the length of the foot, caused by an increase in body
mass, is observed in the Artiodactyla, Perissodactyla and Carnivora (see
Chapter 4).

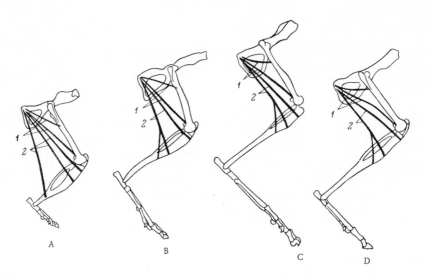

FIGURE 210.  Types of limbs of ricocheting animals, built for force and for speed:

A) force type in Mesocricetus brandti; B) intermediate type in Rattus norvegicus;   C, D) speed
type: C) in Allactaga williamsi, D) Macropus rufus. 1) short postfemoral, 2) long postfemoral
muscles.


    The appreciable difference in the absolute dimensions of the animals
studied, kangaroos and rodents, means that a simple comparison of the
relative size of the limb segments becomes inadequate when assessing their
adaptive changes.   The difference in the absolute running speeds and sizes
of jumps in these animals is not at all proportional to the increase in their
absolute size (weight and the length of the legs and of the body), as is also
well illustrated by the jump index (Table 26).   However, the influence we
have noted of an increased body mass on the relative size of the foot is
also confirmed when we take the absolute indexes into account.   A kangaroo
weighs about 350 times more than a jerboa (Table 26), but there is only a
fivefold difference in the absolute length of the jump in these animals and
a tenfold difference in the length of the legs.   Remembering that the force
developed by the limb in jumping increases in proportion to the increase in
the length of the limb and increases in proportion to the body mass, the
loads on the limb must be at least 100 times greater in the kangaroo than
in the Dipodidae.
    In rodents adapted to the primitive ricocheting jump the exclusively
shock-absorbing role of the forelimbs creates special conditions which
have an effect on the change of the length of the segments.   The average
force of shock absorption (p. 70) changes in inverse proportion to the

duration of the phase of landing on the forefeet.  The duration of this phase
depends partly on the length of the leg, as with an increase in the size of the
segments of the forelimb skeleton there is an increase in the length of the
path traversed by the animal's center of gravity in the phase of landing on
the forefeet.  All the rodents studied which are adapted to the primitive
ricocheting jump land either on the ends of the digits or, much more often,
on the base of the forepaw with the forefeet stretched forward.  In these
conditions the yielding extension of the wrist joint and flexion of the elbow
joint allow the animal to smoothly change the direction of movement of the
center of gravity from sloped down to horizontal and even upward.  The
lowering of the body between the legs upon the stretching of the pectoral
muscles and m. serratus ventralis is also of shock-absorbing value.
Because the body descends between the shoulder joints, its flexion is use-
less for cushioning the impact.  We may therefore expect that specialization
for the primitive ricocheting jump will be reflected in a certain lengthening
of the hand (in animals which land on the ends of the digits) and a lengthening of the forearm (in species which land on the forepaw).  Indeed, if we
compare Calomyscus bailwardi with the other Cricetinae, we see
that it is the hand which is primarily lengthened and then the forearm.  In
Conilurus sp. in comparison with other Muridae the forearm is enlarged,
and in Gerbillus pyramidum in comparison with other gerbils both
the forearm and the hand are larger.  The relative size of the forelimb segments in mammals adapted to the bipedal ricochet (rigid and supple) in fact
differ very little from those in rodents little specialized for running
(Table 28).  Hence, the widespread belief that the forelimbs are reduced in
connection with the transition to the bipedal ricochet must be considered
to be based on misunderstandings, which apparently stem from formal comparisons of the forelimb skeleton with the skeleton of the overdeveloped
hind limbs.

### Adaptation to the dorsostable and dorsomobile modes

Adaptation to fast running in Dasyprocta agouti (up to 60 km/hr) and
Spermophilopsis leptodactylus (up to 40 km/hr) on the one hand
and to large jumps in squirrels (to 3.0 m along the horizontal) on the other
leads to an increase in the length of the limb segments (Table 29).

Jumping onto a far-away branch enables the squirrel to shake off pursuers quickly, in connection with which it has adapted to large springs.
Preservation of this type of locomotion during running along the ground
leads to progression by means of large leaps with a steep trajectory* of
movement of the center of gravity.  Running in the open, on the other hand,
to which Spermophilopsis leptodactylus is adapted, results in
gently sloping movement of the center of gravity during each jump.  Still,
both during high-speed running with a little-inclined trajectory, and as a
result of adaptation to one-time jumps with a steep trajectory, elongation of
the limbs arises; in both cases it is profitable (formulas (6) and (9)).

* According to the observations conducted by O.V. Egorov (oral communication), bounds through loose snow
  are performed along a flat trajectory in squirrels in Yakutia.

TABLE 29. Changes in the length ratios of the limb segments upon specialization for the dilocomotory dorso-stable and dorsomobile modes of running in rodents (% of the total length of the lumbar and thoracic regions of the spine)

| Species | Relative size | | | | | | | | Modes and forms of dilocomotory running, degree of cursorial specialization |
| --- | --- | --- | --- | --- | --- | --- | --- | --- | --- |
| | femur (a) | tibia (b) | foot (c) | total (a+b+c) | shoulder (b) | forearm (e) | hand (f) | total (d+e+f) | |
| Dasyprocta agouti | 37.1 | 43.4 | 43.8 | 124.3 | 31.4 | 27.6 | 18.0 | 77.0 | Dorsostable, cursorial |
| Sciurus vulgaris | 46.5 | 52.3 | 51.8 | 150.6 | 34.4 | 32.0 | 29.3 | 95.7 | Dorsomobile, saltatorial, — highly specialized |
| S.persicus | 43.5 | 47.0 | 48.0 | 138.5 | 32.3 | 30.2 | 28.7 | 91.2 | |
| Spermophilopsis leptodactylus | 44.8 | 46.7 | 48.2 | 139.7 | 32.2 | 29.4 | 28.2 | 89.8 | Dorsomobile, cursorial |
| Citellus undulatus | 35.0 | 35.8 | 34.6 | 105.4 | 28.5 | 24.0 | 19.4 | 71.9 | |
| C.pygmaeus | 34.7 | 34.4 | 32.0 | 101.1 | 31.0 | 29.2 | 18.8 | 79.0 | |
| C.relictus | 34.5 | 33.8 | 32.8 | 101.1 | 26.8 | 23.9 | 19.2 | 69.9 | Dorsomobile, little specialized |
| C.citellus | 34.5 | 33.1 | 30.4 | 98.0 | 27.6 | 24.0 | 21.0 | 72.6 | |
| Marmota bobac | 37.6 | 37.2 | 34.2 | 109.0 | 37.0 | 29.6 | 22.0 | 88.6 | |
| M.camtschatica | 37.8 | 37.6 | 35.4 | 110.8 | 34.1 | 27.8 | 24.9 | 86.8 | |

The relative length of the limbs is greater and the running speed lower in Sciurus vulgaris than in Spermophilopsis leptodactylus (Table 27). Analogous cases have been examined above taking examples of mammals in other groups. For example, ungulates and lagomorphs which live in forest and shrub do not run as fast as closely related species which inhabit the open spaces. However, the relative length of the limbs is smaller in the species which live in the open. The reason for this difference in the relative length of the limb segments in the Ungulata and Lagomorpha is the same as that governing the ratio of the segments in the Sciuridae. In the forest-dwelling ungulates and lagomorphs, just as in the squirrels, jumps on the run are performed along a steeper trajectory than in forms living in the open.

TRANSFORMATIONS OF THE LIMB MUSCLES IN
CONNECTION WITH SPECIALIZATION OF
RODENTS AND KANGAROOS FOR DIFFERENT
MODES AND FORMS OF RUNNING

Adaptation to different forms of the metalocomotory mode

In general, the work of muscles for flexing and extending the joints in the hind limbs is similar in all mammals. Specialization for the primitive and bipedal ricochet is expressed primarily in changes in the ratios and structure of the hind limb muscles. These muscles take the main load

during the phase of support, and those which bear the greatest load are the
extensors of the hip, knee and talocrural joints.   The single-jointed muscles
of all groups (short postfemoral and gluteal of the hip and knee joint extensors) undergo yielding stretching in the preparatory period and actively
contract in the starting period of the phase of hind support.   The activity of
the double-jointed muscles is more complicated.   The work of the long postfemoral muscles depends on the places of their insertion onto the sacrum
and pelvis on the one hand and on the tibia on the other, and also on the
nature of flexor-extensor movements in the knee and talocrural joints.
Unfortunately, the lack of data on movement in the phase of support for all
the marsupials and rodents studied forced us to limit ourselves to representatives adapted to each of the forms of metalocomotory running.   As
typical examples of animals adapted to the ricochet we took the Norway rat
(Rattus norvegicus) — primitive ricocheting jump, William's jerboa
(Allactaga williamsi) — rigid ricochet, and the red kangaroo
(Macropus rufus) — supple ricochet.   Having drawn a diagram of the
movement of the appendicular skeleton of these animals in the phase of
support, we plotted on the straight line corresponding to the ischium three
points at different distances from the hip joint (Figure 211, a, b, c).   Three
points were also plotted on the tibia, descending distally from the knee joint
(Figure 211, d, e, f).

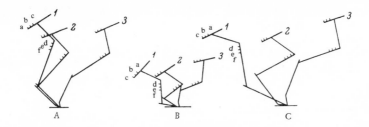

FIGURE 211. Diagram showing the movement of the hind limb skeleton of ricocheting mammals during the
phase of support:

A) kangaroo; B) Norway rat; C) jerboa. 1) beginning,  2) middle,  3) end of the phase of support.
a, b, c) hypothetical points of origin of muscles on the ischium;  d, e, f) hypothetical points of ending of
muscles on the tibia.

The distribution of these points represents three hypothetical variants
of insertion of the long postfemoral muscles on the ischium on the tibia.
If we join up these points in all the variants we can see how the working
conditions of the model muscles depend on the position of their origin and
termination in the animals adapted to the three forms of the metalocomotory mode of running.   Taking the size of the nominal long postfemorals
at the end of the preparatory period to be 100% (Figure 211, 2), we can
trace the relationship between the position of the points of their insertion on
the ischium and tibia and the changes of the length of these muscles during
the phase of support (Table 30).

TABLE 30. Length of "nominal long postfemoral muscles" in different moments of the phase of support (% of their length in the middle position 2 on the diagram in Figure 211)

| Species | Position of skeleton | Combination of points | | | | | | | | |
|---|---|---|---|---|---|---|---|---|---|---|
| | | a | | | b | | | c | | |
| | | d | e | f | d | e | f | d | e | f |
| Rattus norvegicus | 1 | 109 | 122 | 134 | 107 | 120 | 133 | 106 | 117 | 130 |
| | 3 | 100 | 112 | 123 | 90 | 102 | 112 | 83 | 92 | 101 |
| Allactaga williamsi | 1 | 111 | 124 | 133 | 113 | 125 | 135 | 114 | 125 | 138 |
| | 3 | 98 | 106 | 112 | 92 | 98 | 103 | 87 | 92 | 97 |
| Macropus rufus | 1 | 103 | 109 | 111 | 100 | 106 | 108 | 100 | 103 | 108 |
| | 3 | 95 | 100 | 101 | 88 | 93 | 95 | 84 | 88 | 90 |

In the preparatory period the length of the muscles in all the variants is reduced or remains the same in all three species (Table 30, *1*). The most considerable shortening of the muscles is observed in those which end relatively distally on the tibia (Table 30, a—f, b—f, c—f). They become 30—34% shorter in the rat, 33—38% shorter in the jerboa and 8—11% shorter in the kangaroo. In the same period the muscles ending proximally on the tibia (Table 30, a—d, b—d, c—d) are shortened by only 6—9% in the rat, by 11—14% in the jerboa and by 0—3% in the kangaroo.

In the starting period of the phase of support in all the species in Table 30, the muscles which end proximally on the tibia continue to shorten. Those which end distally on the tibia become either longer or else very slightly shorter. The muscles beginning closer to the hip joint (a—f, b—f) become 1—23% longer in the rat and 3—12% longer in the jerboa, while those which begin farther from it (c—f) become 3% shorter. In the kangaroo, only the muscles which originate the closest to the hip joint (a—f) become just 1% longer, whereas the other muscles are shortened by 5—10%.

Hence, by using models to examine the elongation and contraction of the long postfemoral muscles in the phase of support in relation to the points of their insertion on the ischium and tibia, we can assess the most favorable conditions for the work of the various components of this group in animals adapted to the three forms of ricochet. As we noted above (p. 138), favorable working conditions for these muscles are slight changes in their length in the preparatory period and continued contraction in the starting period of the phase of support.

As seen from Table 30, for all three species the muscles which end proximally on the tibia (a—d, b—d, c—d) optimally meet these conditions. We should therefore naturally expect that with cursorial specialization the endings of the long postfemoral muscles will be shifted proximally along the tibia. And in fact, when comparing little-specialized and highly specialized forms in different groups of rodents (Cricetinae, Gerbillinae, Muridae, Dipodidae, Dipodomyidae), we see that the endings of the long postfemorals do become shifted proximally along the tibia as cursorial specialization is enhanced (Figures 212 and 213).

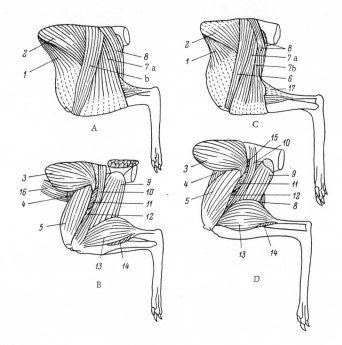

FIGURE 212. Hind limb muscles of R a t t u s   n o r v e g i c u s  (A and B) and  C o n i l u r u s  sp. (C and D), lateral:

A, C) superficial layer; B, D) deep layer. 1) tensor fasciae latae; 2, 3) m. gluteus: 2) superficialis, 3) medius; 4) m. rectus femoris; 5) m. vastus lateralis; 6, 7a, 7b) m. biceps femoris: 6) anterior, 7a) posterior, 7b) posterior pars caudalis; 8) m. semitendinosus; 9) m. semimembranosus; 10) m. praesemimembranosus; 11–12) m. adductor: 11) brevis, 12) maximus; 13) m. gastrocnemius lateralis; 14) m. soleus; 15) m. gemelli; 16) m. iliopsoas; 17) band to m. gastrocnemius.

Table 30 shows that in the kangaroo the most favorable action is that of the muscles beginning caudally along the pelvic girdle (c–d, c–e, c–f), since they undergo the lest contraction in the preparatory period and continue to contract in the starting period. In the jerboa these conditions are answered rather by muscles a–d, b–d, b–e (Table 30; Figure 211).

In rodents and marsupials the muscle which ends the farthest distally in the group of long postfemorals is m. semitendinosus, which is weakened in all small mammals adapted to ricocheting jumps. It is thus weaker in C o n i l u r u s  sp. than in R a t t u s   n o r v e g i c u s,  weaker in G e r b i l l u s   p y r a m i d u m  than in most of the other gerbils, weaker in C a l o m y s c u s   b a i l w a r d i  than in P h o d o p u s   s u n g o r u s  and C r i c e t u l u s   m i g r a t o r i u s  and, finally, it is weaker in the majority of the specialized Dipodidae than in the Sicistinae (Table 31). In all these animals, m. semitendinosus occupies more or less the position c–f (Figures 212 and 213); only in the kangaroo is the position rather c–e, since in this animal all the postfemoral muscles are shifted much farther proximally along the tibia.

TABLE 31. Relative weight of the hind limb muscles in rodents and the kangaroo (% of the total

| Muscles | Nesokia indica | Rattus norvegicus | Conilurus sp. | Meriones vinogradovi | Meriones blackleri | Meriones persicus | Meriones meridianus | Gerbillus pyramidum |
|---|---|---|---|---|---|---|---|---|
| m. gluteus medius . . . . . | 4.1 | } 7.3 | 11.3 | 6.8 | 6.2 | 6.9 | 6.8 | 6.1 |
| m. gluteus minimus [1] . . . . | 1.2 | | 1.4 | 1.3 | 1.0 | 1.1 | 1.1 | 1.1 |
| Gluteal . . . . . . . . | 5.3 | 7.3 | 12.7 | 8.1 | 7.2 | 8.0 | 7.9 | 7.2 |
| m. quadratus femoris . . . . | 0.9 | 0.7 | 0.5 | 0.7 | 0.6 | 0.7 | 0.8 | 0.8 |
| m. pectineus . . . . . . . | 0.8 | 0.9 | 0.4 | 0.4 | 0.4 | 0.6 | 0.3 | 0.5 |
| m. adductor brevis . . . . . | 2.1 | 1.9 | 3.2 | 2.1 | 1.8 | 2.1 | 2.4 | 3.2 |
| m. adductor maximus . . . . | 1.2 | 2.5 | 2.3 | 2.9 | 3.0 | 3.0 | 3.5 | 3.6 |
| m. adductor longus . . . . . | 0.4 | 0.5 | 0.2 | 0.2 | 0.2 | 0.2 | 0.3 | 0.3 |
| m. praesemimembranosus . . . | 0.4 | 0.7 | 1.2 | 0.6 | 0.8 | 0.8 | 0.8 | 0.5 |
| Short postfemoral . . . . . | 5.8 | 7.2 | 7.8 | 6.9 | 6.7 | 7.4 | 8.1 | 8.9 |
| m. biceps anticus [2] . . . . . | 1.4 | 1.4 | 2.1 | 2.8 | 3.8 | 2.9 | 2.8 | 3.2 |
| m. biceps posticus [3] . . . . . | 4.0 | 5.7 | 1.3+12.9 | 7.0 | 6.5 | 6.7 | 6.2 | 7.0 |
| m. semitendinosus . . . . . | 2.8 | 2.9 | 2.8 | 2.1 | 3.1 | 2.4 | 2.5 | 2.5 |
| m. semimembranosus [4] . . . . | 5.2 | 5.1 | 4.9 | 6.1 | 5.9 | 5.5 | 5.4 | 6.3 |
| m. gracilis . . . . . . . . | 0.5 | 0.6 | 1.8 | 1.3 | 1.1 | 1.1 | 0.8 | 0.9 |
| Long postfemoral . . . . | 13.9 | 15.7 | 25.8 | 21.3 | 20.4 | 18.0 | 17.8 | 19.9 |
| Hip joint extensors . . . . | 25.0 | 30.2 | 46.3 | 36.3 | 33.8 | 33.4 | 32.8 | 36.0 |
| m. vastus lateralis . . . . . | 3.4 | 3.5 | 5.8 | 4.3 | 4.7 | 5.0 | 4.5 | 5.1 |
| m. vastus medialis [5] . . . . . | 1.9 | 2.0 | 2.8 | 2.2 | 2.0 | 2.3 | 2.4 | 3.2 |
| Knee joint extensors . . . . | 5.3 | 5.5 | 8.6 | 6.5 | 6.7 | 7.3 | 6.9 | 8.3 |
| m. gastrocnemius . . . . . | 3.5 | 4.3 | 9.0 | 6.3 | 5.6 | 5.7 | 6.4 | 7.8 |
| m. psoas minor . . . . . . | — | — | — | — | — | — | — | 0.3 |
| m. iliopsoas . . . . . . . | 4.5 | 7.0 | 4.0 | 5.9 | 5.7 | 4.8 | 4.8 | 3.5 |
| m. gluteus superficialis [6] . . . | 3.9 | 3.4 | 2.3 | 3.0 | 3.1 | 3.0 | 2.7 | 2.6 |
| m. popliteus . . . . . . . | 0.3 | 0.2 | 0.2 | 0.3 | 0.3 | 0.3 | 0.3 | 0.3 |
| m. rectus femoris . . . . . | 2.3 | 2.7 | 2.9 | 3.7 | 3.9 | 4.4 | 3.7 | 4.6 |
| m. plantaris . . . . . . . | 0.8 | 1.0 | 1.4 | 1.6 | 1.4 | 1.6 | 1.9 | 1.9 |
| m. obturator[7] . . . . . . . | 1.0 | 1.0 | 1.1 | 0.6 | 0.8 | 0.7 | 0.5 | 0.7 |
| m. gemelli . . . . . . . . | 0.4 | 0.3 | 0.1 | 0.6 | 0.2 | 0.2 | 0.2 | 0.7 |
| m. tibialis anterior . . . . . | 1.4 | 1.6 | 2.8 | 2.1 | 2.2 | 2.3 | 2.0 | 3.4 |
| Extensors of digits . . . . | 1.3 | 0.9 | 1.3 | 1.8 | 1.5 | 1.4 | 1.4 | } 2.7 |
| Flexors of digits . . . . | 1.8 | 2.3 | 1.7 | 1.4 | 1.6 | 1.7 | 1.7 | |

[1] m. m. gluteus minimus + piriformis.
[2] For Macropus rufus — + m. caudofemoralis.
[3] s. str. + pars anterior.
[4] m. m. semimembranosus anticus + posticus.
[5] m. m. vastus medialis + intermedius.
[6] m. m. gluteus superficialis + tensor fasciae latae + sartorius.
[7] m. m. obturator externus + internus.

weight of the fore and hind limb muscles)

| Gerbillus gleadovi | Phodopus sungorus | Cricetulus migratorius | Calomyscus bailwardi | Sicista caudata | Pygerethmus platyurus | Allactaga jaculus | Allactaga saltator | Allactaga williamsi | Eremodipus lichtensteini | Macropus rufus |
|---|---|---|---|---|---|---|---|---|---|---|
| 7.8 | } 5.6 | 4.2 | 5.4 | 3.5 | 4.3 | 5.2 | 5.5 | 4.1 | 6.0 | 8.1 |
| 1.3 | | 2.7 | 2.3 | 0.9 | 3.8 | 2.2 | 1.6 | 1.5 | 2.7 | 1.3 |
| 9.1 | 5.6 | 6.9 | 7.7 | 4.4 | 8.1 | 7.4 | 7.1 | 5.6 | 8.7 | 9.4 |
| 1.3 | 0.5 | 0.4 | 0.6 | 0.5 | 0.3 | 0.3 | 0.9 | 0.9 | 0.3 | — |
| 0.8 | 0.5 | 0.5 | 0.6 | 0.9 | 0.3 | 0.5 | 0.2 | 0.3 | 0.3 | 0.2 |
| 2.8 | 1.8 | 1.1 | 2.6 | 1.9 | 1.3 | 0.8 | 0.6 | 1.2 | 5.6 | 1.3 |
| 4.1 | 1.5 | 1.5 | 2.9 | 1.9 | 1.5 | 2.2 | 7.2 | 1.6 | 4.0 | 3.9 |
| 0.2 | 0.4 | 0.2 | 0.5 | 0.2 | 0.2 | 0.3 | 1.1 | 0.9 | 0.7 | 0.4 |
| 0.7 | 0.9 | 0.5 | 0.8 | 0.9 | 0.7 | 0.4 | — | — | — | — |
| 9.9 | 5.6 | 4.2 | 8.0 | 6.3 | 4.3 | 16.1 | 17.2 | 15.7 | 15.5 | 9.8 |
| 2.2 | 0.8 | 1.7 | 1.7 | 0.9 | 8.0 | 14.2 | 22.3 | 21.6 | 14.8 | 3.9+4.9 |
| 6.0 | 3.0 | 3.8 | 4.5 | 3.0 | 6.3 | 8.9 | 1.6 | 1.8 | 6.3 | 1.9+8.9 |
| 3.1 | 2.4 | 3.6 | 2.2 | 2.3 | 3.1 | 1.6 | 1.8 | 2.8 | 1.9 | 3.8 |
| 6.1 | 3.0 | 4.2 | 5.5 | 3.5 | 4.2 | 11.6 | 7.1 | 10.8 | 4.6 | 5.1 |
| 1.5 | 1.4 | 1.2 | 1.3 | 1.1 | 0.5 | 1.0 | 0.9 | 1.6 | 1.0 | 1.6 |
| 18.9 | 10.6 | 14.5 | 15.2 | 10.8 | 22.1 | 25.7 | 26.6 | 27.8 | 24.4 | 26.1 |
| 37.9 | 21.8 | 25.6 | 30.9 | 21.5 | 34.5 | 49.2 | 50.9 | 51.6 | 48.6 | 45.3 |
| 5.7 | 2.4 | 2.6 | 5.1 | 3.2 | 8.4 | 8.0 | 8.1 | 10.1 | 8.1 | 7.6 |
| 1.9 | 0.9 | 1.3 | 3.0 | 1.4 | 2.2 | 3.5 | 3.3 | 3.3 | 3.2 | 2.7 |
| 7.6 | 3.3 | 3.9 | 8.1 | 4.6 | 10.6 | 11.5 | 11.4 | 13.4 | 11.3 | 10.3 |
| 7.3 | 2.5 | 2.7 | 4.6 | 4.2 | 9.8 | 10.1 | 10.1 | 9.2 | 10.3 | 5.7 |
| — | 0.4 | 0.9 | 0.1 | 0.5 | 1.2 | 0.6 | 0.6 | 0.3 | 0.2 | 2.6 |
| 5.3 | 3.9 | 4.0 | 5.1 | 5.5 | 5.2 | 2.1 | 1.9 | 2.0 | 2.9 | 1.9 |
| 2.2 | 3.1 | 5.7 | 3.2 | 2.8 | 1.5 | 2.2 | 1.9 | 2.2 | 2.1 | 3.1 |
| 0.4 | 0.1 | 0.3 | 0.2 | 0.5 | 0.5 | 0.4 | 0.4 | 0.5 | 0.3 | 0.4 |
| 4.1 | 2.3 | 2.4 | 3.7 | 3.2 | 2.2 | 2.6 | 3.1 | 2.8 | 3.5 | 3.8 |
| 2.2 | 0.7 | 0.2 | 0.7 | 1.6 | 1.9 | 2.1 | 2.1 | 2.5 | 2.0 | 4.7 |
| 1.3 | 0.9 | 1.9 | 1.6 | 1.1 | 2.7 | 1.2 | 0.5 | 0.5 | 1.1 | 1.4 |
| 0.3 | 0.4 | 0.1 | 0.2 | 0.2 | 0.2 | 0.1 | 0.1 | 0.1 | 0.2 | 0.1 |
| 0.5 | 1.0 | 1.3 | 1.9 | 2.3 | 3.9 | — | 3.8 | 3.3 | 3.4 | 3.5 |
| 2.5 | 0.3 | 1.3 | 1.4 | 2.1 | 1.7 | } 7.2 | 0.8 | 0.8 | 0.4 | 0.9 |
| 3.1 | 1.9 | 1.5 | 1.4 | 1.9 | 2.0 | | 1.3 | 1.9 | 1.6 | 1.8 |

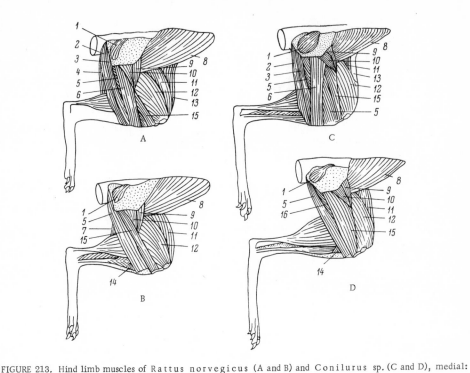

FIGURE 213. Hind limb muscles of Rattus norvegicus (A and B) and Conilurus sp. (C and D), medial:

A, B) superficial layer; B, D) deep layer. 1) m. obturator internus; 2) m. semitendinosus; 3) m. semimembran osus posterior; 4) m. biceps femoris posterior; 5) m. semimembranosus anterior; 6) m. gracilis; 7) m. biceps femoris; 8) m. iliacus; 9) m. pectineus; 10) m. adductor longus; 11) m. rectus femoris; 12) m. vastus medialis; 13) m. tensor fasciae latae; 14) m. popliteus; 15, 16) m. adductor: 15) maximus, 16) brevis.

Its relatively strong development in the kangaroo is explained by the fact that the lever of application of its force to the hip joint is fairly large, and this, as we said above, is especially important for the kangaroo.   The most interesting of the long postfemoral muscles in the mammals under consideration is m. biceps femoris, and in the kangaroo also m. caudofemoralis, which is absent in all the rodents studied.   M. caudofemoralis begins (Figure 214) on the transverse processes of the first two or three caudal vertebrae and the last sacral vertebra and ends on the straight ligament of the knee.   Its position corresponds approximately to c—d.   In the kangaroo an analogous position is occupied by the anterior part of m. biceps femoris posterior, which originates on the proximal tuber ischiadicum and ends on the proximal end of the tibia and the straight ligament of the knee (Figure 214). From the above analysis of the model of maximally profitable activity of the kangaroo's long postfemoral muscles it is clear that the strengthening of m. caudofemoralis and m. biceps posticus pars anterior fully meets the theoretically optimal working conditions for them.   The sum of the indexes of m. caudofemoralis and m. biceps posticus pars anterior therefore expectedly makes up 13.8% of the weight of the fore and hind limb muscles in the kangaroo, while the sum of m. biceps anticus and m. biceps posticus pars posterior only 5.8% (Table 31).

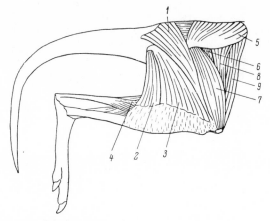

FIGURE 214. Hind limb muscles of the kangaroo:

1) m. caudofemoralis;  2, 3) m. biceps femoris posterior: 2) pars caudalis,  3) pars cranialis;  4) tendinous band to m. gastrocnemius;  5) m. gluteus medius;  6) m. rectus femoris;  7) m. vastus lateralis;  8) m. sartorius; 9) m. tensor fasciae latae.

The process of progressive development of the primitive ricocheting jump leads to a strengthening in rodents sometimes of m. biceps femoris anterior and sometimes of m. biceps posticus pars anterior.  In the latter case there is a shift of the proximal tuber ischiadicum closer to the center of movement in the hip joint.  Therefore, m. biceps posticus pars anterior occupies a position close to a—d or b—d in rodents (Table 30).  The strengthening of this part of the muscle sometimes goes so far that it almost becomes differentiated into a separate muscle (Figure 212).  A good index of the reorganization of m. biceps femoris posterior is its cross section taken at about the level of the distal third of the venter.  On this section (Figure 215) we see that in Nesokia indica, Phodopus sungorus and Cricetus cricetus, which are little specialized for the ricochet, and in Allactaga saltator and A. williamsi, highly specialized species, the thickness of the posterior biceps muscle is more or less uniform its whole length.  In Rattus norvegicus a certain thickening of the anterior bundles is observed, while in Conilurus sp. not only are they thickened, but they are even separated from the other portion into an independent muscle.  Something of the sort is seen in Calomyscus bailwardi, in which the anterior bundles of this muscle are markedly thicker than in other species of hamsters.  The same is observed in the three ricocheting jerboas and in the kangaroo (Figure 215).  It is interesting to note that the sum of the indexes of the anterior and posterior biceps muscles is practically uniform in all the highly specialized bipedal Dipodidae, constituting 23.1—23.9% of the weight of the limb muscles.  However, in two of them almost the entire mass of both biceps muscles refers to the anterior muscle, while in two others part of the mass refers to m. biceps posticus anterior.

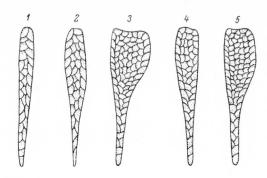

FIGURE 215.  Cross sections of m.biceps femoris in rodents:

1) Nesokia indica; 2) Rattus norvegicus; 3) Conilurus sp.;
4) Meriones persicus; 5) Gerbillus gleadovi.

Finally, of the long postfemoral muscle it remains to discuss the changes in force and in points of attachment characteristic for m. m. semimembranosus.  On the whole it may be said that with specialization for the ricocheting jump their ending is shifted proximally and in most cases they are strengthened.  Only in Conilurus sp. are they not strengthened, and here their relative weight is even lower than in Rattus norvegicus and Nesokia indica.  These muscles are strengthened in the gerbils, hamsters and jerboas (Table 31).  Interestingly, in the Dipodidae the insertion of m. semimembranosus anterior is shifted from the tibia onto the femur.  Because of this its activity is similar to that of the femoral adductors.  It is therefore understandable that a similar effect may be achieved by a strengthening either of the semimembranosus muscles or of the femoral adductors (cf. Allactaga jaculus, A. saltator, A williamsi and Eremodipus lichtensteini in Table 31).

Adaptation to all forms of the ricochet leads to a strengthening of the hip and knee joint extensors, while the talocrural joint extensors are strengthened in the rodents and to a lesser degree in the marsupials.  Thus, in Conilurus sp. as compared with Nesokia indica the hip joint extensors are 1.85 times, the knee joint extensors 1.62 times and the talocrural joint extensors 2.57 times stronger.  In Gerbillus pyramidum as compared with Meriones blackleri these extensors are strengthened respectively 1.08, 1.24 and 1.39 times.  In Calomyscus bailwardi as compared with Phodopus sungorus the hip joint extensors are strengthened 1.43 times, the knee joint extensors 2.46 times and the gastrocnemial muscles 1.84 times.  In representatives of the genera Allactaga and Eremodipus in comparison with Sicista betulina the hip joint extensors are strengthened more than 2.27—2.42 times, the knee joint extensors 2.46—2.92 times and the talocrural joint extensors (gastrocnemial muscles) 2.20—2.46 times.  The indexes of the hip and knee joint extensors in the kangaroo are close to the indexes of the most highly specialized Dipodidae.  As for m. gastrocnemius, its index is much lower in the kangaroo than in the highly specialized ricocheting rodents (Table 31).

From the above analysis we see more or less similar trends of change of the main muscle groups bearing loads in the phase of support in all types of specialization for the ricocheting jump. The degree to which these groups are strengthened also more or less coincides with the size of the difference in the depth of specialization for the ricochet. Thus, C o n i l u r u s sp. is markedly more specialized than N e s o k i a   i n d i c a, and in degree of specialization for the primitive ricochet G e r b i l l u s   p y r a m i d u m has departed very little from the other Gerbillinae. C a l o m y s c u s   b a i l w a r d i shows quite a high degree of adaptation to the primitive ricochet, while P h o d o - p u s   s u n g o r u s has hardly yet learned to move by the primitive ricocheting jump. Whereas the Sicistinae are weakly adapted to the primitive ricochet, the Dipodidae are the most highly specialized of the ricocheting animals which employ the bipedal ricochet.

Despite the similar patherns of change of the hind limb muscles in the Rodentia, the ratio of the different muscles differs markedly in different forms. Thus, in the family Muridae we see that adaptation to the ricochet is manifested in a strengthening of the gluteal group, which is 2.4 times stronger in C o n i l u r u s sp. than in N e s o k i a   i n d i c a, while in other groups of rodents this strengthening is not even twofold.* Along with this, a lesser degree of strengthening of the knee joint extensors is observed in the family Muridae than in other groups of rodents adapted to the ricocheting jump. For example, the knee joint extensors are only 1.62 times stronger in C o n i l u r u s sp. than in N e s o k i a   i n d i c a, but more than 2 times in the Dipodidae and Cricetinae (Table 31). Even in G e r b i l l u s   p y r a m i d u m the knee joint extensors are markedly more strengthened than in other gerbils, particularly if we consider that the difference in specialization of the various representatives of this subfamily is very minor. An interesting point of comparison is that there is a similar ratio in the development of the gluteal muscles and the extensors of the knee joint among the Ungulata (compare them in the Equidae and the gnu on the one hand and in the other Artiodactyla on the other, p.147).

In the Dipodidae we note a very considerable strengthening of the short postfemoral group of muscles, which comprises m. m. semimembranosus whose endings are shifted onto the femur.

The difference in the relative development of the gastrocnemial muscles in kangaroos and rodents adapted to the ricochet was mentioned earlier. While they are markedly strengthened in rodents in connection with specialization for the ricochet, in the kangaroo their level of development is much lower than in the Dipodidae, Gerbillinae and C o n i l u r u s sp. The difference is due to the different position of the origins of the gastrocnemial muscles. They begin in rodents on the Vesalius ossicles, each of which is attached to the femur by ligaments. In the kangaroo the gastrocnemial muscles originate directly on the femur. Because of this, the direction of pull of the bundles passes side by side with the center of movement in the knee joint in rodents, while in the kangaroo, on the contrary, at a considerable distance from this (Figure 66). In rodents, therefore, tension of the gastrocnemial muscles has scarcely no effect on flexor movements in the knee joint on the one hand, while on the other, the flexor-extensor movements

* Even in the most highly specialized jerboas the gluteal group is strengthened about 1.3—1.9 times in comparison with S i c i s t a   b e t u l i n a (Table 31).

in the knee joint also have no effect on movements in the talocrural joint. But in the kangaroo, the flexor-extensor movements in the knee joint exert a direct effect on the talocrural joint, while tension of the gastrocnemial muscles affects flexion of the knee joint.


### Adaptation to the dilocomotory dorsomobile and dorsostable modes

The dilocomotory dorsomobile mode of running was investigated in the Sciuridae and the dorsostable mode in the agouti. To compare the pattern of progress in the cursorial adaptation of some species of Sciuridae let us analyze the data in Table 27 using additional indexes of body structure. For further calculations we compute the angle of departure of the center of gravity $\alpha$ (according to formula (1)) during maximum running speeds and determine the average force for running from formula (9) and for jumping from formula (6). Multiplying these indexes by the speed of motion, we obtain comparable data on the power of the animals (Table 32). For all the species mentioned the pace angle $\beta$ is 80°. whereby $\sin \beta/2$ is equal to 0.65. In calculating the average force of the squirrel's jump from formula (6), we proceed from the fact that the length of the jump is 2.9 m for the red squirrel and 2.8 m for the Persian squirrel; the angle of departure is taken to be 30°, and all the other indexes are given in Table 32. With this as a point of departure, we can calculate the average power, which will be 8.2 kg· m/sec for S. vulgaris and 7.1 kg·m/sec for S. persicus.

TABLE 32. Description of the average force and power developed during running and jumping in the Sciuridae

| Species | Weight of body ($m$), kg | Leg length($h$), m | sin $\alpha$ | Speed ($v$), m/sec | Length of jump ($l$), m | Force ($F$), kg | Power ($W$), kg·m/sec |
|---|---|---|---|---|---|---|---|
| Sciurus vulgaris .. | 0.4 | 0.15 | 0.23 | 5.5 | 1.4 | 1.49 | 8.2 |
| S.persicus ....... | 0.4 | 0.15 | 0.20 | 5.5 | 1.2 | 1.29 | 7.1 |
| Spermophilopsis leptodactylus ... | 0.6 | 0.15 | 0.04 | 10.0 | 0.8 | 1.25 | 12.5 |
| Citellus undulatus | 0.6 | 0.15 | 0.17 | 5.5 | 0.85 | 1.37 | 6.8 |
| C.citellus ....... | 0.5 | 0.12 | 0.15 | 5.0 | 0.6 | 0.98 | 4.4 |

Apart from being adapted to running, some representatives of this family are climbers and diggers. In the Sciuridae adaptation to running is associated with leaving the forest for the open spaces, where the predator problem is tackled by digging burrows, the distance between which depends on the frequency of occurrence of the main food items. These intervals are small in the steppe, but in the desert are of several kilometers. The level of fossorial specialization differs little among the different species because the Sciuridae do not dig food galleries in the burrows, and it is precisely this action which leads to deep fossorial specialization (Gambaryan, 1960).

Therefore, in comparing a number of ground-dwelling Sciuridae, the animals which will be more attached to their burrows will be those which are less adapted for running. As a result, a comparison of them enables us to picture the process of adaptation to running within this group. However, in the work of the hind limbs the necessary forces developed during digging and climbing by no means exceed the force and power developed during running, especially in the case of one-time jumps. During digging operations the hind limbs act as a support, providing better working conditions for the forelimbs and incisors that are breaking up the earth. In addition, the hind legs throw aside the loosened earth. This is done in small batches (not more than 25–40 g even in marmots) and over short distances (no more than one meter). Naturally, there is no need for the hind limbs to develop a great deal of force for this. For support the position of the limbs is more important than muscle strain (Gambaryan, 1960). A climbing squirrel has to have great mobility in the distal and partly the proximal joints of the limbs and it has to develop the best techniques for seizing branches. All this is reflected in the details of the structural changes in the claws and limb joints rather than in the specific nature of the limb muscles.

We established that Spermophilopsis leptodactylus develops the most power during running (Table 32), followed by the squirrels and then the sousliks of the genus Citellus. Adaptation to one-time jumps from one spot is expressed in squirrels in that they have to develop a much greater force than during running along the ground. However, reorganizations of the muscles making for endurance during sustained, high-speed running may be more pronounced than with saltatorial specialization. We may therefore expect that despite the different trends of specialization, the long-clawed ground squirrel and the squirrels will show similar morphological changes in the musculature, promoting sustained, swift running in the one animal and large leaps in the others. These species are very similar in terms of the relative weight of the hind limb muscles (Sokolov's data on the markedly lower relative weight of the hind limb muscles in S. persicus probably stem from some misunderstanding: according to our data, the ratio of the hind limb muscles to the weight of the body is not 4.7 but 8.4).

The champion runner among the Sciuridae is the long-clawed ground squirrel (Spermophilopsis leptodactylus), whose hip joint extensors are much more strongly developed than in the other members of the family. An even greater relative weight of the hip joint extensors is observed in the well cursorially adapted agouti (Table 33). In both species of Sciurus and in Spermophilopsis the short postfemoral muscles are relatively better developed than in the other Sciuridae and the gluteal muscles more weakly developed, whereas in the agouti the opposite is true. The relative weight of the gluteals is least in S. vulgaris, which is characterized by the greatest relative weight of the knee joint extensors. The high speeds developed by the agouti are promoted by the development of not only the gluteal muscles but also the knee joint extensors. However, the relative weight of the knee joint extensors in this animal is nowhere near that in the mammals specialized for ricocheting (Tables 31 and 33).

The gastrocnemial muscles begin in all the Sciuridae and in the agouti on the Vesalius ossicles. However, the attachment of the latter on the femur is of a somewhat different nature than in other rodents. In the agouti the Vesalius ossicles are attached to the lateral and medial epicondyles of the femur by ligaments which are directed both parallel and perpendicular to the axis of the bone. Strengthening of the gastrocnemial muscles in connection with cursorial specialization is therefore less strongly expressed in the agouti than in the rodents adapted to the ricochet.

Flexion of the hip joint when the hind limbs are being transported in the air takes place with increasingly lesser acceleration the greater the jump. Therefore, specialization for the ricocheting jump leads to a lesser relative weight of m. iliopsoas (Table 31), the main flexor of the hip joint. During the gallop the hind limbs are brought forward after the animal has landed on the forefeet, and the acceleration of their transfer is proportional to the speed of running. Hence, in the Sciuridae and the agouti the relative weight of m. iliopsoas increases in proportion to the increase in cursorial specialization (Table 33). Something of an exception to this rule is Calomyscus bailwardi, which is adapted to the ricocheting jump and in which this muscle is much better developed than in the other hamsters. Analyzing Table 31, we see that in Rattus norvegicus this muscle has a greater weight than in Nesokia indica, which is less specialized for the ricocheting jump. In the gerbils its relative weight is very similar to that in the mouselike hamster. Specialization for the metalocomotory mode originally probably led to a strengthening of m. iliopsoas similarly to what happens during specialization for the gallop. This is because with a low degree of specialization for metalocomotory running the process of forward deflection of the hind limbs takes place mainly in the phase of front support, when acceleration of the transfer of the hind limbs forward is certainly advantageous. With enhanced specialization for metalocomotory running the hip joint is fully flexed already in the air, and acceleration of its flexion proves inversely proportional to the size of the jump. The distinctive development of m. iliopsoas in C. bailwardi is explained by the relatively low degree of specialization for the metalocomotory mode of running; the mouselike hamster and the gerbils differ little from each other in the degree of specialization for this.

Comparing Table 33 with Tables 8, 18 and 31, we see that m. tibialis anterior differs widely in various mammals in terms of its relative weight. In rodents and marsupials this muscle is well developed, and its development is enhanced with specialization for running. Thus, in Conilurus sp. it is twice as large as in Nesokia indica and in Gerbillus pyramidum 1.4—1.7 times larger than in the genus Meriones. In Calomyscus bailwardi it is developed 1.5—1,9 times more than in the other Cricetinae, and in the specialized Dipodidae 1.5—1.7 times more than in the marmots. The relative weight of this muscle is more or less the same in the kangaroo as in the specialized jerboas. In Sciurus and Spermophilopsis the relative weight of m. tibialis anterior is 1.5—2.7 times greater than in the other Sciuridae. On the other hand, in carnivores and ungulates, its relative weight is negligible and decreases with increasing cursorial specialization (Tables 8 and 18). Thus, this muscle is 1,8—2.5 times weaker in the cheetah than in other Felidae. It is also weaker in the wolf than in the Canidae less specialized for running. It is weaker in the Equidae than in the tapir, and so on.

TABLE 33. Relative weight of the hind limb muscles in the Sciuridae and Dasyproctidae (% of the total weight of the fore and hind limb muscles)

| Muscle, group of muscles | Sciurus vulgaris | Sciurus persicus | Spermophilopsis leptodactylus | Citellus undulatus | Citellus citellus | Marmota sibirica | Marmota baibacina | Dasyprocta agouti |
|---|---|---|---|---|---|---|---|---|
| m. psoas minor | 0.5 | 0.5 | 0.9 | 0.6 | 1.0 | 1.0 | 0.6 | 0.2 |
| m. iliopsoas | 4.5 | 4.8 | 5.4 | 4.3 | 3.9 | 2.4 | 3.0 | 6.3 |
| m. gluteus superficialis[1] | 2.6 | 3.4 | 3.1 | 2.8 | 3.9 | 3.0 | 2.6 | 2.3 |
| m. gluteus medius[2] | 3.6 | 4.3 | 4.3 | 5.1 | 4.6 | 4.0 | 3.2 | 9.9 |
| m.m. obturatores | 1.3 | 1.1 | 1.3 | 1.2 | 1.3 | 1.1 | 0.9 | 1.2 |
| m. gemelli | 0.2 | 0.2 | 0.2 | 0.3 | 0.3 | 0.2 | 0.2 | 0.3 |
| m. pectineus | 0.3 | 0.3 | 0.2 | 0.4 | 0.3 | 0.2 | 0.4 | 0.2 |
| m. quadratus femoris | 0.4 | 0.3 | 0.5 | 0.4 | 0.6 | 0.5 | 0.4 | 0.3 |
| m. adductor brevis | 1.0 | 0.8 | 1.1 | 1.5 | 1.3 | 1.0 | 1.0 | 1.2 |
| m. adductor maximus | 2.6 | 1.6 | 3.5 | 1.9 | 1.6 | 1.7 | 2.3 | 0.5 |
| m. adductor longus | 0.2 | 0.1 | 0.2 | 0.1 | — | 0.2 | 0.2 | 0.5 |
| m. adductor accessorius | 0.3 | 0.8 | 0.9 | — | — | 0.7 | 0.6 | 0.6 |
| m. biceps anterior | 2.0 | 3.3 | 2.9 | 1.4 | 1.9 | 1.1 | 1.0 | 5.2 |
| m. biceps posterior[3] | 2.9+ 0.6 | 2.4 | 4.0+ 1.5 | 4.2 | 4.2 | 1.3+ 1.9 | 3.6 | 3.8+ 1.6 |
| m. semitendinosus | 2.1 | 2.4 | 2.4 | 1.9 | 1.7 | 1.7 | 1.8 | 3.6 |
| m. semimembranosus anterior | 2.5 | 2.4 | 2.1 | 3.9 | 3.0 | 1.7 | 1.6 | 0.5 |
| m. semimembranosus posterior | 3.2 | 4.1 | 5.6 | 4.5 | 3.9 | 2.9 | 3.5 | 3.7 |
| m. praesemimembranosus | 1.3 | 0.9 | 0.6 | — | 0.3 | 0.4 | 0.5 | — |
| m. gracilis anterior | 0.1 | } 2.4 | 0.2 | 0.1 | } 1.2 | 0.1 | 0.1 | 2.0 |
| m. gracilis posterior | 2.3 | | 1.8 | 1.4 | | 1.4 | 1.7 | 1.5 |
| m. popliteus | 0.4 | 0.3 | 0.3 | 0.3 | 0.3 | 0.2 | 0.2 | 0.3 |
| m. tenuissimus | 0.1 | 0.3 | 0.3 | 0.2 | — | 0.1 | 0.2 | — |
| m. rectus femoris | 3.6 | 3.3 | 3.4 | 3.0 | 3.0 | 2.2 | 2.6 | 3.2 |
| m. vastus lateralis | 6.6 | 4.4 | 4.3 | 4.1 | 3.8 | 3.3 | 3.5 | 4.2 |
| m. vastus medialis + intermedius | 2.2 | 1.8 | 1.7 | 1.5 | 1.3 | 1.3 | 1.5 | 2.0 |
| m. gastrocnemius | 3.7 | 4.3 | 3.4 | 3.1 | 2.5 | 2.3 | 3.2 | 3.9 |
| m. soleus | 0.3 | 0.2 | 0.5 | 0.6 | 0.5 | 0.5 | 0.6 | 0.3 |
| m. plantaris | 1.7 | 1.2 | 3.2 | 1.7 | 1.6 | 1.4 | 1.6 | 1.1 |
| m. tibialis anterior | 2.5 | 2.5 | 2.7 | 1.6 | 1.3 | 1.0 | 1.5 | 1.1 |
| m. m. extensorius digitorum[4] | 1.8 | 1.7 | 2.5 | 1.3 | 1.4 | 1.1 | 1.4 | 1.0 |
| m. m. flexores digitorum[5] | 2.0 | 1.8 | 1.6 | 1.7 | 1.7 | 1.5 | 1.7 | 1.7 |
| Long postfemoral | 11.3 | 11.6 | 15.8 | 12.1 | 11.0 | 9.3 | 10.7 | 21.4 |
| Short postfemoral | 10.6 | 10.5 | 12.0 | 9.6 | 9.0 | 7.5 | 8.0 | 3.8 |
| Gluteal | 3.6 | 4.3 | 4.3 | 5.1 | 4.6 | 4.0 | 3.2 | 9.9 |
| Hip joint extensors | 25.5 | 26.4 | 32.1 | 26.8 | 24.6 | 20.8 | 21.9 | 35.1 |
| Knee joint extensors | 8.8 | 6.2 | 6.0 | 5.6 | 5.1 | 4.6 | 5.0 | 6.2 |
| Talocrural joint extensors | 4.0 | 4.5 | 3.9 | 3.7 | 3.0 | 2.8 | 4.0 | 4.2 |
| Total for the hind limbs | 59.4 | 57.7 | 66.6 | 55.1 | 50.4 | 43.4 | 47.4 | 64.2 |

[1] m. gluteus superficialis + m. tensor fasciae latae.
[2] m.m. gluteus medius + minimus.
[3] m. biceps posticus pars anterior + pars posterior.
[4] m.m. ext. digitorum longus + ext. hallucis longus + peroneus longus + peroneus brevis + peroneus digiti quinti.
[5] m. m. fl. hallucis longus + fl. digitorum longus + tibialis posterior.

Beginning on the anterior margin of the proximal end of the tibia and ending on the proximal end of the first metatarsal bone, m. tibialis anterior is the flexor of the knee joint. With ventral support (the phase of support) and simultaneous tension of the gastrocnemial muscles it pushes the proximal end of the tibia forward. In rodents and marsupials, when the animal lands on the ground, the hind limbs are brought forward far beyond the level of the perpendicular from the center of gravity, and tension of m. tibialis anterior imparts a thrust to the center of gravity forward and upward. When a carnivore or ungulate lands on the hind feet, the perpen- ↖ dicular from the center of gravity lies in front of the place of their landing, and tension of m. tibialis anterior acts to lower the center of gravity further rather than to raise it, in connection with which this muscle is weakened. Interestingly, in the agouti, the gallop of which is very similar to that of the ungulates, the perpendicular from the center of gravity, as in ungulates and carnivores, comes to be placed, when the animal lands, in front of the place where the hind feet touch down. At the same time, Table 33 shows that the relative weight of m. tibialis anterior is markedly lower in the agouti than in other rodents specialized for running.

During running, in the Sciuridae and Dasyprocta, the forelimbs not only absorb the shock but also promote the crossed stage of flight. Remembering that crossed flight is weakly expressed in the Sciuridae (details above), we can understand why there is no particular intensification of the load on the muscles with cursorial specialization. In the agouti, although the stage of crossed flight is better expressed, the main locomotory role still falls to the hind limbs, as in other mammals in which the basic high-speed gait is the gallop.

The digging of burrows, which is characteristic for some ground-dwelling Sciuridae, is reflected in the strengthening of a number of forelimb muscles which bear loads during running (the pectoral muscles, the extensors of the shoulder and elbow joints and the flexors of the digits). Therefore, the difference in the length of the segments in the forelimbs between the squirrels and the long-clawed ground squirrel and between the sousliks and the marmots is quite clearly expressed, whereby we can trace its connection with cursorial specialization (see above); on the other hand, the differences in the development of the muscles are much more weakly expressed.


PATHS AND CAUSES OF REORGANIZA-
TION OF THE LOCOMOTORY ORGANS
IN RODENTS AND KANGAROOS


The ancestors of the rodents branched off at an early stage from the common stem of the Mammalia; they were probably small animals living in forests. Any kind of temporary refuge (cracks in bark, niches under fallen trees, shrub thickets, etc.) served for protection from predators and partly also from unfavorable environmental conditions. The main high-speed asymmetrical gait of these creatures must have been the primitive ricocheting jump, which was preserved and progressively developed in a number of recent groups; in the most highly specialized of these it became the bipedal ricochet. In other groups of rodents a more progressive gait, the gallop, appeared for various reasons.

During the primitive ricocheting jump in the stage of flight the hind limbs are deflected forward in proportion to the size of the jump. The forelimbs play almost exclusively the role of shock absorbers. As speed increases the jumps get larger, at first enhancing the shock-absorbing role of the forelimbs. Then, with the transfer of part of the shock-absorbing function onto the increasingly forward deflected hind limbs, the forelimbs come to bear less and less the force of the impact. The secondary decrease of the shock-absorbing role of the forelimbs continues until in connection with further specialization for swift movement the animal goes over to the bipedal ricochet, the stage of forelimb support being canceled. Hence, cursorial specialization in these groups proceeds synchronously with the change of the gait and the enhancement of the role of the hind limbs not only for furthering propulsion of the body but also for shock absorption, which is carried out by two different mechanisms. In the one case, this mechanism acts to diminish the amplitude of flexor-extensor movements in the sacro-lumbar region as far as possible, and in the second, on the contrary, to increase the mobility of this region. We have therefore isolated three forms of running associated with progressive specialization for the ricochet: *primitive, rigid, and supple ricocheting jumps.*

The primitive ricocheting jump formed the basis for both the rigid and the supple forms of the ricochet. Deepened specialization for swift running in animals adapted to the primitive ricocheting jump led to the appearance of either the supple form or, what is observed more frequently, the rigid form of the ricochet.

The gallop was worked out in the Sciuridae in connection with the specialization of their ancestors for the climbing mode of life. Leaping from branch to branch led to an inhibition of the forward deflection of the hind limbs, and the animal always grabbed the next branch with the forepaws, actively pulling up the body and hind limbs onto the branch or letting them down smoothly on account of flexion of the spine and the work of the forelimbs. During such jumps it proved better for the hind limbs to be brought closer to the forelimbs in the stage of flight in jumps of various size, since then hardly any energy would be wasted in drawing up the body. When this type of leg movement is retained for running along the ground, the result is that in tracks of squirrels an increase in the stage of extended flight (the distance from the hind footprints to the front footprints) brings about hardly any change in the stage of crossed flight (from the front prints to the hind prints). Only when the running speed is markedly accelerated does the stage of crossed flight begin to increase. Curiously, tracks of this type are preserved also in the typically ground-dwelling sousliks (C i t e l l u s  u n d u-l a t u s,  C . c i t e l l u s), which may indicate that the ancestors of the sousliks went through a stage of adaptation to the arboreal mode of life. It is only in the long-clawed ground squirrel (S p e r m o p h i l o p s i s  l e p t o d a c t y l u s), which is highly specialized for running swiftly along the ground, that the tracks become similar to those of typical galloping animals (ungulates, carnivores, lagomorphs). In this animal, an increase in the length of a full stride causes an increase in the stages of both crossed and extended flight. Characteristic for all the Sciuridae is the bound in the presence of locomotory hind limbs and forelimbs and flexor-extensor movements of the spine. Hence, we call their mode of running *dorsomobile dilocomotory*.

A study of motion pictures of the gallop of the agouti reveals a close resemblance in the mechanics of motion to that of the Ungulata, proving that a number of rodents are also adapted to the dorsostable dilocomotory mode of running.

Adaptation to the above modes and forms of running is reflected primarily in the specific features of the skeleton of the spine. Specialization for the primitive ricocheting jump leads to the formation of a peculiar organ of shock absorption of the spine, consisting of a strongly developed spinous process of the second thoracic vertebra. To its apex is attached a movable triangular plate from which a number of tendinous bands stretch to the spinous processes in front. When these enter into action one after the other, the head is smoothly lowered. With increasing specialization for the primitive ricocheting jump this organ is at first perfected, and then with an increasing transference of the role of shock absorption to the hind limbs it is secondarily simplified until it disappears altogether in animals which have gone over to the bipedal ricochet. The deep convergent similarity of structure of the organ of shock absorption of the spine, which is achieved independently in various rodents, does not prevent the mechanism causing the triangular plate to revert to its initial position and the supraspinous ligament to extend beyond the second thoracic vertebra from appearing independently in different groups.

Along with the perfection and secondary loss of this organ of shock absorption progressive development takes place of the organ of rigidity, that is, the sacrolumbar region. The working principle of this organ is based on the utilization of the spinous processes of the last lumbar and first sacral vertebrae and the supraspinous ligament as lattice girders. The increase in the size of the last lumbar vertebrae and the reduction of the spinous processes of the first sacral vertebrae, from the region of which the fanwise widening tendinous plates stretch to the supraspinous ligament, permit only very restricted flexion of the sacrolumbar region.

In the kangaroo, adapted to the supple ricochet, the spinous processes of the lumbar and sacral vertebrae are weakly developed and directed parallel to each other, and instead of the supraspinous ligament the interspinal muscle stretches between them, making for smooth flexor-extensor movements of the sacrolumbar region in the cycle of running.

Spinal rigidity in the agouti is achieved by strong growth of the articular processes of the lumbar vertebrae, from which tendinous attachments of m. semispinalis stretch to the spinous processes. In the Sciuridae the sacrolumbar region is very similar in structure to that in the Carnivora.

Specialization for the metalocomotory mode of running leads to progressive elongation of the hind limb segments, while the forelimb skeleton becomes lengthened as specialization for the primitive ricocheting jump is enhanced. Their further growth ceases in animals which go over to the bipedal ricochet.   Whereas with specialization for the primitive and rigid ricochet, it is the foot which is primarily lengthened, in the larger kangaroo the relative size of the foot is secondarily reduced and the tibia undergoes the most considerable elongation.   Specialization for the dilocomotory dorsomobile and dorsostable modes of running also leads to a marked lengthening of the limb segments.

As is the case with other mammals, in rodents and marsupials cursorial specialization goes along with a strengthening of the hip, knee and talocrural joint extensors.

In an analysis of the hip joint extensors the most complicated problem is to figure out the causes of the change in the relative power of the individual components of the long postfemoral muscles.   We therefore set up models of the movement of the hind limb skeleton in animals adapted to the three forms of metalocomotory running.   Using these models, we were able to show that for the kangaroo, among the long postfemoral muscles which encounter the most favorable working conditions are those which begin caudally on the ischium and end proximally on the tibia, while in the Dipodidae, this applies to the muscles which originate cranially on the ischium and proximally on the tibia.   It is these muscles which are the most strengthened in the kangaroo and jerboas.

As noted for other mammals, whereas it is the gluteal muscles which of the three groups of hip joint extensors are strengthened with cursorial specialization, their indirect influence for extending the knee joint leads to lower rates of reinforcement of the knee joint extensors proper.   Such a division of labor between the extensors of the hip and knee joints is observed in the Muridae among the rodents.

In the rodents and marsupials studied we see two types of attachment of m. gastrocnemius on the femur: in the kangaroo directly from the femur; in the Sciuridae, Cricetidae, Gerbillinae and Dipodidae, on the Vesalius ossicles linked to the femur by ligaments.   Hence, cursorial specialization in the kangaroo did not involve a strengthening of the gastrocnemial muscle, since flexion of the knee joint affects extension of the talocrural joint.
M. gastrocnemius is markedly strengthened in the Sciuridae, while in the other rodents it is strengthened in proportion to the degree of specialization for swift running.

The enormous diversity of trends and degrees of cursorial specialization among the recent rodents allows us to select groups which contain both little-specialized and highly specialized forms and also groups where similar forms of running arise independently.   As a result, the comparative biomechanical and comparative morphological methods of investigation reveal both the general trends of change of the locomotory organs and the distinctive pathways of their change for solving the same tasks.   We should also mention the similarity in the changes of the organs of movement when performing essentially different functions. Analysis of these relationships can be used directly for determining the reliability of the signs of similarity and dissimilarity used in phylogenetic schemes.   For instance, the similarity in the lengthening of the limb segments in squirrels and the long-clawed ground squirrel comes about through different trends of specialization, and cannot serve as a basis for grouping them as related animals. These examples show clearly that they promise to be useful in a further in-depth analysis of a large number of representatives of any group of animals.

*Chapter 9*

## MODES AND FORMS OF RUNNING IN MAMMALS

### CLASSIFICATION OF THE MODES OF RUNNING

Cursorial adaptation has played an important role in the phylogeny of many groups of mammals. The ways in which speed of land locomotion is increased are very diverse, and therefore we may associate adaptation to high-speed running with various changes in the organs of movement. Hence, different modes and forms of running must be distinguished, characterized by morphological and biomechanical features of the locomotory apparatus.

Cursorial adaptation is specialization for swift motion. A classification of the modes of running may be based on the means by which speed is achieved. Acceleration may be imparted to the body by a thrust of the hind and forelimbs and also by extensor-flexor movements of the spine. Since these elements may work separately or together in different combinations, this is clearly the key to isolating the different modes of running observed in mammals.

The hind limbs are most advantageously placed for lending propulsion to the body (Kovalevskii, 1873; Bekker, 1955; Gray, 1961; Rashevsky, 1961; etc.), and therefore it is to be expected that any specialization for running will proceed along the lines of an increasing enhancement of their role. The part played by the spine and the forelimbs is always less significant.

We have called *metalocomotory* the mode of running in which only hind limb thrust is used for imparting acceleration to the body. If the forelimbs also have some part in this, we propose the term *dilocomotory*. We may distinguish two trends of adaptation of the spine. With the *dorsostable*, the spine remains rigid throughout the cycle, while with the *dorsomobile*, flexion and extension of the spine actively participate in transmitting acceleration to the body.

We may conclude from the material presented in the preceding chapters that four modes of cursorial specialization are observed in mammals: metalocomotory, metalocomotory dorsomobile, dilocomotory dorsostable and dilocomotory dorsomobile (Table 34; Figure 216).

It is advisable to examine the conditions under which these four modes arose in different groups of mammals in order to assess their significance. Mammals show a multitude of gaits, many of which are typical only for this class of animals, being absent in reptiles and amphibians. A review of the opinions on why this diversity of gaits appeared obliged us to consider the question of the primary mode of life of the Mammalia.  As a result, we propose a new hypothesis on the primary semifossorial mode of life of the common ancestors of the Marsupialia and placental mammals.

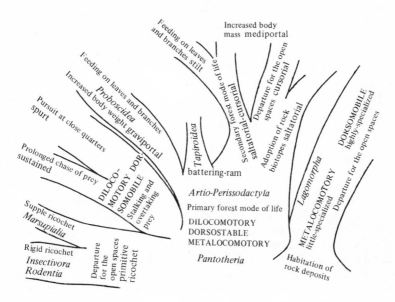

FIGURE 216.  Diagram showing the origin and evolution of the modes and forms of running in mammals

Digging and raking the upper soil layers or the forest litter had the result that the primary mammals used the forelimbs either simultaneously or alternately while the hind limbs supported the body.  This broke down the coordination of the motor mechanism inherent in the lower tetrapods and based on a symmetrically diagonal sequence of leg movement, when movement of a forelimb was followed by that of the hind limb on the opposite side.  In addition, when digging it is more profitable for the femur to be placed parallel to the plane of symmetry of the trunk (parasagittally), so as to provide optimum conditions for support while the legs are working, since in this case the pressure is diffused via the bone-ligament apparatus and does not affect active work of the muscles.  Hence, the appearance of an asymmetrical coordinating mechanism and reorganized locomotory apparatus, as occurred with the transition from reptiles to mammals, may be attributed to the original semifossorial mode of life of the ancestors of the marsupials and placentals.*

* Many workers accept another hypothesis, that the climbing mode of life was the original one for the ancestors of the mammals.  We have criticized this view (pp. 52–58).

TABLE 34. Adaptation to various modes of running in animals of different groups

| Mode of running | Our material | Presumably (according to published data) |
|---|---|---|
| Metalocomotory | Macropodidae, Muridae, Gerbillinae, Cricetinae, Dipodidae, Capromyidae | Dasyuridae, Peramelidae, Soledontidae, Macroscelidae, Pedetidae, Heteromyidae, Octodontidae |
| Metalocomotory dorsomobile | Lagomorpha | Ctenodactylidae, Chinchillidae |
| Dilocomotory dorsomobile | Sciuridae, Carnivora | |
| Dilocomotory dorsostable | Proboscidea, Artiodactyla, Perissodactyla, Dasyproctidae | Caviidae, Hydrochoeridae, Hyracoidea |

The primary semifossorial mode of life of these animals is associated with their relatively low degree of mobility, but progress in the order Mammalia was to lead to acceleration of motion. The acquired parasagittal position of the limbs and the asymmetrical sequence of movement made saltation the most profitable means of speeding up movement. The primitive Tetrapoda used only symmetrical gaits, in which directly after the phase of support was over the limbs begin to move forward to promote the next thrust. When this primary neuromuscular rhythm of reflex chains is preserved during an asymmetrical jump, the hind limbs are deflected forward in the air in proportion to the speed of motion. We have called this gait the *primitive ricocheting jump*. It is primarily progressive, because it permits the propelling limbs to get ready for the next thrust while they are still in the air.

Once animals moving by means of the primitive (quadrupedal) ricochet (Marsupialia, Insectivora, Rodentia) moved out into the open spaces, they began to move faster, mainly on account of increasing the length of jumps, that is, the distance covered during free flight. On the one hand this called for progressive development of the hind limbs and on the other led to the result that the hind limbs came to play an increasingly important part in locomotion. They became responsible not only for propelling the body but also for cushioning the impacts arising upon landing after each jump, in other words, they usurped the role formerly fulfilled by the forelimbs. This was the simple outcome of increasing deflection of the hind limbs when the duration of free flight was increased. The peak of specialization for primitive ricocheting saltation is therefore the true (bipedal) ricochet (Marsupialia, Insectivora, Rodentia). There was no other way of accelerating motion in animals employing the primitive ricocheting jump. At the same time it seems that, being based on a primitive neuromuscular mechanism of limb motion, expressed in propulsion immediately upon leaving the ground, the primitive ricochet should have been the initial gait of mammals. It is no accident that this is how animals of the primitive groups mostly moved (Table 34). The only mode of running which was able to evolve on the basis of the primitive ricochet was the metalocomotory.

But the primitive ricochet was not the universal gait.  As we have already said, it was particularly suitable for the open spaces.  Where other habitats were concerned, in which obstacles had to be overcome or the animals had to jump from stone to stone or from branch to branch,  the coordination apparatus had to be organized in a new way.  In these conditions it proved best to delay somewhat the deflection of the hind limbs after pushing away.  The advantages of this are as follows.  First, it lessened the steepness of jumps over obstacles, and second, it made the jumps more stable.  The latter is clearly especially important during jumps from one object to another with a variable distance between them.  Of course, with the primitive ricochet the fact whether the animal lands on the forefeet or on the hind feet depends not on its preference but on the size of the jump.  Thus appeared the gallop.  During the gallop, forward movement of the hind limbs after landing on the forefeet may take place owing to: 1) flexion of the spine, 2) active pulling up of the trunk by the forelimbs, 3) both these together.  All these three means also promote acceleration and can thereby serve as a basis on which specialization for different modes of running evolves.  It is interesting that all three possible modes of cursorial specialization are in fact used by galloping mammals.  The development of each of them has a direct link with the habits and mode of life of the ancestors of the group under consideration.

The use of flexor-extensor movements of the spine for acceleration is widespread in many mammals, even though such movements are uneconomical.  The reason is that movement of the spine is achieved owing to mobility between the segments of the vertebral column, and friction in the intersegmental joints and the work of segmented muscles lead to considerable losses of energy.

For the ancestors of the Lagomorpha, living in stone deposits, jumping from rock to rock resulted in that the hind limbs were brought close to the forelimbs which had landed due to active flexion of the spine.  Meanwhile, the forelimbs could not promote the appearance of a second stage of free flight in the cycle, that is, they had mainly a shock-absorbing role.  When this type of work of the spine and forelimbs was preserved after the Lagomorpha had emerged into the open, the metalocomotory dorsomobile mode of running evolved, where acceleration is promoted in the main by a thrust of the hind limbs and active flexion of the spine (Table 34).

The evolution of the Carnivora saw the working out of various techniques of hunting.  Stalking prey, or lying in wait were inherent in all carnivores to some extent or other.  Utilizing any unevennesses of ground or hiding in ambush, the predator creeps up to its victim until it makes the final, sudden bound.  It is here that flexion and extension of the spine are useful, since the wastefulness of these movements is compensated by the attainment of speed over a short distance.  With further specialization for swift running, even if the main hunting strategy becomes a prolonged chase (family Canidae) during which there is no special need for spinal mobility, flexion and extension of the spine continue to be used for increasing the speed of motion.  The pattern is as follows.  The predator makes a leap by thrusting off with the hind limbs and at the same time by actively extending the spine.  After it has landed, the hind limbs are brought forward owing to the impact of the forelimbs and flexion of the spine.  The thrust at the front is so great that the animal finds itself in the air once again.  Hence, characteristic for carnivores is the dilocomotory dorsomobile mode of running, in which speed is achieved both by thrusts of the hind and fore-limbs and by flexor-extensor movements of the spine.

In the Sciuridae, the climbing ancestors of which probably jumped from branch to branch, an adaptation was developed to pulling up the body and hind limbs onto the next branch due to the work of the forelimbs and active flexion of the spine.  When the animals came down to the ground, the gallop developed and the mode of running became analogous to that in the Carnivora.

If the features of the mode of life did not demand special development of spinal mobility, adaptations to fixation of the spine appeared.  The high-speed running of many plant-feeding forms did not call for a mobile spine, and therefore the Ungulata, Proboscidea, and Dasyproctidae developed special rigid structures lending stability to the spine during the running cycle.  Whence the dilocomotory dorsostable mode of running, in which the forelimbs and hind limbs are active and the spine remains rigid.

We do not know of one species among the recent mammals whose type of motion could be considered as transitional between any of these four modes of running.  On the contrary, specialization for any mode of running leads to ever more clearly expressed morphological differences.  Moreover, adaptation to a particular mode of running is characteristic for large groups of animals — orders and, more rarely, superfamilies and families.  This indicates that the modes of running originated at the dawn of evolution of any particular group and that they became perfected according to the path chosen by the given group.  However, each of the principal modes of running could have arisen in different groups independently.  In this case we have to expect that the same mechanisms of motion will be implemented by dis-similar morphological features of the skeleton and muscles.


CLASSIFICATION OF THE FORMS OF RUNNING

In the course of specialization for each mode of running, specific vari-ants of these modes, called here forms, appeared in dependence on the build of the animals and the conditions under which this specialization took place. Their characterization takes into account speed, endurance, type of shock absorption, nature of the movement of the center of gravity, and so on.

Metalocomotory mode.  In specialized bipedal mammals propulsion is imparted to the body in two ways.  In the first case, active flexion and ex-tension of the sacrolumbar region take place (the supple ricochet of the Macropodidae) and in the second, special arrangements develop which re-strict movement in this region (the rigid ricochet of the Dipodidae, Macro-scelidae, Pedetidae, etc.).  The rodents, little specialized for running, cu-shion the impact with the forelimbs (the primitive ricochet of the Muridae and Cricetidae).  Hence, the character of shock absorption allows us to distinguish in the metalocomotory mode of running *primitive, supple and rigid ricochets* (Table 35; Figure 216).

TABLE 35. Adaptation to various forms of the metalocomotory mode of running in animals of different groups

| Form of running | Our material | Presumably (according to published data) |
|---|---|---|
| Primitive ricochet | Muridae, Cricetinae, Gerbillinae, Zapodinae, Capromyidae | Metachirops, Marmota, Dasyurus, Selevinia, Octodontidae, Soledontidae |
| Rigid ricochet | Dipodidae | Macroscelidae, Antechinomys, Macrotis, Perameles, Dypodomys, Pedetes |
| Supple ricochet | Macropodidae | |

**Metalocomotory dorsomobile mode.** Only the Lagomorpha and, possibly, the Chinchillidae, are adapted to this mode. The mechanics of running in these groups of animals is not so different that we can isolate special forms of running. It is best simply to speak of various degrees of specialization for it.

**Dilocomotory dorsostable mode.** Adapted to this mode are the Ungulata, Proboscidea and Dasyproctidae. The diverse modes of life of these animals, which are highly specialized for running, makes it essential to distinguish the different forms of running. We can isolate three basic types of habitats where the ungulates are concerned, these determining the specific nature of the motion mechanics: forest biotopes, open spaces and rocky areas.

TABLE 36. Adaptation to various forms of the dilocomotory dorsostable mode of running in animals of different groups

| Form of running | Our material | Presumably (according to published data) |
|---|---|---|
| Battering-ram | Tapiroidea | Suidae |
| Cursorial | Equidae, Rangifer, Taurotragus, Connochaetes, Gazella gutturosa, Saiga | Strepsiceros, Gazella grandti, Oryx |
| Saltatorial-cursorial | Dama, Cervus, Capreolus, Ovis, Gazella subgutturosa, Dasyprocta | Tragulidae, Moschinae, Antilocapra, Elaphurus, Hydrochoeridae, Hyracoidea |
| Saltatorial | Capra, Rupicapra | Ammotragus, Oreamnos |
| Stilt | Giraffa | Okapia, Litocranius |
| Mediportal | Bos, Bison, Phöephagus | Anoa, Ovibos |
| Graviportal | Proboscidea | |

Forest habitats display the most varied conditions, yet they have one feature in common which is directly reflected in the mechanics of motion. In the forest, ungulates are forced to overcome obstacles all the time. The tapirs, which have evidently retained the original mode of life, move by means of the *battering-ram* form of running (Table 36).

The trajectory of the tapir's common center of gravity is relatively steep during the gallop than during the trot. Also characteristic for the battering-ram form of running is a large amplitude of flexor-extensor movements in the limb joints. This is what makes this form of running uneconomical. More economical is the *saltatorial-cursorial* form, in which obstacles are hurdled mainly by increasing the size of leaps. As with the battering-ram form, the center of gravity in animals using the saltatorial-cursorial form (Table 36) moves along a steeper trajectory during the gallop than during the trot or rack, but the amplitude of flexor-extensor movements in the joints is smaller in this case.

Penetration of the open spaces was a milestone in the evolution of the Ungulata. From these habitats they could now pass into other biotopes. In all open biotopes the only means of protection from predators are swiftness and endurance. The form of running of ungulates living in the open may therefore be called *cursorial* (Table 36). In animals adapted to this form the center of gravity moves along a gently sloping trajectory during the gallop and a steeper one during the trot and rack. The speed of running depends mainly on the number of paces per unit time, not on the length of each stride.

The rock-dwelling ungulates adapted to special conditions of movement. Their distinctive way of fleeing from predators became short bounds from refuge to refuge and leaps from one rock to another. Under these conditions the capacity for large leaps was a substitute for specialization for swift running, and thus we call their form of running *saltatorial*. This form is characterized by slow speeds (in rocky areas high speeds may be dangerous) and little stamina (a few leaps spare the ungulate from being pursued further). The trend of specialization in the saltatorial form of running is diametrically opposite to that in the cursorial form, since for large leaps there must be a steep trajectory of movement of the center of gravity, whereas for high-speed running a little-inclined trajectory is more economical.

For a number of groups of Ungulata and Proboscidea, the latter especially, an increase in the size of the body in the course of phylogenesis is characteristic. At first an increased body size plays a subordinate role in working out the features of motion mechanics, but it gradually becomes more and more important, and finally, in large forms it has more influence on the mechanics of motion than the nature of the habitat. Thus the *mediportal* and *graviportal* forms appeared (Table 36). Animals adapted to the mediportal form show a marked decrease in the pace angle and diminished vertical movements of the center of gravity. In large ungulates adapted to the mediportal form, and living in various biotopes, the cursorial mechanics has its own specific nature in each particular case. Thus, for example, the gallop of the steppe-dwelling bisons can be distinguished from that of the

European bison even visually.  Still, the overall character of motion, governed by the size of the animal, prevails over these particular differences.

The main trend of phylogenetic development of the Proboscidea, whose roots, like those of the Ungulata, evolve into the group Condylarthra, is associated with a progressive increase in the size of the body.  The recent elephants as a result are the largest of the terrestrial animals.  The ancient small predecessors of the Proboscidea lived in reed swamps and probably ran similarly to the recent tapirs (battering-ram form).  It was in these biotopes that the group's main development took place.  The by now much larger proboscidians departed into the forest-steppe, where they began to feed on branches and leaves from tree crowns.  As a result, apart from the body's being larger, the legs now grew longer.  An enormously heavy body and long legs led to a pillarlike arrangement of the limbs and the appearance of a specific graviportal form of dilocomotory dorsostable running (Table 36).  The typical features here are extremely small fluctuations of the center of gravity, due to the negligible vertical movements of the hip joint and the apex of the scapula.  In addition, the elephants began to lose the ability to move by asymmetrical gaits, and their main high-speed gait became the fast walk, or amble, during which at least one foot was always on the ground.

Some ungulates became adapted to feeding chiefly on leaves and branches of trees.  This led to elongation of the legs and neck, which in turn resulted in changes in the mechanics of motion during running.  They became able to move on legs which were straightened, especially their distal segments.  This can aptly be termed the *stilt* form of running (Table 36).

Hence, from the battering-ram form of the dilocomotory dorsostable mode of running the cursorial form developed in connection with the transition to life in the open; the secondary transition to forest biotopes led to the saltatorial-cursorial form of running.  Some descendants of this branch of the Ungulata adapted to the stilt form when they went over to feeding on branches.  Animals adapted to the cursorial form developed saltatorial running when they came to inhabit rock biotopes.  An increase in body mass, on the other hand, caused the appearance of the mediportal and, in elephants, the graviportal forms of running.

**Dilocomotory dorsomobile mode.**  In the Carnivora, cursorial adaptation is closely bound up with the techniques of hunting, and therefore the different forms of running are governed by different mechanics of motion related to the means used to capture prey.  Specialization for swift running developed in two ways in these animals: short-distance spurts and prolonged chases.  The mustering of considerable speed, exceeding that of the prey, becomes of importance only once the predator has been detected.  This is why the spurt form of running is calculated for speed but not for endurance.  On the other hand, a long chase has to be such that it will exhaust the prey.  We thus define these forms of running in the Carnivora as the *spurt* form and the *sustained* form (Table 37; Figure 216), which differ in the demands made on economy and speed.  The type of habitat does not have any appreciable influence on the development of the motion mechanics in carnivores.  Although the spurt form of running is special to those carnivores which live in the open, there are other carnivores which also live in the open and are specialized for sustained running. Thus, while for the ungulates the nature of the habitat ranks first in determining the form of running, in carnivores it is irrelevant.

TABLE 37. Adaptation to various forms of the dilocomotory dorsomobile mode of running in animals of different groups

| Form of running | Our material | Presumably (according to published data) |
|---|---|---|
| Spurt | Acinonyx | |
| Sustained | Canis, Lycaon, Vulpes | Otocyon, Megalotis, Cyon, Lyciscus, Lupulella |

Adaptation to running in the Mustelidae and Ursidae is very distinctive, since their speed of movement is determined by very strong development of the forelimbs (bears) and of the spine (Mustelidae) caused by characteristic habits that have no relation to running along the ground.

## GAITS AND MODES AND FORMS OF RUNNING

Detailed investigations of mammalian gaits were begun in the 19th century. More recently, substantial progress has been made in their classification (Howell, 1944; Hildebrand, 1963, 1966; Sukhanov, 1963, 1967, 1968; Gambaryan, 1967a, 1967b). It was found that the gaits are extremely diverse. Naturally, some researchers found a connection between the structural features of the locomotory organs of animals and the gaits they prefer (Sokolov et al., 1964). It is true that most animals show some sort of selectivity of gaits, but the question whether there is a relation between the structure of the organs of locomotion and gaits calls for special study. It appears more worthwhile to distinguish on the one hand gaits and on the other, modes and forms of running, and where the latter are concerned to seek the causes of the changes occurring in the locomotory apparatus.

We may consider two aspects in the locomotion of the Tetrapoda: the mechanics of motion and the mechanism of the pattern of movement. The mechanics of motion is defined as the physical parameters of motion itself: the speed at which the body advances, the size and angle of the pace, the angles in the limb joints, the acceleration of the body and its parts, etc. The mechanism of the pattern of movement is characterized by the particular sequence of movement of the limbs and the rate of alternation of the different phases and stages performed by the neuromuscular apparatus in the cycle of motion.

Only the mechanism of the pattern of movement has any bearing on gaits. Very instructive here is an experimental study of the mechanism of this pattern (Arshavskii et al., 1965;  Shik and Orlovskii, 1965;  Orlovskii et al., 1966a, 1966b;  Shik et al., 1966).  Experiments showed that a change of gaits (a change in the sequence of movement and the sequence of stages in the cycle) may be provoked by regulating the intensity of stimulation of the motor centers of the mesencephalon.

The relationship between the motor apparatus and the mode of running may be partly colored by the presence of selectivity for gaits.  It often happens that animals of any one species display only one gait.  However, also here it may be shown that the leading role is played by specialization for just this mode of running.  In a group of animals it is sometimes possible to select a number of forms with a various degree of cursorial specialization, which illustrates the synchronous change of the mode of running and the gait.

The original asymmetrical mammalian gait was probably the primitive ricocheting jump.  Cursorial specialization in animals adapted to this jump proceeded at the same time as the change of their typical gait and gradually brought about the transition to the bipedal ricochet.  In many groups of mammals we can find all the links in the chain of gradually mounting specialization for running, accompanied by the switch from the typical primitive to the bipedal ricochet.  Furthermore, the little-specialized forms move only by means of the primitive ricocheting jump, the intermediate forms by both the primitive and the bipedal ricochet, and the specialized ones, even during slow motion, only by the bipedal ricochet (Muridae, Gerbillinae, Cricetinae, Dipodidae, Dipodomyidae, Macroscelidae, etc.).  In this case we see almost complete synchronization of the changes in the modes of running with selectivity for the gait characteristic for the given species, which moreover does not change even with a change in the speed of movement.  The changes refer only to the size of the pace and the pace frequency per unit time.  An analogous synchronization of switches in the types of gaits is observed in the lower Tetrapoda in the course of specialization for land locomotion (Sukhanov, 1967, 1968).

A more complicated picture presents itself when we study selectivity for gaits in animals which have adopted various forms of the gallop.  In the Sciuridae the gallop was worked out in connection with the adaptation of their ancestors to leaping from branch to branch.  With each jump the forefeet landed on the next branch, and then the hind limbs were drawn up to them.  When this type of movement was preserved during running along the ground, the bound evolved as the characteristic high-speed gait of all the Sciuridae.  In studying the tracks of Spermophilopsis leptodactylus, Citellus undulatus and Citellus citellus and also of Sciurus persicus and S. vulgaris, we see that they use only the bound for practically all speeds of motion.  Hence, the bound is the typical gait of the Sciuridae, just as the primitive and bipedal ricochets are the typical gaits for all the mammals discussed above.  However, no changes in the depth of cursorial specialization among the Sciuridae are in any way reflected in changes in the gait selected by them.  The squirrels highly specialized for saltation and the long-clawed ground squirrels highly specialized for swift running use the same bound, despite the profound differences in the mechanics of motion in these forms.  The larger representatives of this family characteristically employ symmetrical gaits (trot, walk, etc.) when movement is slowed down.

A study of tracks of the weasel, stoat, marten and other small Mustelidae showed that they, too, often use the bound with any speed of motion. As with the Sciuridae, changes in the depth of specialization are not expressed in changes of the typical gait. The larger forms (badger, honey-badger, wolverine, etc.) freely go over to other gaits (trot, various forms of the walk, and so on) when moving slowly.

In the Lagomorpha, large or small, the typical gait is the half-bound, which does not change whatever the speed. This type of gallop remains constant with various depths of cursorial specialization.

We described most of the known gaits of mammals when studying movement in the horse, which shows a very large variety of gaits, as is the case with most of the Ungulata, the large Carnivora, many large rodents, the Hyracoidea and the terrestrial Edentata. But this variety of possible gaits for each species certainly does not mean that selectivity is absent. Of course, there are trotters and amblers among the horses; it is also known that most of the artiodactyls and carnivores use the lateral gallop as the high-speed gait and horses the diagonal gallop, while elephants adopt the fast walk, and so on. However, in all these mammals the gait always changes as the speed does, so that cursorial specialization is not associated with selectivity for any one gait.

Hence, the mutual relationship between perfection of land locomotion and gaits may have three variants in the Tetrapoda:

1. Each species is characterized by one particular gait which does not change with the speed of motion. But in the systematic group to which this species belongs the gait changes with enhanced specialization for running (the lower Tetrapoda, adapted to ricocheting jumps, the mammals of the groups Muridae, Cricetinae, Gerbillinae, Dipodidae, Dipodomyidae, Macroscelidae, Macropodidae, etc.).*

2. The use of one particular gait and its irrelevance to the speed of motion are preserved, but the connection between the change of the gait with the degree of cursorial specialization disappears (Lagomorpha, small Mustelidae, small Sciuridae, etc.).

3. Acceleration of motion has a direct relationship to the change of gait. The animals characteristically use many gaits, but in animals of the same species and even in individuals in different periods of life, selectivity for a gait may change markedly. Along with this, no relation whatsoever is observed between the depth of cursorial specialization and the change of gait.

For mammals adapted to ricocheting jumps and for the Lagomorpha, the gaits which do not depend even on the size of the body are very specific. Where the Mustelidae and Sciuridae are concerned, the specificity of gaits breaks down only for the large representatives of these groups. The specificity of gaits for the first two groups of mammals is apparently connected with different tempos of development of the forelimbs and hind limbs and their considerable functional nonequivalence. The result of all this is that it is difficult for them to make the transition from their typical asymmetrical gaits to symmetrical ones, in which the role of the forelimbs and hind

* The gait does not change within the main speed limits characteristic for the animal. With very slow movement, however, the gait does change.

limbs becomes equalized.  Although there is little difference in the role of the fore and hind limbs in the Sciuridae and Mustelidae, jumps are not a problem for small forms, even when moving slowly, and various inconsiderable obstacles can easily be overcome.  But for the larger species jumps are profitable only when movement is swift, and at low speeds these animals go over to symmetrical gaits.

The mechanism of motion may be highly specific in various mammals even in the presence of selectivity for identical gaits.  The modes and forms of running are what actually characterize the features of this mechanics. Special attention must be paid to the mechanics of motion with the fastest gaits, as they hold the key to the changes in the skeleton and muscles.  This does not exclude the fact that the features of locomotion at lesser speeds can yield additional material for a description of the modes and forms of running.

MORPHOLOGICAL CHARACTERISTICS OF THE LOCOMOTORY
ORGANS RELATED TO THE MODES AND FORMS OF RUNNING

**Structural features of the spine**

The above classification of the modes and forms of running shows that the work of the spine is essentially different with different paths of cursorial specialization.  We shall therefore naturally expect there to be considerable reorganizations of the spine connected with its specific function. With the metalocomotory mode of running the function of the spine is directly associated with shock absorption.  The actual type of shock absorption is taken as a criterion for isolating forms in this mode of running.  There is, of course, a specific nature of development of the spine which morphologically determines its different functions for each form of this mode of running.  The active influence of spinal mobility in imparting acceleration to the body, which is generally characteristic for the dorsomobile metalocomotory and dilocomotory modes of running, leads to special adaptations promoting vertical flexures of the vertebral column in the running cycle. The dilocomotory dorsostable mode of running necessarily leads to rearrangements making the spine rigid.

A uniform role of the spine appearing in different groups of animals independently may result in typical changes in its structure, without deep convergent similarity.

Hence, the structure of the spine may show specific features either directly depending on the modes and forms of running or related to the evolution of the group.

The various principles of shock absorption in animals adapted to the metalocomotory mode of running are associated with distinctive changes in the vertebral column.  In rodents adapted to the primitive ricocheting jump there is convergent development of an organ of shock absorption of the spine (Muridae, Cricetinae, Gerbillinae, Myocastoridae), which is represented by the strongly developed spinous process of the second thoracic vertebra, on the apex of which a triangular plate is fixed.  From this plate to the spinous processes of the first thoracic and last cervical vertebrae stretch a number of tendinous bands which are tensed one after the other after the animal has landed on the forefeet.  This organ becomes perfected as specialization is enhanced for the primitive ricocheting jump, and then it is secondarily simplified until it disappears with the transition to the bipedal ricochet.  Despite the deep convergent similarity of the organ of shock absorption of the spine in different rodents, there are specific features in it which are characteristic for individual groups.  For instance, for the same smoothness of extension of the region of the spine situated behind the second thoracic vertebra the spinous processes of thoracic vertebrae III and IV are lengthened and their slope backward is increased in the Muridae, while they are shortened in the Gerbillinae.  The return of the triangular plate to its original position is achieved by m. semispinalis in the Cricetinae and Muridae, by a system of ligaments in the Myocastoridae, by a multisegmented muscle in the Gerbillinae, and so on.

In animals adapted to the primitive ricochet the spinous processes of the last lumbar vertebrae are sloped forward while those of the first sacral vertebrae are sloped backward.  In connection with this the sacrolumbar region of the supraspinous ligament is lengthened.  In addition, in the more specialized forms C o n i l u r u s and G e r b i l l u s the spinous process of the first sacral vertebra is markedly reduced, which makes the part of the supraspinous ligament which runs from the spinous process of the last lumbar vertebra to sacral vertebra II even longer.  Lengthening of this part of the little-extensible ligament in the presence of its overall rigidity leads to improved shock-absorbing properties of the whole region.  The sacrolumbar region of the spine in animals adapted to the rigid ricochet smoothly changes its configuration in very restricted limits with a quick return to the initial position.  This takes place due to the strong growth of the spinous process of the last lumbar vertebra and that of sacral vertebra III or IV.  Between them is the supraspinous ligament to which fanlike tendinous plates extend from the rudimentary spinous processes of the first sacral vertebrae. These formations act as braces when the supraspinous ligament is greatly strained.

Finally, in the kangaroo, which is adapted to the supple ricochet, the spinous processes of the lumbar and sacral vertebrae are parallel and weakly developed, and between them is an interspinal muscle; the supraspinous ligament characteristic for animals adapted to the rigid ricochet is absent here.

The Ungulata, Proboscidea and Dasyproctidae adapted independently to the dorsostable dilocomotory mode of running.  In ungulates, spinal rigidity is achieved owing to the considerable growth of the spinous processes, the anteroposterior widening of their apexes, which are connected up by the poorly elastic supraspinous ligament, and the formation of hinges on the articular processes.  Furthermore, in these animals the transverse processes of the lumbar vertebrae are arranged strictly in the horizontal plane and are wider toward the ends, on which intertransversal ligaments are attached.  The Proboscidea show an arcuate structure of the thoracic and lumbar regions of the spine.  Transverse processes and hinges on the articular processes are virtually absent.  However, lateral rigidity of the spine is assured by the marked shortening of the lumbar region* and the union between the last rib of the tendinous plate and the wing of the ilium. In the Dasyproctidae, spinal rigidity stems from the growth of the articular processes, which are connected with the nearest spinous processes by the bundles of m. semispinalis, that has well-expressed tendinous attachments.

In carnivores adapted to the dorsomobile dilocomotory mode of running, unlike the ungulates, proboscidians and D a s y p r o c t a, the flexor-extensor movements of the spine actively promote acceleration of motion.  The apexes of the spinous processes are not widened but narrowed, the supraspinous ligament is replaced by a series of fanlike tendinous plates which make for a large amplitude of movement, and in the Mustelidae, for which spinal mobility is especially important, the spinous processes are still further shortened and, in the total absence of interspinous ligaments, the interspinal muscles attain a high degree of development.  Cursorial specialization in carnivores is also connected with marked strengthening of the extensors of the spine.

In the Lagomorpha, adapted to the dorsomobile metalocomotory mode of running, active flexion of the spine is even more important for speed than extension.  The flexor muscles are extremely well developed, constituting 7.8—9.5% of the weight of the limb muscles as against 2.0—4.4% in other mammals.  Apart from this, in the most cursorially specialized forms, ventral spinous processes appear on the last thoracic and first lumbar vertebrae, these enhancing the force levers when the spine is flexed.

As we see from the above, each mode of running, and for the metalocomotory mode also each form of running, can be clearly defined according to the structure of the spine.  At the same time it was found that whereas animals have developed a certain form or mode of running independently of each other, in the structure of the spine we find highly specific morphological features of complex functional types of work.

## Structural features of the appendicular skeleton

With specialization for all modes of running it is the hind limbs which give the main thrust for imparting acceleration to the body.  The system of imparting this acceleration is always the same.  The thrust is given in the phase of support, in the preparatory period of which the knee and talocrural joints and sometimes also the hip joint are flexed, while in the starting period they are extended.  Under these conditions long legs are useful for speed and for increasing the size of jumps.  The outcome of all this is that the degree of convergence in the length ratios of the hind limb segments is very high with specialization for the various modes and forms of running. In connection with the varying role of the forelimbs in animals adapted to different modes of running, considerable differences are observed in the size

---

* The elephant has only two or three lumbar vertebrae.

ratio of the segments in the forelimbs and hind limbs.  In addition, the size of the body, the characteristics of the motion mechanics, etc., affect the length of the limb segments.  It is therefore more useful to consider them within the limits of adaptation to the various modes of running.

Most of the primitive mammals are adapted to the metalocomotory mode: Marsupialia, Insectivora, Rodentia.  In all these groups cursorial specialization is expressed in an increasing enhancement of the role of the hind limbs, gradually leading the way to the bipedal ricochet.  The varying level of specialization is clearly detected, first, from the length ratio of the fore and hind limbs and second, from the ratio of the length of the limb segments to the length of the spine.  A comparison of the percentage length ratios of the limb segments to the total length of the lumbar and thoracic regions of the spine shows that it is markedly smaller in little-specialized than in specialized forms.  This can be seen in any group which contains both more and less specialized forms.  Thus, in C a l o m y s c u s   b a i l w a r d i  in comparison with P h o d o p u s   s u n g o r u s  the foot is 1.8 times, the tibia 1.3 times, and the femur 1.2 times larger.  In G e r b i l l u s  as compared with M e r i o n e s, the foot is 1.8 times, the tibia 1.5 times and the femur 1.1 times larger.  In C o n i l u r u s  sp. in comparison with N e s o k i a   i n d i c a, the foot is 2.1 times, the tibia 2.0 times and the femur 1.5 times larger.  Finally, in the Dipodidae, in which the difference in the degree of specialization for the ricochet is especially marked, in A l l a c t a g a   e l a t e r, which is more highly specialized cursorially than S i c i s t a   b e t u l i n a, the foot is 2.2 times, the tibia 2.1 times and the femur 1.5 times larger.

The kangaroo is very highly specialized for the bipedal ricochet.  Yet the relative length of the foot (52—72% of the length of the thoracic and lumbar regions of the spine) is smaller than in the specialized jerboas (98—112%).  This phenomenon finds its explanation in the weight of these animals, which is 350—700 times greater in the kangaroo.

The length change of the forelimb segments with specialization for the metalocomotory mode of running differs significantly from that in the hind limb segments.  The forearm and hand at first become longer, then more slowly longer, and finally, with the bipedal ricochet, this process of elongation ceases altogether.

Hence, specialization for metalocomotory running leads to certain changes in the length ratios of the limb segments.  Still, these changes are nonspecific for each of the three forms of running.  All that can be said is that the relative length of the foot in animals adapted to the supple ricochet increases less than in animals adapted to the rigid ricochet.  But this is related to the increase in the weight of the body rather than with the specific nature of running.

It is also hard to grasp the specific character of the changes occurring in the ratios of the limb segments within different systematic groups, since the process of specialization for metalocomotory running leads to a convergently similar change in these ratios in all the groups studied.

In the highly cursorially specialized Lagomorpha adapted to the dorsomobile metalocomotory mode of running the length of the hind limb segments is much greater than in the less specialized forms.  Thus, in the Leporidae in comparison with the Lagomyidae the percentage length ratio of the foot to the length of the lumbar and thoracic regions of the spine is 1.4—1.5 times larger, while the tibia and femur are only 1.1 times larger.  Elongation of the forelimbs in connection with cursorial specialization is much more weakly expressed in the Lagomorpha.

When we compare the increase in the length of the limb segments arising due to adaptation to the metalocomotory and the metalocomotory dorso-mobile modes of running, we see that the rate of elongation of the forelimb segments is markedly higher in the second than in the first case, while in animals adapted to the dilocomotory dorsostable and dorsomobile modes the forelimb segments lengthen at the same rate as the hind limb segments.

The dorsostable dilocomotory mode of running develops independently in the Artiodactyla, Perissodactyla, Proboscidea and Dasyproctidae differing in size and in mode of life.  All the animals which have worked out this mode of running are highly specialized runners, and the length ratio of the limb segments is characteristic for the various forms of running.

Calculations show that with an increase of the angle of departure in each jump during running, the average forces developed during the phase of support are markedly increased.  Therefore, in inhabitants of forest and shrub adapted to saltatorial-cursorial running, in spite of their relatively low speeds, the segments of the limbs are elongated to a relatively greater extent than in inhabitants of the open spaces adapted to cursorial running. These differences come out especially clearly when comparing animals of about the same weight.  Thus, for example, in the saiga and Mongolian gazelle the percentage ratio of the length of the foot to the length of the thoracic and lumbar regions of the spine is 54—55%, whereas in the forest roe deer it is 70% and in the goitered gazelle, which also lives in shrub thickets and is adapted to the saltatorial-cursorial form of running, it is 66%. Analogous differences are noted when comparing the reindeer and forest deers.  The length of the reindeer's foot is 60% of that of the thoracic and lumbar retions, while in the red deer, which is similar in size, it is 66%.  Animals adapted to these two forms of running are characterized by a relative shortening of the limb segments with an increase in body mass.  For instance, in the cursorially adapted gnu the length of the foot is 52% of that of the spine, while in the saiga and Mongolian gazelle it is, as stated,  54—55%. Similarly, in the heavy maral it is 60%, while in the red deer 66%, and in the wild ass 58%, but in the larger zebra only 53%.  A further relative shortening of the foot is seen in animals with the mediportal form of running.   The foot of the European and American bisons is equal to 43—46% of the length of the spine.

These differences in the length of the limb segments in animals adapted to the cursorial and saltatorial-cursorial forms of running would seem to indicate that there is a subsequent increase in the length of the foot in animals with the saltatorial form of running, since they have to develop still greater forces in the phase of support.  However, specialization for the saltatorial form proceeds by two paths in the Artiodactyla.  With the first path, extension of the talocrural joint in the phase of support occurs a good deal later than that of the knee and hip joints (goats) and with the second path, all these joints are extended synchronously (chamois).  Because of this, in all goats the load tending to break the foot is increased to a considerable extent in the phase of support, while in the chamois the load on the foot is distributed parallel to its length.  As a result, the goat's foot is secondarily much shortened (49% of the length of the spine) while that of the chamois is just as long as that of the saltatorial-cursorial forms (68% of the length of the spine).

The relative length of the limb segments in the tapir, which uses the battering-ram form of running, is very similar to that in the ancestors of the Artiodactyla and Perissodactyla.  This group of perissodactyls apparently branched off from their common stem early and has always lived in little changing biotopes, so that it has preserved the very primitive length ratios of the segments.  For instance, the length of the tapir's foot is about 39% of the length of the lumbar and thoracic regions of the spine, that is, much less than in ungulates with the mediportal form of running.

It is interesting to note that in natural phylogenetic series of mammals which have progressed by way of increasing the size of their body, a shortening of the foot is also observed.  Thus, comparative anatomical and paleontological investigations indicate that the foot became secondarily shorter as the body mass increased.

The interactions we have described between the mechanics of motion, the weight of the body and the relative dimensions of the limb segments cannot embrace the whole diversity of the ratios in the skeletal elements.  A graphic example of an opposite development of these ratios in the presence of a need for an overall lengthening of the limbs is the size of the limb segments in the elephant and giraffe.  In terms of weight, the giraffe approaches the largest representatives of the family Bovidae which have the mediportal form of running, but the relative size of its foot and hand (106 and 94% of the length of the thoracic and lumbar regions of the spine) exceeds that in even the most specialized saltatorial-cursorial ungulates.  In elephants, on the other hand, the distal segments are shortened to 15—16% of the length of the spine, while the femur is lengthened, reaching 56—59% of the length of the spine (as against 31—48% in ungulates).  Long legs help the giraffe and elephant, which have similar modes of feeding, to pluck food from the tops of trees.  The flexing moments of the limb joints are diminished in elephants owing to the shortening of the distal segments and the pillarlike arrangement of the limbs, expressed in the straightening of the proximal joints.*  In the giraffe, it is the distal segments which are straightened, and this also somewhat diminishes the flexing moments.  The opposite patterns of change in the proportions of the skeleton and straightening of the joints in the giraffe and elephant can probably be explained by different phylogenetic pathways. The small forest-dwelling ancestors of the giraffe probably adapted to the saltatorial-cursorial form of running, during which the distal segments of the limbs became maximally elongated.  This trend was maintained when the body mass increased, since the specific nature of feeding required an overall lengthening of the limbs.  The negative consequences of the ever increasing loads for flexing the distal joints were compensated by their straightening, which was also useful for enhancing tallness.  The small ancestors of the elephants lived in littoral swamp thickets and had no need to

---

* A shortening of the distal segments in the presence of a phylogenetic increase in the body mass is observed not just in elephants but in many groups of mammals (Gregory, 1912; Simpson, 1951; and others).

adapt to running fast.  The ratio of their limb segments was similar to that
in the recent tapirs (Osborn, 1936).  The gradually increasing weight of the
animals led to a decrease in the size of the distal segments and to a straightening of the proximal joints.  When the animals entered sparse forests,
branches and leaves became an increasingly important part of the diet, and
this resulted in the need for longer legs.  The solution was found in a pillarlike arrangement of the limbs and a lengthening of their proximal segments.

For all animals adapted to the dorsostable dilocomotory mode of running
the work of the forelimbs has almost the same importance for speed as that
of the hind limbs.  Therefore, in all the groups studied the ratio of the skeletal elements of the forelimbs differs little from that of the hind limbs.

Adaptation to the dorsomobile dilocomotory mode of running is observed
in the Carnivora and Sciuridae, the trend of cursorial specialization of which
is determined mainly by the hunting habits in the former and by the arboreal
mode of life in the latter.  The one or two jumps that all the Felidae characteristically make before seizing the prey are replaced in the cheetah by a
series of jumps which represent the spurt form of running.  A sustained
form of running was worked out in the wolf and Cape hunting dog as a result
of the need for prolonged pursuit of fleet-of-foot ungulates.  Long legs are
useful for both these forms.  The elongation of the limb segments differs
little in the cheetah from that in the other species of the family, so that the
load during the cheetah's spurt is almost the same as that during the jumps
of other cats.  It is particularly hard to trace this process due to the very
different sizes of cats, for in the small forms we should expect there to be
a greater relative elongation of the limbs.  This elongation may be seen in
the cheetah only when comparing its indexes with those of Felidae more or
less its own size (puma, jaguar, leopard).

The opposite is observed in the Canidae.  Here the hunting habits are
more varied than in the Felidae: the raccoon-dog, for example, is a form
little specialized for running.  Therefore, the elongation indexes of the limbs
are much better expressed in dogs than in cats with cursorial specialization.


## Structural features of the limb muscles and their connection with the modes and forms of running

It is difficult to perceive a specific link between the structure and strength
of development of the limb muscles and the modes and forms of running.
The ratio of muscle strength of the fore and hind limbs is often similar with
both dilocomotory modes of running.  For the metalocomotory dorsomobile
and metalocomotory modes, specialization for swift land locomotion is always associated with an increasingly pronounced strengthening of the hind
limb muscles and a weakening (metalocomotory) or slight strengthening
(metalocomotory dorsomobile) of the forelimb muscles.  This is linked
with the exclusively shock-absorbing role of the forelimbs which is characteristic for both these modes of running.

In the phase of support the action of the muscles causes the body to move
forward, while in the phase of transit positive and negative accelerations of
the distal parts of the limbs develop.  The muscles working in the phase of

support are naturally subjected to an especially heavy load. At the moment of the transition from support to transit the load on the muscles abruptly alters, and the flexors of the leg joints go into action toward the end of support to diminish the harmful consequences of this switch of the load. The flexor muscles are now working in a yielding regime, and therefore the transition from extension of all the joints to their flexion at the end of the phase of support proceeds automatically. The active work of the extensors lends acceleration to the body. Since cursorial specialization is connected with increased speed, the muscles promoting this acceleration are strongly developed. The load on the hip joint extensors is particularly great, as the hip joint is chiefly responsible for imparting propulsion to the body. These extensors comprise three groups of muscles: gluteal and the short and long postfemoral. The first two groups are single-jointed muscles, and their work schedule is directly connected with the schedule of flexor-extensor movements in the hip joint.

The function of the long postfemorals changes in relation both to the flexor-extensor movements going on simultaneously in the hip and knee joints and the distance of their origin and termination from the center of movement in these two joints. Due to the simultaneous extension of the hip and knee joints, the shortening of the corresponding muscles may be less expressed than in the single-jointed muscles, while in those which end distally on the tibia there may be a lengthening toward the end of the phase of support, despite the continuing extension of the hip joint. The strengthening of the various components of the hip joint extensors is very diverse with specialization for the various modes and forms of running. Thus, adaptation to the metalocomotory mode leads to a strengthening of the hip joint extensors in all the groups studied. In Calomyscus bailwardi, in comparison with Phodopus sungorus, all three groups of hip joint extensors are 1.4 times enlarged. The strengthening of the different groups of hip joint extensors is not uniform in the family Muridae in the process of cursorial specialization. The group of short postfemorals is only 1.3 times stronger in Conilurus sp. than in Nesokia indica, whereas the long postfemorals are 1.9 times and the gluteal muscles even 2.4 times stronger. In Gerbillus, as compared with Meriones, the hip joint extensors are 1.1—1.2 times strengthened as a whole, while of their individual components only the group of short postfemoral muscles is markedly strengthened, and the other groups may even be more weakly developed in Gerbillus than in Meriones. The hip joint extensors show various types of strengthening in the jerboas as compared with the marmots. In Pygerethmus and Eremodipus the gluteal muscles are markedly strengthened, but in the other Dipodidae these muscles are developed more or less to the same extent as in the marmots. In Pygerethmus the group of short postfemorals is even weaker than in the marmots, while in the other jerboas both the long and the short postfemorals are strengthened in comparison with those in marmots by a factor or 2 or more. The hip joint extensors are developed more or less to the same extent in the kangaroo as in the Dipodidae. Among the three groups the gluteal muscles are considerably developed, even more strongly than in the jerboas.

In the Lagomorpha, specialization for the metalocomotory dorsomobile mode of running causes strengthening of the long and short postfemoral muscles, which are 1.6—2.5 times larger in hares than in pikas; the gluteal group is developed practically to the same extent in these animals.

Specialization for the dilocomotory dorsomobile mode of running is expressed mainly in a strengthening of the short postfemoral muscles.

All the Ungulata, Proboscidea and Dasyproctidae are highly specialized for the dorsostable dilocomotory mode of running. The weight of the hip joint extensors in these animals, as in all other highly specialized runners, constitutes 48—55% of that of the hind limb muscles.

The strength ratio of the three groups of hip joint extensors is not uniform in these animals. The gluteal muscles are particularly strengthened in the Perissodactyla. Thus, in the tapir they account for 14.3% of the weight of the hind limb muscles, in the Equidae 21.3—23.5%, while in most of the Artiodactyla only 7.5—11.8%. Considering that the relative development of the gluteal group in the tapir is similar to the initial degree of development for the whole group, we see a progressive development of these muscles in the phylogeny of the Perissodactyla. Among the Artiodactyla the gluteal muscles show the greatest relative weight in the gnu (13.1% of the weight of the hind limb muscles). Several groups of artiodactyls show special development of the short postfemoral muscles (saiga, reindeer), while in other groups the long postfemorals are more developed (kudu, gnu, bison, yak, giraffe, and others).

It is hard to understand the causes of the differences observed in the development of the various components of the hip joint extensors. Some clues as to their interactions may be provided by the different ratios in the development of the skeleton and muscles. In all cases where the gluteal muscles are strengthened there is simultaneous growth of the trochanter major of the femur. Strengthening of the short postfemoral muscles goes along with a reduction in the size of the ischium. This process is readily perceived by examining a diagram of the pelvic girdle and femur in various animals when the femur is always drawn the same length (Figure 217). The figure clearly shows that the ischium is relatively markedly smaller in the cheetah than in rats.

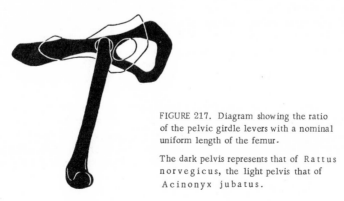

FIGURE 217. Diagram showing the ratio of the pelvic girdle levers with a nominal uniform length of the femur.

The dark pelvis represents that of Rattus norvegicus, the light pelvis that of Acinonyx jubatus.

The strengthening of the long postfemorals is apt to vary most considerably and is the most complicated to explain.  As the muscles contract, the mechanical effect (the force developed) gradually diminishes, and when they contract by more than 30% of their initial length the efficiency of contraction decreases sharply (Hill, 1948, 1953).  Hence, an analysis of the length changes of the muscles during the phase of support during high-speed gaits of various mammals can show which positions of the origin and ending of the different components of the long postfemorals are most advantageous.  Thanks to such an analysis it became clear why in the large Bovidae, the Equidae and the gnu,  m. semitendinosus is greatly strengthened, whereas in the other ungulates it is not so strongly developed.  The same analysis explains the strengthening of the short postfemoral muscles in animals with a relatively short ischium, since with uniform extension of the hip joint the process of contraction of the short postfemorals depends directly on the length of the ischium: the longer it is, the greater the degree of contraction of these muscles.

The short postfemoral and the gluteal muscles indirectly act to extend the knee joint, and therefore when they are strengthened, the knee joint extensors proper, that is, the three femoral capita of m. quadriceps femoris, are often less intensively strengthened than in animals in which the long postfemoral muscles attain the strongest development in the process of cursorial specialization.  Thus, for instance, in the Muridae, Perissodactyla and the gnu, the gluteal muscles are strengthened and at the same time the knee joint extensors have a lesser relative weight than in other cursorially specialized mammals.  In the Asiatic wild ass, for example, the femoral capita of m. quadriceps constitute 8.3—8.9% of the weight of the hind limb muscles, but in the tapir 11.6%.  In C o n i l u r u s the knee joint extensors are hardly any stronger than those in N e s o k i a   i n d i c a, while in the Cricetinae, Gerbillinae and Dipodidae they are even stronger than the hip joint extensors.  An analogous influence on the relative strengthening of the knee joint extensors proper may be exerted by strongly developed short postfemoral muscles.  Thus, in the cheetah in comparison with other Felidae and in the wolf and Cape hunting dog in comparison with the raccoon-dog, the short postfemorals are markedly strengthened, while the relative weight of the three femoral capita of m. quadriceps is not increased.

The strength ratios of the different muscle groups are of a distinctive nature in elephants.  Of the hip joint extensors it is the gluteal and the short postfemoral muscles which are strongly developed.  At the same time, in the Proboscidea the knee joint extensors proper attain a high degree of development, being strengthened due both to an increase of their mass and to the transformation of the lateral caput of m. quadriceps femoris into a pentapinnate muscle.

Extension of the talocrural joint is one of the most important functions in the work of the hind limbs lending propulsion to the body. However, actual extension of the talocrural joint cannot be deeply connected with the work of the other limb joints.  It is interesting to note that two absolutely differently acting moments associated with specialization for swift running may influence the reinforcement of talocrural joint function.  In the first instance the extensors of the joint (m. soleus) are strengthened independently.  Under these conditions the gastrocnemial muscles are inserted directly alongside the center of movement in the knee joint on the sesamoid Vesalius ossicles, these being attached to the femur by ligaments the fibers of which are directed along the axis of the femur. This type of origin of the gastrocnemial muscles makes for almost full independence of extending movements in the knee and talocrural joints (Muridae, Gerbillinae, Criceti-nae, Dipodidae). In the second instance the gastrocnemial muscles are inserted directly onto the femur, and extension of the knee joint automatically affects extension of the talocrural joint (Ungulata, Carnivora, Leporidae, kangaroo, Dasyproctidae, etc.).

This fundamental difference in the working conditions for the talocrural joint extensors had the result that in the first group of animals cursorial specialization took the path of independent strengthening of the gastro-cnemial muscles, while in the second group, on the contrary, it proceeded along the lines of diminished strength of the gastrocnemial muscles. Thus, in the rodents specialized for the metalocomotory mode of running the gastro-cnemial muscles are 2—4 times stronger than in species with a low degree of cursorial specialization. On the other hand, in the cheetah in comparison with the other Felidae and in the wolf and Cape hunting dog in comparison with the raccoon-dog the gastrocnemial muscles are 2—2.5 times weaker. These muscles are relatively weakly developed in the Ungulata, Lagomorpha, kangaroo, etc.

Apart from the strengthening and weakening of individual groups of muscles, cursorial specialization is reflected in deep-seated reorganizations of the structure of the muscles and of the morphological relations in their work and in their movements in the different joints.

The diversity of movements of the primitive mammals led to the need for autonomous work of the different muscles and their groups. During loco-motion, on the other hand, movements are restricted to just one chain of strictly determined movements in the joints and in primitive mammals are regulated only by the coordinating nerve center. In this sense we can speak of a reduction in the freedom of the motor apparatus in the performance of a specific task (Bernshtein, 1935, 1947). A neurocoordinating means of achieving locomotion can apparently develop also in the presence of a cur-sorial specialization in which the morphological links in the work of the dif-ferent joints are weakly developed (adaptation to the metalocomotory mode of running in the rodents). Progressive specialization for running, however, usually goes along with profound morphological changes which limit the num-ber of degrees of freedom without direct connection with the nerve-coordinating centers. An example of this is the development of a link be-tween the extension of the knee and talocrural joints on account of the shift of the origin of the gastrocnemial muscles onto the femur.

The morphological features of structure of the locomotory organs in the cursorially specialized kangaroo, carnivores, lagomorphs and, especially, the ungulates and proboscidians restrict almost all movements except for those directly connected with land locomotion. This is caused by an in-creased number of double-jointed and in particular triple-jointed muscles, a system of ligaments which limits abduction and adduction, and closely interrelated work of all three joints of the hind limbs. The appearance of tendinous bands on the long postfemoral muscles which form a connection with the calcaneal tubercle has the result that extension of all three joints is automatically interlinked.

The structural reorganizations of the muscle complexes are of great interest. For instance, m. biceps femoris, m. gluteus superficialis and m. tensor fasciae latae are united into a single functional complex in a number of mammals. The main working component in this complex is m. biceps femoris anterior (Ungulata) or m. biceps posticus par anterior (Felidae), while all the other components work to improve the working conditions for the main muscle. M. biceps posticus pars posterior (Felidae) or m. biceps femoris posterior (Ungulata) keep the posterior tendon of m. biceps anterior tensed. This muscle ends on the tibia, descending along it sometimes by more than half the length of the bone. M. biceps femoris anterior is maximally strained at the end of the phase of support. At this time the bundles of m. biceps femoris posterior are working in a yielding regime, that is, they are extended despite the active tension, as was confirmed by direct experiments on the domestic cat. Under these conditions there is a sharp increase in the pull on the posterior tendon of the anterior biceps muscle and it is brought into a straightened position. M. tensor fasciae latae and m. gluteus end on the anterior tendon of m. biceps femoris anterior (Ungulata) or on m. biceps posticus pars anterior (Felidae). Extension of the hip joint leads to a lengthening of the bundles of these muscles, and toward the end of the phase of support these also work in a yielding regime. The result is that the anterior tendon is also straightened. The fibers of the working part of the muscles stretch from the anterior to the posterior tendon, and in this case, if these tendons are not strained, they cause them to sag instead of acting to extend the hip joint. Hence, tension of the anterior and posterior tendons by the muscles working in a yielding regime is a necessary condition for successful work of the most strongly developed part of the biceps muscle.

Similar rearrangements are seen in the structure of m. gracilis and m. sartorius. At the end of the phase of support, m. sartorius (Carnivora) or m. gracilis anterior (Ungulata, Rodentia) are extended and work in a yielding regime. These muscles end on the anterior margin of m. gracilis posterior, enhancing its tension. To its posterior margin is attached a tendinous band which ends on the calcaneal tubercle and links up the work of the extensors of the hip and talocrural joints.

For both dilocomotory modes of running, the dorsomobile and the dorsostable, the forelimbs play almost the same role as the hind limbs. For the forelimbs, as for the hind limbs, the greatest load is borne in the phase of support. In this phase the trunk is actively pulled between the legs. The acceleration thus developed engenders a stage of flight with crossed legs, and this stage is usually no shorter than the stage of extended flight, which arises under the influence of the thrust given by the hind limbs.

Contributing to drawing the body between the legs are the pectoral muscles, m. serratus ventralis pars thoracalis, m. latissimus dorsi, and partly m. spinotrapezius. The first two of these muscles are also mainly responsible for keeping the trunk between the legs. Performance of both these functions in the phase of support is possible only in the presence of active work of the shoulder and elbow joint extensors.

Animals employing the dilocomotory dorsomobile mode of running are characterized by a considerably greater mobility of the forelimbs with respect to the trunk than animals using the dilocomotory dorsostable mode of running.  One of the manifestations of this difference which is most readily seen is the position of the shoulder at the moment the animal lands on the forefeet.  In the Carnivora and Sciuridae (dilocomotory dorsomobile mode) the shoulder at this time is positioned almost vertically (Figures 142, 145, 147 and 196–199), while in the Ungulata and Dasyproctidae (dilocomotory dorsostable mode) it is at an angle to the vertical plane (Figures 70–74 and 78–86).  As a result, cursorial specialization in the first group leads mainly to a strengthening of the pectoral muscles and of m. latissimus dorsi, while in the second group m. serratus ventralis is strengthened. The extensors of the shoulder and elbow joints undergo marked development in both groups.

The fixation of the proximal margin of the scapula in proboscidians and the giraffe reduces the load for extension of the shoulder joint and for keeping the trunk between the legs.  The latter takes place due to passive resistance to extension of the ligaments connecting the scapula with the trunk, not as a result of active contraction of m. serratus ventralis.  There is therefore a marked reduction in the strength of m. serratus ventralis and the pectoral muscles, and also of the shoulder joint extensors (m. supraspinatus and m. infraspinatus) in these animals.

As when investigating the hind limb muscles, it was found in the forelimbs that despite their common character, the flexor-extensor movements in the joints are achieved by different ratios in the development of the muscles. To illustrate this we may point out the fact that whereas in most ungulates extension of the shoulder joint is achieved primarily on account of the development of m. supraspinatus and m. infraspinatus, in the gnu, yak, and American and European bisons these muscles are weakly developed,  but m. rhomboideus is strengthened instead.

Summing up the above, we may draw a number of general conclusions.

1.   Specialization for swift land locomotion in mammals proceeded by various paths, enabling us to distinguish a number of modes and forms of running.

2.   Unlike gaits, for a characterization of which it is enough to know the sequence of movement of the limbs and the sequence of supporting stages in the cycle of motion, each mode and form of running is characterized by a specific cursorial mechanics which is displayed in uniform gaits.

3.   Specialization for each mode and form of running leads to increasingly clearly defined morphological differences in the structure of the locomotory organs.

4.   With the convergent appearance of uniform modes and forms of running, when it is difficult to perceive any differences in the mechanics of motion, the highly specific patterns of change are always revealed on a background of parallelly developing features of the skeleton and muscles.

COMPARATIVE DESCRIPTION OF VARIOUS TYPOLOGICAL SYSTEMS
OF LOCOMOTION AND LOCOMOTORY APPARATUS

Any typification is introduced for convenience of analyzing manifold and multifaceted processes, and to make subsequent work easier, these are distributed according to certain signs into simpler and more uniform groups. We had a choice of two ways of working out a classification of the modes and forms of running. One of them consists in selecting the most clear-cut, homogeneous criteria (morphological, biological) with subsequent distribution of all the species studied under the corresponding headings. In this case the typification could have been more logically sustained, but at the same time it would have lost its flexibility: diverse trends of evolution are hard to perceive according to previously and formally prepared measuring rods. The second way, which was the one we chose, consists in the following. The point of departure is the concept of the diversity of factors affecting the phylogenesis of animals, and the basis for isolating the different modes and forms of running are the principal mechanisms of motion governing the trend of changes in the skeleton and muscles in the evolution of a group. To select criteria in this case we have to choose the main ones from this huge diversity of factors which permit us to discover the causes of the appearance of the different motion mechanics. This is of course a very complicated business and calls for research from many different angles. It is all the more complicated that there are very few data on the motion mechanics of mammals, so that this work was based almost entirely on original comparative biomechanical studies, which were necessarily performed with a varying degree of completeness for various groups.

It is desirable to evaluate the proposed classification from different aspects. First, from the point of view of full coverage of the possible pathways of specialization for swift land locomotion, and second, in comparison with other typologies of locomotion and the support-motor apparatus. In addition, it is advisable to assess the possibility of using this kind of classification for applied purposes and its fruitfulness for allied fields of theoretical study.

Regarding the modes of running, we can safely say that all the possible paths of specialization for swift land locomotion have been embraced, as any other means of attaining speed is hardly profitable and therefore scarcely likely. This is confirmed very clearly by those instances when running speed is achieved in certain groups of animals by another means. An actual comparison of different representatives of these groups which are variously adapted to running shows that the more specialized forms gradually come to adopt one of the four modes of running we have described. It is much more difficult to speak of a full set of forms of running. Further specialized research may show that there is a need for a more detailed classification of these. However, it seems that in this case it is more worthwhile to distinguish forms of running according to the characteristic features of the motion mechanics appearing as a result of conditions of phylogenetic development which are specific for each group of mammals.

For an evaluation of the proposed classification we may draw a comparison with the typologies of the support-motor apparatus which have been

most widely accepted in the literature.  In the USSR, the classification of
the support-motor apparatus, which was originally used for quite different
purposes, enjoys the widest recognition.  At the beginning of this century,
Osborn (1910), studying the evolution of the Ungulata, put forward the theory
that these animals descend from the five-toed plantigrade forms.  Later on
the ungulates went through a stage of digitigrade movement and finally became unguligrades, in which the lateral digits were gradually reduced.
These three positions, and also the structure of the foot, were used by
Severtsov (1928) to illustrate the "phase fixation principle."  Severtsov
showed that in walking, plantigrade forms support themselves at first on
the whole foot, then on the digits and at the end of the phase of support only
on the distal phalanges.  In digitigrade forms landing is on the toes, and
support on the foot is excluded from the process of movement, which indicates that a phase of digitigrade walking is fixed.  Finally, in ungulates
support on the toes is also lost and only a phase of support on the last phal-
anx, sheathed in the hoof, remains in the cycle.

These three types of support and structure of the foot were later discussed in numerous summaries, including those of Böker (1935) and Krüger
(1948) on comparative and ecological morphology, in which they were already
credited as the basic types of movement.

The use of this typology of the support-motor apparatus gradually became widespread.  Plantigrade, digitigrade and unguligrade movement originally characterized stages in the evolution of the Ungulata.  For Severtsov
they served merely to illustrate the "phase fixation principle."  Böker and
Krüger attributed to them a partial functional significance.  At present,
many anatomists (Glagolev, 1941, 1952, 1954;  Artemenko, 1949;  Lebedev,
1951;  Manzii, 1953;  Kas'yanenko, 1956;  and many others) have begun to
believe that any feature of the bone-muscle system can be explained by the
fact that the animal belongs to a group showing one of these three types of
foot structure.  This made the initially useful classification, showing
the type of structure of the foot in the ancestors of the Ungulata and
demonstrating the pattern of change of the limbs in the phylogeny of
the group, unequivocally harmful.  The absurdity of its extreme application is shown by the following.  Plantigrade, digitigrade and unguligrade movement, each by itself, arise as a morphological expression of
adaptation to very diverse mechanisms of motion, and therefore it is wrong
to use these types as being functionally uniform.  Cursorial specialization
may proceed along the lines of a transition from plantigrade to unguligrade
movement and even have in its development a stage of digitigrade forms,
but it may also proceed quite differently.  The actual structure of the foot
does not always have decisive significance for the motion mechanics of the
animal as a whole.  And taking into account that similar mechanisms of
movement may be achieved by various morphological systems, we have to
discount the usefulness of the structural features of any organ of locomotion
for determining the work and structure of the foot.

We believe that morphological typology cannot be discussed in the same
breath with functional typology, as the same functions may be fulfilled by
different structures and, vice versa, similar structures are not always

associated with the same function. This is seen particularly clearly when such diffuse concepts as plantigrade, digitigrade and unguligrade movement are taken to be a similar structure. Whereas the appearance of the last two is probably in most cases indeed associated with specialization for swift running (which is achieved by extremely various means in mammals), plantigrade movement appears in the most different animals, including both very highly specialized forms adapted to running, digging, climbing and swimming and animals which show little specialization for any form of locomotion.

Still less success may be expected from attempts to use this typology for studying phylogenetic changes in the structure of the locomotory organs if a rough comparative anatomical series is taken as the basis. Thus, investigations of the structural features of the shoulder region (Glagolev, 1941, 1952) or the tibia (Lebedev, 1951) in the plantigrade bear, digitigrade dog and unguligrade horse can hardly give us an idea of the changes which take place in these organs in the phylogeny of the Perissodactyla. Studying the phylogenetic changes in the appendicular skeleton of the Ungulata and Proboscidea, Gregory (1912) justifies the need to distinguish four types of running (subcursorial, cursorial, mediportal and graviportal). Each of these is characterized by speed, endurance and the size of the animal. A happy choice of the factors basically influencing the mechanics of motion, and thereby also the structure of the animals, enabled Gregory to single out the main trends of adaptive changes of the skeletal ratios in the phylogeny of these groups of mammals. We have, therefore, with some additions and modifications, used Gregory's grouping of the paths of cursorial adaptation in the discussion of the forms of running in the Ungulata and Proboscidea.

The adaptive nature of phylogenesis makes it very important to accurately define the features of the motion mechanics which are responsible for reorganizations of the skeleton and muscles. Each mode and form of running is characterized precisely by these features. This determines the possibility of using the proposed classification of the modes and forms of running. Examination of the adaptations to each mode and form showed that independent adaptation to them in all cases took place in different ways. Evidence of independent adaptation to similar mechanisms of movement can introduce corrections into the existing systems of individual mammalian groups. Similar purposes may be served by explaining similar features in the structure of animals with different mechanisms of motion, as in this case the absence of a genetic closeness in the appearance of similar characters has been proved.

Since the author has not specifically intended to study the process of phylogenesis of any particular group of animals, these corrections bear a negative character, in other words, they may indicate the invalidity of certain conclusions but they do not determine the true position of animals within a group. Thus, for example, proof of the independent paths of adaptation to life among rocks in Capra and Rupicapra enables us to decide against uniting these animals in the same tribe as Sokolov (1953) did, but it does not permit us to establish to which of the groups of Bovidae the chamois is most similar. As our studies showed, adaptations both to the saltatorial form of running (C a p r a) and to the cursorial form (Saiga tatarica, Gazella gutturosa, Rangifer tarandus) are reflected in a reduction of the relative size of the foot and hand.  But adaptation to the saltatorial-cursorial form of running (Gazella sub-gutturosa, Cervus elaphus, Capreolus capreolus, etc.), on the contrary, leads to an increase in the size of the foot and hand.  We therefore have doubts about the hypothesis of Sokolov, Klebanova and Sokolov (1964) on the genetic relationship between the genera C a p r a and S a i g a: they tried to prove that the saiga's ancestors passed through a mountain-dwelling stage of life only according to the difference in the rela- tive size of the hand and foot in the goitered gazelle and saiga.

This example is interesting from the point of view of a broader than usual interpretation of the concept on the convergent development of characters. Convergence in the development of characters is probably often found in ani- mals which have adapted to different life conditions with a profound difference in the original and definitive functions.  With such a broad interpretation of convergence it seems advisable to dwell on this question in more detail. From Darwin on the causes of convergent development of characters have been treated very similarly by all investigators, and in fact Shmal'gauzen's formulation sums up the general theme: "Convergence means the indepen- dent acquisition of similar characters by unrelated organisms.  In the pres- ence of convergence, similarity is explained by adaptation to a similar en- vironment, and differences are attributed to a different origin, that is, the absence of close affiliation" (Shmal'gauzen, 1946,  p. 433).

The convergent development of relatively short distal segments in the limbs of goats adapted to the saltatorial form of running is connected with diametrically opposite requirements with regard to the mechanics of running. With the large single jumps characteristic for goats, a steep trajectory of the center of gravity is more profitable, while with high-speed running a less inclined one is best.  The cursorial form of running is associated with adap- tation to maximum speed, whereas for animals adapted to saltation speed be- comes a danger.  The factors bringing about the appearance of shorter dis- tal segments are also different in this case.  With a gently sloping trajec- tory of the center of gravity the individual jumps during a gallop are small and speed of motion is achieved on account of the pace frequency, not the size of each stride.  On the other hand, when the limbs become longer, steeper trajectories of each jump might appear, and this would be unprofit- able.  In goats, however, owing to the delayed extension  of the talocrural joint relative to the schedule of extension of the other joints  in the hind limbs in the phase of support, the load on the foot is increased, and there- fore the foot is shortened.  The similar relative size of the foot in goats and saigas is therefore not connected with adaptation to similar conditions of existence or to a similar mode of life.

We described an analogous instance for the family Sciuridae. In squirrels, which are adapted to jumping from branch to branch, the legs are lengthened, as occurs in animals adapted to running swiftly in the open:  Spermophil- opsis leptodactylus.  The mechanics of the basic movements of these forms is diametrically opposite.  Squirrels are adapted to one-time large leaps with a steep trajectory, but the ground squirrel is adapted to frequent, gently inclined leaps, enabling great speed to be mustered.  Long legs are useful both for single jumps and for speed, and therefore despite the different mode of life and conditions of existence, these forms show a convergent elongation of the limbs.

A theory of running is proposed as a result of a detailed study of terrestrial locomotion in mammals.  The theory is based on: 1) general laws of economy relating to the attainment of high speed and large jumps; 2) specific investigations of the typical features of the biomechanics of animals and the causes of their appearance; 3) an analysis of the role of individual organs in providing the basic patterns of movement, allowing for high speeds or large jumps to be achieved, plus a study of endurance, taking into account the influence exerted by the different sizes of animals on the mechanics of motion; 4) an analysis of the structural features of these organs.

Data of this kind revealed various paths of specialization for swift land locomotion.  Each of these paths led to the appearance of different modes and forms of running, a classification of which is given in this work.  They are characterized not only by the mechanical features of motion during high-speed, often identical gaits but also by adaptive features of the skeleton and muscles.  However, in the life activity of each species of mammals slow movements may be just as important as fast ones.  The use of different gaits introduces different variations into the speed of motion.  The overall diversity of the gaits typically employed by mammals is so great that the need arose to revise the classification of gaits and probe the reasons for their appearance.

In investigating the causes and paths of development of the modes and forms of running and also the factors in the appearance of the different gaits, it proved necessary to give a review of the mode of life of the ancestors of all the mammals and different groups of them.  Therefore, the general theory and classification of the modes and forms of running which we propose may serve as a basis for a study of the phylogenetic changes of the skeleton and muscles in groups whose evolution is associated with adaptation to swift land locomotion.

# BIBLIOGRAPHY

Adamson,Joy. Die Geschichte eines zahmen Gepardin, die ihre Jungen in der Wildnis aufzog. – Das Tier 12 (1966), 4–7.

Akaevskii,A.I. Anatomy of the Reindeer. – Moscow, 1939. (Russian)

Akaevskii,A.I. Anatomy of Domestic Animals, Vol.1. – Moscow, 1961. (Russian)

Aleksander,R. Biomechanics. – Moscow, 1970. (Russian)

Andrews,R.C. Living animals of the Gobi Desert. – Nat. Hist. (Am. Mus.) 24 (1924), 150–159.

Annandales,N. Bipedal locomotion in lizards. – Nature 56 (1902), 577–578.

Arshavskii,Yu.I., Ya.M. Kotz, G.N. Orlovskii, I.M. Rodionov, and M.L. Shik. Study of the biomechanics of running in the dog. – Biofizika 10, No.4 (1965), 665–672. (Russian)

Artemenko,B.A. Kinematic principle in the limb structure of land vertebrates. – Tez. Dokl. V Vses. S"ezda Anat. Gistol. Embriol., pp.134–136, Leningrad, 1949. (Russian)

Badoux, D.M. Some notes on the functional anatomy of Macropus giganteus Zimm. with general remarks on the mechanics of bipedal leaping. – Acta Anat. 62 (1965), 418–433.

Bartholomew,G.A. and H.H.Caswell. Locomotion in kangaroo rats and its adaptive significance. – J. Mammal. 32, No.2 (1951), 155–169.

Bartholomew,G.A. and C.G. Reynolds. Locomotion in pocket mice. – J. Mammal 35, No.3 (1954), 386–392.

Bashenina,N.V. Ecology of the migratory hamster (Cricetulus migratorius Pall.) in the European USSR. Data on the Fauna and Flora of the USSR. – Mosk. Obshch. Ispyt. Prirody 27 (1951), 157–183. (Russian)

Bekker,M.G. Theory of Land Locomotion in the Mechanics of Vehicle Mobility. – Birmingham, 1955.

Benninghoff,A. and H. Rollhäuser. Zur inneren Mechanik des gefiederten Muskels. – Pflügers Arch. Ges. Physiol. 254 (1952), 527–548.

Bensley,B.A. On the question of an arboreal ancestry of the Marsupialia and the interrelationships of the mammalian subclass. – Am. Nat. 35 (1907a), 117–138.

Bensley,B.A. A theory of the origin and evolution of the Australian Marsupialia. – Am. Nat. 35 (1907b), 245–270.

Bentley,P. The drama of Serengeti: Two men and a "flying zebra." – Courier UNESCO 9 (1961), 19–22.

Bernshtein,N.A. The problem of coordination in relation to locomotion. – Arkh. Biol. Nauk 38, No.1 (1935), 1–34. (Russian)

Bernshtein,N.A. Relation between speed, angle of departure, thrust force, and distance of flight during running jumps. – In: Issledovaniya po Biodinamike Khod'by, Bega i Pryzhka, pp.284–288, Moscow, 1940. (Russian)

Bernshtein,N.A. Movement Patterns. – Moscow, 1947. (Russian)

Bohmann,L. Die grossen einheimischen Nager als Fortbewegungstypen. – Zeitschr. Morphol. Ökolog. der Tiere 35, No.3 (1939), 317–388.

Böker,H. Die Enstehung der Wirbeltiertypen und der Ursprung der Extremitäten. – Zeitschr. Morphol. Anthropol. 26 (1926), 559–602.

Böker,H. Vergleichende biologische Anatomie der Wirbeltiere, Vol.1, Jena, 1935.

Böker,H. and R.Pfaff. Die biologische Anatomie der Fortbewegung auf dem Boden und ihre phylogenetische Abhängigkeit vom primären Baumklettern bei den Säugern. – Gegenbaurs. Morph. Jb. 68 (1931), 496–540.

Borellus,J.A. De motu animalium, Vols.1, 2, 1710.

Bourdelle. Les allures de la giraffe, en particulier le galop. – Bull. Mus. d'hist. natur., sér. 2, 6, No.4 (1934), 329–339.

Brovar,V.Ya. Biomechanics of the withers (in connection with the role of the spinous processes in vertebrates). — Tr. Mosk. Zootekhn. Inst. 2 (1935), 42—58. (Russian)

Brovar,V.Ya. Analysis of the ratios of the weight of the head to the length of the spinous processes of the vertebrae. — Arkh. Anat. Gistol. Embriol. 24 (1940), 54—75. (Russian)

Brovar,V.Ya. Forces of Gravity and Animal Morphology. — Moscow, 1960. (Russian)

Brull,E.L. du. The general phenomenon of bipedalism. — Am. Zool. 2 (1962), 205—208.

Bryant,M.D. Phylogeny of the nearctic Sciuridae. — Am. Midl. Nat. 33, No.5 (1945), 257—390.

Camp,C.L. and A.F. Borell. Skeletal and muscular differences in the hind limbs of Lepus, Sylvilagus and Ochotona.— J. Mammal. 18, No.3 (1937), 15—326.

Casamiquela,R.M. Estudias incológicos. — Buenos Aires, 1964.

Clark le Gros,W.E. History of the Primates. — London, 1949.

Croix,P.M. de la. Filogenia de las locomociónes cuadrupedal y bipedal en los vertebratos y evolutión de la forma consecutiva de la evolutión de la locomoción. — An. Soc. cient. Argentina 108 (1929), 383—406.

Croix,P.M. de la. Evolutión del gallope transverso. — An. Soc. cient. Argentina 113 (1932), 38—45.

Croix,P.M. de la. Les modes de locomotion des vertébrés terrestres. — Nature, Paris 2922 (1934), 97—100.

Croix,P.M. de la. The evolution of locomotion in mammals. — J. Mammal. 17, No.1 (1936), 51—54.

Dagg,A.I. and A. de Vos. The walking gaits of some species of Pecora.— J. Zool. 155, No.1 (1968), 103—110.

Denny-Brown,D. On the nature of postural reflexes. — Proc. Roy. Soc. London 104 (1929), 252—301.

Dollo,L. Les ancêstres des marsupiaux étaient-ils arboricoles? — Trav. Stn. Zool. Wimereux 7 (1899a), 188—204.

Dollo,L. Le pied du Diprodon et l'origine arboricole des Marsupiaux. — Bull. Sci. Belgie 33 (1899b), 275—280.

Dombrovskii,B. Anticlinal vertebrae in mammals and its functional explanation. — Zool. Zh. 14, No.1 (1935), 37—44. (Russian)

Dondogin,Ts. Comparative ecological-morphological analysis of the organization of Mongolian Lagomyidae. Author's Summary of Candidate Thesis. — Moscow, 1950. (Russian)

Donskoi,D.D. Biomechanics of Physical Exercises. — Moscow, 1958. (Russian)

Druzhinin,A.N. Morphological-functional analysis of the shoulder girdle muscles in the Indian elephant.— In: Pamyati Akademika A. N. Severtsova, pp. 209—277, 1941. (Russian)

Eble,H. Funktionelle Anatomie der Extremitätenmuskulatur von Ondatra zibethica. Beiträge zur Anatomie der Bisamratte (II). — Wiss. Martin-Lüther-Univ.    Halle-Wittenb. Math.-naturwiss. Reihe 4, No.5 (1955), 977—1004.

Egorov,O.V. Ecology of the Siberian ibex (Capra sibirica Meyer). — Tr. Zool. Inst. AN SSSR 27 (1955), 7—134. (Russian)

Ellenberger,W. and H.Baum. Topographische Anatomie des Pferdes, Vol.1. — Berlin, 1893.

Engberg,J. Reflexes to foot muscles in the cat. — Acta Physiol. Scand. 62, No.235, suppl.(1964), 1—64.

Engberg,J. and A.Lundberg. An electromyographic analysis of stepping in the cat. — Experientia 18 (1962), 174—176.

Flint,V.E. and A.N. Golovkin. Outline of the comparative ecology of the Cricetinae in Tuva. — Byull. Mosk. Obshch. Ispyt. Prirody, Otd. Biol. 66, No.5 (1961), 57—77. (Russian)

Fokin,I.M. Features of running in the Dipodidae. — Byull. Mosk. Obshch. Ispyt. Prirody. Otd. Biol. 68, No.5 (1963), 22—28. (Russian)

Gabuniya,L.K. Present status of knowledge on fossil Equidae. — In: V.O.Kovalevskii. Sobranie Nauchnykh Trudov, Vol.2, pp.241—279, Moscow, 1956. (Russian)

Gambarjan,P.P. and V.S.Karapetjan. Besonderheit im Bau des Seelöwen (Eumetopias californianus), der Baikalrobbe (Phoca sibirica) und Seeotters (Enhydra lutris) in Anpassung an die Fortbewegung im Wasser. — Zool. Jahrb. Anat. 79, No.1 (1961), 123—148.

Gambaryan,P.P. Biomechanics of the ricocheting jump in rodents. — Zool. Zh. 34, No.3 (1955), 621—630. (Russian)

Gambaryan,P.P. On the function of plumose muscles. — DAN Arm. SSR 25, No.2 (1957), 87—91. (Russian)

Gambaryan,P.P. Adaptive Features of the Locomotory Organs in Fossorial Mammals. — Erevan, 1960. (Russian)

Gambaryan,P.P. Morphofunctional analysis of the limb muscles in the tapir (Tapirus americanus). — Zool. Sb. AN Arm. SSR 13 (1964), 5—50. (Russian)

Gambaryan,P.P. Origin of the diversity of mammalian gaits. — Zh. Obshch. Biol. 28, No.3 (1967a), 289—305. (Russian)

Gambaryan,P.P. Methods of studying gaits from tracks.— Zool. Zh. 46, No.8 (1967b), 1224—1228. (Russian)

Gambaryan,P.P. and N.M. Dukel'skaya. The Rat. — Moscow, 1955. (Russian)

Gambaryan,P.P. and B.A. Martirosyan. Ecology of the mouselike hamster (Calomyscus bailwardi Thomas). — Zool. Zh. 39, No.9 (1960), 1408—1413. (Russian)

Gambaryan, P.P. and R.O. Oganesyan. Biomechanics of the gallop and the primitive ricocheting jump in mammals. — Izv. AN SSSR, No.3 (1970), 441—447. (Russian).

Gambaryan,P.P., G.N. Orlovskii, T.G. Protopopova, F.V. Severin, and M.L. Shik. Muscle work during different forms of locomotion in the cat and adaptive changes of the locomotory organs in the family Felidae. — Tr. Zool. Inst. AN SSSR 42 (1970), 270—298. (Russian)

Gambaryan,P.P., S.B. Papanyan, and B.A. Martirosyan. Data on the biology of the midday gerbil (Meriones meridianus dahli Schidl.) in Armenia. — Byull. Mosk. Obshch. Ispyt. Prirody Otd. Biol. 65, No.6 (1960), 17—22. (Russian)

Gambaryan,P.P. and R.G. Rukhkyan. Morphofunctional analysis of the limb muscles in Elephas indicus and Loxodonta africana. — Tr. Zool. Inst. AN SSSR (Manuscript), 1972. (Russian)

Gans,C. and W.J. Bock. The functional significance of muscle architecture — a theoretical analysis. — Rev. Anat. Embryol. and Cell. Biol. 38 (1965), 115—142.

Gasparyan,K.M. Hind limb muscles of ungulates. — Biol. Zh. AN Arm. SSR 20, No.12 (1967), 87—98. (Russian)

Gasparyan,K.M. Ecology of the wild goat and morphofunctional features of the locomotory organs of some Bovidae. Author's Summary of Candidate Thesis. — Erevan, 1969. (Russian)

Gindtse,B.K. Animal Anatomy. — Moscow, 1937. (Russian)

Glagolev,P.A. The shoulder joint of mammals. — Tr. Voenno-Vet. Akad. Krasnoi Armii 3 (1941), 30—45. (Russian)

Glagolev,P.A. Structural features of the shoulder girdle in the horse. — Izv. Mosk. Zootekhn. Inst. 1 (1952), 90—110. (Russian)

Glagolev,P.A. Evolution of the thoracic limb of mammals on the path to unguligrade movement. — Izv. Timiryazev. Sel'skokhoz. Akad. 1 (1954), 12—25. (Russian)

Goiffon and Vincent. Mémoire artificielle des principes rélatifs à la fidèle représentation des animaux tant en peinture, qu'en sculpture. I partie concernant le cheval. — Alfort, 1779.

Goubaux, A. and G. Barrier. De l'extérieur du cheval. 1884.

Gray,J. Studies in the mechanics of the tetrapod skeleton. — J. Exp. Biol. 20 (1944), 88—116.

Gray,J. How Animals Move. — Cambridge, 1953.

Gray,J. General principles of vertebrate locomotion. — Symp. Zool. Soc. London 5 (1961), 1—11.

Gray,J. Animal Locomotion. — Cambridge, 1968.

Green,E.E. Bipedal locomotion of a Ceylonese lizard. — Nature 66 (1902), 492—493.

Gregory, W.K. Notes on the principles of quadrupedal locomotion and of the mechanism of the limbs in hoofed animals. — Ann. N. J. Acad. Sci. 22 (1912), 267—294.

Gregory,W.K. Relationships of the Tupaiidae and Eocene lemurs, especially Notharctus. — Bull. Geol. Soc. Am. 24 (1913), 241—251.

Gregory,W.K. Evolution Emerging, Vols.1,2.— New York, 1951.

Gromova,V.I. (Ed.). Fundamentals of Paleontology: Mammals, Vol.13. — Jerusalem, IPST, 1968.

Grzimek,B. Elefanten graben sich Brunnen wie Menschen. — Das Tier 9 (1966), 19—25.

Grzimek,M. and B. Grzimek. A study of the game of the Serengeti Plains. — Z. Säugetierk. 25 (1960), 45—96.

Gureev,A.A. Lagomorpha. — Fauna SSSR 3, No.10, Leningrad, 1964. (Russian)

Haines,R.W. Arboreal or terrestrial ancestry of placental mammals. — Quart. Rev. Biol. 33, No.1 (1958), 1—23.

Hájek,K. Weidmannsheil. — Praga, 1954.

Hatt,R.T. The vertebral column of ricochetal rodents. — Bull. Am. Mus. Nat. Hist. 63 (1932), 599—738.

Hesse, R. Die Tierkörper als selbständiger Organismus. — Jena, Fischer, 1935.

Hildebrand,M. Motion of the running cheetah and horse. — J. Mammal. 40, No.4 (1959), 481—738.

Hildebrand,M. How animals run. — Am. Sci. 202 (1960), 148—157.

Hildebrand,M. Further studies on locomotion of the cheetah. — J. Mammal. 42, No.1 (1961), 84—91.

Hildebrand,M. Walking, running and jumping. — Am. Zool. 2, No.2 (1962), 151—155.

Hildebrand,M. The use of motion pictures for the functional analysis of vertebrate locomotion. — In: Proc. XVI. Intern. Congr. Zool. 3 (1963), 263—268.

Hildebrand,M. Analysis of the symmetrical gaits of tetrapods. — Folia Biotheoret. 13, No.6 (1966), 9—22.

Hildebrand,M. Symmetrical gaits of Primates. — Am. J. Phys. Anthropol. 26, No.2 (1967), 18—27.

Hildebrand,M. Symmetrical gaits of dogs in relation to body build. — J. Morphol. 124, No.3 (1968), 320—330.

Hill,A.V. The pressure developed in a muscle during contraction.— J. Physiol. London 107 (1948), 518—526.

Hill,A.V. The mechanics of active muscle. — Proc. Roy. Soc. London 141 (1953), 104—117.

Hirsch,W. Zur physiologischen Mechanik des Froschsprunges. — Z. Vergleich. Physiol. 15, No.1 (1931), 1—50.

Howell,A.B. The saltatorial rodents Dipodomys; functional and comparative anatomy of its muscular osseous system. — Proc. Am. Acad. Arts Sci. 67 (1932), 377—536.

Howell,A.B. Speed in Animals. — Chicago, 1944.

Huxley,J. Arboreal ancestry of the marsupial. — Proc. Zool. Soc. London (1880), 655—668.

Huxley,J. Africa's wild life in peril. — Courier UNESCO 9 (1961), 8—14.

Huxley,J. Elephants and giraffes. — Animals 3, No.16 (1964), 422—427.

Kambulin,E.A. Data on the ecology of the great gerbil (Rhombomys opimus) in Kazakhstan, and control measures. — In: Gryzuny i Bor'ba s Nimi, Vol.1, pp.43—80, Alma-Ata, 1941. (Russian)

Kas'yanenko,V.G. Motor-Support Apparatus of the Horse (Functional Analysis). — Kiev, 1947. (Russian)

Kas'yanenko,V.G. Patterns of adaptive transformations of the limb joints in mammals. — Zool. Zh.35, No.3 (1956), 321—344. (Russian)

Kirchshofer,R. Freiland- und Gefangenschafsbeobachtungen an der nord-afrikanischen Rennmaus Gerbillus nanus garamantis Letaste 1981. — Z.Säugetierk. 23, No.1—2 (1958), 33—49.

Klapperstück,J. Vergleichend-anatomische Untersuchungen am Achsen- und Extremitätenskelett von Myocastor coypus Mol.— Wiss. Z. Martin-Luther-Univ. Halle-Wittenb. Math.-naturwiss.Reihe 5, No.2 (1955), 251—284.

Klebanova,E.A. Microscopic structure of the substantia compacta of the long bones in some representatives of the Sciuridae. — Tr. Zool. Inst. AN SSSR 33 (1964), 256—282. (Russian)

Klimov,A.F. The Limbs of Agricultural Animals. Skeleton and Musculature. — Moscow-Leningrad, 1927. (Russian)

Klimov,A.F. Anatomy of Agricultural Animals, Vol.1. — Moscow, 1937. (Russian)

Klimov,A.F. Anatomy of Domestic Animals, Vol.1 — Moscow, 1955. (Russian)

Kolmakov,A. and V. Vasil'ev. The Asiatic wild life. — Konevodstvo 11 (1936), 8—18. (Russian)

Kovalevskii,V.O. Osteology of Anchitherium aurelianense Cuv. as a Form Explaining the Genealogy of Equus. — Kiev, 1873. (Russian)

Kovalevskii,V.O. Osteology of two fossil species of the ungulate group Entelodon and Gelocus Aymardi. — Izv. Obshch. Lyubitelei Estestvozn. Antropol. Etnograf. 16, No.1 (1875), 1—65. (Russian)

Kovalevskij,V.O. Monographie der Gattung Anthracotherium Cuv. und Versuch einer natürlichen Klassifikation der fossilen Huftiere. — Paleontographica 22, No. 3—5 (1874), 131—346.

Koveshnikova,A.K., E.A. Klebanova, and E.S. Yakovleva. Essays on Human Functional Anatomy.— Moscow, 1954. (Russian)

Kozlova,E.V. The relation between the mode of life of birds and the pattern of development of their flight apparatus (order Alciformes). — Zool. Zh. 36, No.6 (1946), 909—921. (Russian)

Krasuskaya,A.A., E.A. Kotikova, A.K. Koveshnikova, and M.V. Lebedeva. Anatomy of the Muscular System. — Moscow-Leningrad, 1938.

Krüger, W. Bewegungstypen. — Handb. Zool. 8, No.15 (1958), 1—56.

Kummer,B. Baumprinzipien des Säugerskeletts. — Stuttgart, 1959a.

Kummer,B. Biomechanik des Säugetierskeletts. — Handb. Zool. 8, No.6 (1959b), 1—80.

Kummer,B. Beziehungen zwischen der mechanischen Funktion und dem Bau der Wirbelsäule bei quadrupeden Säugetieren. — Z. Tierzücht. Zücht. Biol. 74 (1960), 159—169.

Kummer,B. Funktioneller Bau und Anpassung des Knochens. — Anat. Anz. 111 (1962), 261—293.

Lebedev, M.I.  Structural features of the tibia in the bear, dog and horse in relation to the differences in their ways of walking. − Sb. Rabot, Leningr. Vet. Inst. (1951), 58−72. (Russian)

Lesgaft, P.F.  Principles of Theoretical Anatomy, Vol.1. − SPb., 1905. (Russian)

Manter, J.T.  The dynamics of quadrupedal walking. − J. Exp. Biol. 15 (1938), 522−540.

Manzii, S.F.  Evolution of the hand in mammals. − Zool. Zh. 32, No.4 (1953), 756−766.

Marei, E.J.  Mechanics of the Animal Organism. − SPb., 1875. (Russian)

Marey, E.J.  Le mouvement. − Paris, 1894.

Matthew, W.D.  Arboreal ancestry of the Mammalia. − Am. Nat. 38 (1904), 811−818.

Matthew, W.D.  Paleocene faunas of the San Juan Basin, New Mexico. − Trans. Am. Phil. Soc. 30 (1937), 1−372.

Miller, M.E.  Anatomy of the Dog. − Philadelphia, 1964.

Muybridge, E.  Animal Locomotion. − Philadelphia, 1887. Republished as: Animals in Motion. − New York, 1957.

Nestrukh, M.F.  Types of locomotion in apes, and walking upright in man in relation to anthropogenesis. − Sov. Antrop. 2, No.2 (1957), 159−169. (Russian)

Orlovskii, G.N., F.V. Severin, and M.L. Shik.  Influence of speed and load on coordination of movements in a running dog. − Biofizika 10, No.2 (1966a), 364−366. (Russian)

Orlovskii, G.N., F.V. Severin, and M.L. Shik.  Effect of injury of the cerebellum on coordination of movements in a running dog. − Biofizika 11, No.3 (1966b), 509−517. (Russian)

Osborn, H.F.  Proboscidea, Vol.1. − New York, 1936.

Osborn, H.F.  Proboscidea, Vol.2. − New York, 1942.

Otoway, C.W.  Aspects of equine and canine locomotion. − Symp. Zool. Soc. London 5 (1961), 101−113.

Panzer, W.  Beiträge zur biologischen Anatomie des Baumkletterns der Säugetiere. I. Das Nagel-Kralle Problem. − Z. Anat. EntwGesch. 98 (1932), 147−198.

Pauwels, F.  Die Bedeutung der Baumprintzipien des Stütz- und Bewegungsapparates fur die Beanspruchung der Röhrenknochen. − Z. Anat. 114 (1948), 129−166.

Peterka, H.E.  A study of the myology and osteology of tree sciurids with regard to adaptation to arboreal glissant and fossorial habits. − Trans. Kans. Acad. Sci. 39 (1937), 313−332.

Rakov, N.V.  Role of the wolf and other predators in keeping down the saiga population. − Tr. Inst. Zool. AN Kaz. SSR 4 (1955), 56−66. (Russian)

Randall, E.I.  The predatory sequence, with emphasis of killing behavior and its ontogeny, in the cheetah (Acinonyx jubatus Schreber). − Z. Tierpsychol. 27 (1970), 492−504.

Rashevsky, N.  Mathematical Biophysics. − Chicago, 1948.

Rashevsky, N.  Mathematical Biophysics. − New York, 1960.

Rashevsky, N.  Mathemetical Principles in Biology and their Applications. − New York, 1961.

Romer, A.S.  Vertebrate Paleontology. − Chicago, 1945.

Sadovskii, N.V.  Fundamentals of the Topographic Anatomy of Agricultural Animals and a Short Handbook on Operative Surgery. − Moscow, 1953. (Russian)

Schaeffer, B.  The morphological and functional evolution of the tarsus in amphibians and reptiles. − Bull. Nat. Hist. 78, No.6 (1941), 395−472.

Schomber, H.W.  Beiträge zur Kenntnis der Giraffengazelle (Litocranius walleri Brooke, 1878). − Saügetierk. Mitt. 11, No.1 (1962), 1−44.

Schumacher, G.H.  Functionelle Morphologie der Kaummuskulatur. − Jena, 1960.

Schumacher, G.H.  Funktionsbedigter Strukturwandel des m. Masseter. − Gegenbaur Morph. Jb. 102, No.2 (1961), 150−169.

Severtsov, A.N.  Studies on the Theory of Evolution. − Berlin, 1922. (Russian)

Severtsov, A.N.  Morphological Patterns of Evolution. − Moscow-Leningrad, 1939. (Russian)

Shik, M.L. and G.N. Orlovskii.  Coordination of the limbs in a running dog. − Biofizika 10, No.6 (1965), 1037−1047. (Russian)

Shik, M.L., G.N. Orlovskii, and F.V. Severin.  Organization of locomotory synergy. − Biofizika 21, No.5 (1966), 879−886. (Russian)

Shik, M.L., F.V. Severin, and G.N. Orlovskii.  Regulating walking and running by means of electrical stimulation of the mesencephalon. − Biofizika 21, No.4 (1966), 659−666. (Russian)

Shmal'gauzen, I.I.  Problems of Darwinism. − Moscow, 1946.

Shtegman, B.K.  Studies of bird flight. − In: Pamyati Akademika P.P. Sushkina, pp.237−266, Leningrad, 1950. (Russian)

Shtegman, B.K.   Some structural features of the shoulder girdle in pigeons and sand grouse, and the functional significance of the clavicle in birds. — Byull. Mosk. Obshch. Ispyt. Prirody. Otd. Biol. 62, No. 1 (1957), 45—56. (Russian)

Simpson, G.G.   The beginning of the age of mammals. — Biol. Rev. 12 (1937), 1—47.

Simpson, G.G.   The principles of classification and a classification of mammals. — Bull. Am. Mus. Nat. Hist. 85 (1945), 1—345.

Simpson, G.G.   Horses — the Story of the Horse Family in the Modern World and Through Sixty Million Years of History. — New York, 1951.

Sinel'nikov, R.D.   Atlas of Human Anatomy, Vol. 1. — Moscow, 1952. (Russian)

Sludskii, A.A.   Interplay between predator and prey (as illustrated by the antelope and other animals and their enemies). — Tr. Zool. AN Kaz. SSR 17 (1962), 24—143. (Russian)

Smith, J.M. and R.G. Savage.   Some locomotory adaptations in mammals. — Linnean Soc. J. Zool. 42, No. 288 (1956), 603—622.

Snyder, H.G.   The bipedal locomotion of the basilisk lizards. Anat. Rec. 99, No. 4 (1949), 69—70.

Snyder, R.C.   Quadrupedal and bipedal locomotion of lizards. — Copeia 41, No. 1 (1952), 64—70.

Snyder, R.C.   The anatomy and function of the pelvic girdle and hind limb in lizard locomotion. — Am. J. Anat. 13 (1954), 288—328.

Snyder, R.C.   Adaptation for bipedal locomotion of lizards. — Am. Zoologist 2, No. 2 (1962), 191—203.

Sokolov, A.S.   Structure of the hind limb muscles in representatives of the Sciuridae. — Tr. Zool. Inst. AN SSSR 33 (1964), 283—318. (Russian)

Sokolov, I.I.   Experiment on a natural classification of the Bovidae. — Tr. Zool. Inst. AN SSSR 24 (1953), 5—295. (Russian)

Sokolov, I.I., E.A. Klebanova, and A.S. Sokolov.   Morphofunctional features of the locomotory organs in the saiga and goitered gazelle. — Tr. Zool. Inst. AN SSSR 33 (1964), 319—348. (Russian)

Sokolov, S.S.   Data on the ecology of Saiga tatarica L. — Vestn. AN Kaz. SSR 3 (1951), 143—151. (Russian)

Solomatin, A.O.   Reproduction and related behavior of the Asiatic wild ass. — Byull. Mosk. Obshch. Ispyt. Prirody. Otd. Biol. 69, No. 2 (1964), 71—82. (Russian)

Solomatin, A.O.   Feeding and watering of the Asiatic ass in southeastern Turkmenia. — Tr. Inst. Zool. AN Kaz. SSR 20 (1965), 89—130. (Russian)

Sosnikhina, T.M.   The migratory hamster, Cricetulus migratorius Pall., in the conditions of Armenia. — Zool. Sb. AN Arm. SSR 7 (1950), 55—82. (Russian)

Sukhanov, V.B.   Forms of movement (gaits) of land vertebrates (a theory of locomotion and of the evolution of its forms). — Byull. Mosk. Obshch. Ispyt. Prirody. Otd. Biol. 67, No. 5 (1963), 136—137. (Russian)

Sukhanov, V.B.   Data on the locomotion of land vertebrates. 1. General classification of symmetrical gaits. — Byull. Mosk. Obshch. Ispyt. Prirody, Otd. Biol. 72, No. 2 (1967), 118—135. (Russian)

Sukhanov, V.B.   General System of Symmetrical Locomotion and Locomotional Features of the Lower Tetrapoda. — Leningrad, 1968. (Russian)

Trofimov, B.A.   Origin, history, and some patterns of development of the Ruminantia. — In: V.O. Kovalevskii, Scientific Works, Vol. 2, pp. 287—298 (1956). (Russian)

Udovin, G.M. and K.M. Yanshin.   Relative weight of the limb muscles in the horse. — Tr. Chkalovsk. Sel'.-Khoz. Inst. 4, No. 1 (1951), 8—17. (Russian)

Ukhtomskii, A.A.   Collected Works, Vol. 3. — Leningrad, 1952. (Russian)

Whiting, H.P.   Pelvic girdle in amphibian locomotion. — Symp. Zool. Soc. London 5 (1961), 43—58.

Yakovleva, E.S.   Functional characteristics of structure and age-dependent changes of the forearm muscles in man. — Arkh. Anat. Gistol. Embriol. 37, No. 12 (1959), 35—44. (Russian)

Yanushevich, P.A.   Anatomy of Domestic Animals. — Moscow, 1931. (Russian)

Yudin, K.A.   Morphological adaptations of the family Falconidae in relation to problems of systematics. — In: Pamyati Akademika P.P. Sushkina, pp. 135—209, Leningrad, 1950. (Russian)

Yudin, K.A.   Phylogeny and classification of the Charadriidae. — Fauna SSSR 2, No. 1, Leningrad (1965), 1.

Zhukov, E.K., E.G. Kotel'nikova, and D.A. Semenov.   Biomechanics of Physical Exercises. — Moscow, 1963. (Russian)

Zverev, M.D.   On the speed of running of some animals. — Tr. Alma-Atin. Gos. Zapov. 7 (1948), 145—146. (Russian)

# INDEX

# DATE DUE

| | | | |
|---|---|---|---|
| | | | |
| | | | |
| | | | |
| | | | |
| | | | |
| | | | |
| | | | |
| | | | |
| | | | |
| | | | |
| | | | |
| | | | |
| | | | |
| | | | |
| | | | |
| | | | |
| | | | |
| | | | |
| | | | |
| 30 505 JOSTEN'S | | | |